Odors from Stationary and Mobile Sources

COMMITTEE ON ODORS FROM STATIONARY AND MOBILE SOURCES
Board on Toxicology and Environmental Health Hazards
Assembly of Life Sciences
National Research Council

NATIONAL ACADEMY OF SCIENCES
Washington, D.C. 1979

NOTICE: The project that is the subject of this report was approved by the Governing Board of the National Research Council, whose members are drawn from the Councils of the National Academy of Sciences, the National Academy of Engineering, and the Institute of Medicine. The members of the committee responsible for the report were chosen for their special competences and with regard for appropriate balance.

This report has been reviewed by a group other than the authors according to procedures approved by a Report Review Committee consisting of members of the National Academy of Sciences, the National Academy of Engineering, and the Institute of Medicine.

The work on which this report is based was performed pursuant to Contract No. 68-01-4655 with the U.S. Environmental Protection Agency.

Library of Congress Cataloging in Publication Data

National Research Council. Committee on Odors from Stationary and Mobile Sources.
Odors from stationary and mobile sources.

1. Odors. 2. Odor control. I. Title
TD886.N37 1979 614.7'6 79-9068
ISBN 0-309-02877-9

Available from
Office of Publications
National Academy of Sciences
2101 Constitution Avenue, N.W.
Washington, D.C. 20418

Printed in the United States of America

COMMITTEE ON ODORS FROM STATIONARY AND MOBILE SOURCES

AMOS TURK, City College of New York, New York, New York, *Chairman*
LLOYD M. BEIDLER, Florida State University, Tallahassee, Florida
DAVID M. BENFORADO, 3M Company, St. Paul, Minnesota
WILLIAM CAIN, Yale University School of Medicine, New Haven, Connecticut
THOMAS D. CROCKER, University of Wyoming, Laramie, Wyoming
ANDREW DRAVNIEKS, Park Forest, Illinois
RICHARD A. DUFFEE, TRC of New England, Wethersfield, Connecticut
MELVIN W. FIRST, Harvard University, Boston, Massachusetts
DAVID R. FLESH, Copley International Corporation, La Jolla, California
J. RONALD MINER, Oregon State University, Corvallis, Oregon
DAVID G. MOULTON, University of Pennsylvania, Philadelphia, Pennsylvania
EDWARD I. SELIG, Bracken, Selig and Baram, Boston, Massachusetts
JANET T. WITTES, Hunter College, New York, New York

JAMES A. FRAZIER, National Research Council, Washington, D.C., *Staff Officer*
NORMAN GROSSBLATT, National Research Council, Washington, D.C., *Editor*

Acknowledgments

This report was reviewed and discussed by the Committee on Odors from Stationary and Mobile Sources as a whole, under the chairmanship of Dr. Amos Turk, and represents a group effort. Various sections of this report were prepared by the individual Committee members, as follows.

The Executive Summary was drafted by Dr. Turk and Dr. Janet Wittes.

Chapter 1, Introduction, was drafted by Drs. Turk and Wittes and Mr. R. David Flesh.

For Chapter 2, on the structure and function of the olfactory system, Drs. Lloyd M. Beidler and David G. Moulton drafted the section on physiology and anatomy, and Dr. William Cain the section on odor perception.

Chapter 3 discusses the effects of odors on health and welfare, and the drafts were prepared by Drs. Cain, Beidler, Moulton, and Melvin W. First and Mr. Flesh, with the consultation of Drs. Benjamin G. Ferris, Jr., and Robert A. Schneider. Dr. First's views on the effects of odors on public health differ from the views of the other members of the Committee. His statement on the issue appears as Appendix A.

Chapter 4 is a textbook treatise on the techniques and instruments used to measure odors. The draft was prepared by Dr. Andrew Dravnieks in collaboration with Messrs. Flesh, David M. Benforado, and Richard A. Duffee and Dr. Cain.

Chapter 5, dealing with the atmospheric dispersion of odorants and its mathematical modeling, was drafted by Mr. Duffee.

Chapter 6 discusses the control methods used to abate odorous emission. The section on high-temperature oxidation was drafted by Mr. Benforado, that on dispersion in the atmosphere by Mr. Duffee, that on absorption and gas-phase reactions by Dr. First, that on adsorption by Dr. Turk, and that on modification by vaporous additives by Drs. Cain, Turk, and Dravnieks.

Chapter 7 lists some examples of agricultural odor problems and discusses the application of methods to control odorous emission from agricultural activities. The discussions of animal feedlots and the rendering of agricultural byproducts were drafted by Dr. J. Ronald Miner, that on food-processing by Mr. Duffee, and those on fisheries and fish-processing by Dr. First.

Chapter 8 gives examples of selected industrial odor problems and the methods applied to abate odorous emission from industrial activities. The drafts on municipal sewage treatment and the rubber industry were prepared by Dr. First, and Mr. Duffee prepared those on foundries and the steel, pulp and paper, petroleum, and chemical industries.

Chapter 9 is a discussion of the odorous emission from automobiles, aircraft, and other mobile sources; it was prepared by Dr. Dravnieks.

Chapter 10 is a discussion of some of the techniques that might be used in assessing the economic value of odor control. It was drafted by Dr. Thomas D. Crocker, who graciously accepted this task late in the course of the study.

Chapter 11 is a discursive treatment of the regulatory aspects of odor control prepared by Mr. Edward I. Selig, with the assistance of Dr. Crocker.

Chapter 12 summarizes the responses of state and local air-pollution control agencies to the committee's questionnaire concerning odor problems encountered by the agencies. It was prepared by Mr. Benforado in collaboration with staff. The cooperation of the numerous persons in the various air-pollution control agencies that responded is gratefully acknowledged. The detailed data sheets are grouped in Appendix C.

In Appendix B, Mr. Flesh has described some of the methods used in conducting social surveys of odor problems.

Numerous organizations assisted the Committee by supplying information and references to published works used in the preparation of the report: the National Institute of Occupational Safety and Health, which made available the computer printout of its data file on odors; the Illinois Institute of Technology Research Institute; the Air Pollution Control Association; the American Society of Heating, Refrigerating and Air-Conditioning Engineers; the National Renderers Association; the American Society for Testing and Materials; Arthur D. Little, Inc.; and the Environmental Protection Agency. Of the EPA staff, Ms. Libby Smith and

Ms. Eva Metzger in the EPA library at Research Triangle Park assisted in verifying bibliographic references and numerous computer printouts of the data files; and Drs. Robert J. M. Horton, George H. Wahl, Jr., and Donald G. Gillette provided technical information.

Assistance in obtaining information was given by the staff of the Committee on Toxicology, National Reserch Council; the National Agricultural Library; the National Library of Medicine; and the George Washington University Law Library. Other persons assisted in many ways; our appreciation is extended to those not specifically mentioned.

The staff officer for the Committee on Odors from Stationary and Mobile Sources was Mr. James A. Frazier, who acknowledges the generous assistance of Mrs. Jean E. Perrin, secretary. The bibliographic references were verified and prepared for publication by Miss Rebecca E. Curzon. Typing was done in the Manuscript Processing Unit of the National Academy of Sciences, coordinated by Ms. Estelle Miller. The entire report was edited by Mr. Norman Grossblatt.

Contents

APPENDIXES

Odors from Stationary and Mobile Sources

Executive Summary

This report was prepared by the Committee on Odors from Stationary and Mobile Sources of the National Research Council's Assembly of Life Sciences under contract with the U.S. Environmental Protection Agency. Its purpose is to assist the Environmental Protection Agency in responding to the provisions of Section 403(b) of the 1977 Amendments to the Clean Air Act. The report deals with the sources and measurement of odors in the ambient air, with their effects on human health and welfare, and with the methods and costs of controlling them. The following pages briefly summarize the Committee's findings, generally in the order in which the corresponding subjects are discussed in the report.

THE OLFACTORY SYSTEM

What is commonly called the "sense of smell" is a function of two different organs in the nose. One of them, the olfactory epithelium, is a yellow pigmented area of a few square centimeters in the highest part of the nose, remote from the main respiratory airstream. This area contains millions of bipolar receptor cells that connect directly to the olfactory bulbs of the brain. During quiet breathing, only 3% of odorous molecules that enter the nose reach and contact this exquisitely sensitive area. To bring more odorant in, a person sniffs. The other organ of smell in the nose consists of the free endings of the trigeminal nerve distributed throughout the nasal cavity. Odorants that cause irritation, tickling, or burning stimulate the trigeminal receptors. These sensations are sometimes called the "common chemical sense." In many practical situations involving smell, the distinction between the two senses is overlooked; and they are not always easy to separate, because most odorants stimulate both systems. However, the two organs are connected to different regions of the brain, and their effects are different. A major function of trigeminal reception is to initiate protective reflexes, such as sneezing and interruption of inhaling.

The human olfactory system can discriminate among many thousands of different odorous substances and can detect many of them in extremely low concentrations. Odors convey information about their sources and elicit a wide variety of emotional and physical effects. The human memory for odors is retained over long periods—often over much of a lifetime.

ODOR PERCEPTION

Most odorous matter discharged to the atmosphere—from industrial, agricultural, or natural sources—consists of complex mixtures of many components. Human sensory responses to the individual components of such mixtures vary over wide ranges, from component to component and, to some extent, from person to person. Many atmospheric contaminants are odorless, or very nearly so; carbon monoxide is a notorious example. The chemically pure octanes—the major constituents of gasoline—have very little odor. But many other substances are readily detectable in minute concentrations. For example, an organic sulfur compound at a concentration of one molecule per billion molecules of air is likely to be readily detectable.

The magnitude of the human sensory responses to odor (the perceived odor intensity) decreases as the concentration of odorant decreases. This diminution is the basis for the control of indoor odors by ventilation or of outdoor odors by the use of tall stacks. However, the relationship between odor intensity and odorant concentration is by no means a direct proportion. Unfortunately for the objectives of odor control, when odorous air is diluted with odor-free air, the perceived odor decreases less sharply than the concentration; for example, a 10-fold reduction in the concentration of amyl butyrate in air is needed to reduce its perceived odor intensity by half. Nor do all odorants respond by the same ratios; some, like amyl butyrate, show sluggish changes in odor with changes in concentration, and others change more sharply.

Perceived odor intensity decreases rapidly during the course of a continuous exposure; this is the phenomenon of adaptation to odor. The sensitivity to odor is recovered when the exposure is removed. Both these processes, adaptation and recovery, operate over short time scales. Adaptation begins to reduce the perceived odor intensity during the first inhalation. Recovery is also rapid, operating over a span of a few minutes. Habituation to odors, which means "getting used to them" or "becoming tolerant of them," however, operates over much longer periods. The phenomenon reveals itself in industrial situations, where workers report that an initially repulsive odor eventually seems less repulsive. This phenomenon is of uncertain origin, and much about its magnitude, limits, and course remains to be explored.

EFFECTS OF ODORS ON THE HEALTH AND WELFARE OF PEOPLE

Since ancient times, it has been supposed that pleasant aromas preserve health and that unpleasant odors are injurious. These suspicions formed the basis for the use of aromatic eau de Cologne and of pomanders stuffed with balsams and for the attribution of diseases to atmospheric "miasmas." Thus, the word "malaria" is derived from the Italian expression for "bad air," *mala aria*. We have made considerable advances in our concepts of health and disease, but much remains to be learned about the role of odors.

Some effects of odors have been studied and are well known. Odors may affect well-being by eliciting unpleasant sensations, by triggering possibly harmful reflexes and other physiologic reactions, and by modifying olfactory function. Unfavorable responses include nausea, vomiting, and headache; induction of shallow breathing and coughing; upsetting of sleep, stomach, and appetite; irritation of eyes, nose, and throat; destruction of the sense of well-being and of enjoyment of food, home, and external environment; disturbance; annoyance; and depression. Exposure to some odorous substances may also lead to a decrease in heart rate, constriction of blood vessels of the skin and muscles, release of epinephrine, and even alterations in the size and condition of cells in the olfactory bulbs of the brain. However, the relationships between the intensity or duration of the exposure to odor and the magnitudes of these symptoms have not been established.

There is a lack of controlled studies of olfactory sensitivity in human populations exposed to odor sources. Consequently, we do not know whether such populations have a higher incidence of altered olfactory function of a kind attributable to morphologic changes.

Stimulation of receptors in the nasal mucosa can elicit marked respiratory and cardiovascular responses. The reported effects, documented only in various animal species, include sneezing, bronchodilatation, decrease in breathing rate, decrease in heart rate, increase in arterial blood pressure, decrease in cardiac output, and vasoconstriction in various parts of the body. In humans, virtually all the information regarding adverse reactions to environmental odors has come from complaints and surveys. Irrespective of the physiologic mechanism of action, persons who live in malodorous environments report adverse somatic symptoms, such as "odor-induced" nausea and headache. Unfortunately, such symptoms are difficult to verify and measure.

MEASUREMENT OF ODORS

Odorants are chemical substances and can be analyzed by chemical methods. Odors are sensations and must be assessed by measuring human responses to them. If the physical and chemical determinants of odor were fully understood, it would be possible to predict the sensory properties of odorous materials from their chemical analysis—in practical terms, one could construct an "odor meter" analogous to a decibel meter for sound. Such understanding is not yet at hand, nor is any such device available. Nonetheless, various instrumental and sensory methods of measurement have been developed and have been applied to sources of odor and to the ambient atmosphere. However, many of the available techniques are costly and time-consuming, and not all the sensory methods have been validated by interlaboratory testing.

The sensory attributes of odor that are subject to measurement include odor intensity, detectability, character (quality), and hedonic tone (pleasantness-unpleasantness).

Odor intensity (the magnitude of the perceived sensation) can be described by an ordinal categorization, such as faint-moderate-strong. In more precise methods, numbering systems are used to estimate the magnitude of one intensity relative to another. One or more standard substances, in designated concentrations, may serve as references.

The limit of detection, known as the odor threshold, is not a specific property of a substance, like its color or density. Instead, the threshold depends on the mode of presentation of the sample and on the sensitivity and even the expectation of the recipient. When such factors are carefully controlled, reasonably reproducible values can be obtained.

Odor character, or quality, is the property of the odor sensation that permits one to distinguish odors of different substances on the basis of prior exposure. Various systems of description have been proposed, and there have been some unsuccessful attempts to categorize all odors in terms of a small number of "primary" odor types.

The hedonic tone of an odor is the degree to which it is perceived as pleasant or unpleasant. Such perceptions differ widely from person to person. Furthermore, these judgments are strongly influenced by the previous associations that a person brings to the experience and by the emotional context in which the odor is perceived. Hedonic tone can be measured in terms of preference (dislike very much, like slightly, etc.), numbers, or pictorial references to facial expressions (smiling, frowning, etc.). Another approach parallels the estimation of intensity—odors can be numerically rated in accordance with the degree to which they are more pleasant or unpleasant than other specified odors.

All these sensory methods require careful attention to the acquisition and preservation of a representative sample of the atmosphere or emission of interest and to the selection of appropriate human judges.

The chemical analysis of mixtures that contain many different chemical components requires the acquisition of a representative sample and the separation and identification of the components. To relate such information to the odor of the mixture, it is also necessary to determine which of the components are odorous and to assess their contribution to the intensity and character of the mixture. The analysis must be at least as sensitive as human olfaction.

Modern methods of separation, collectively called "chromatography," are adequate. The preferred method of identification, called "mass spectrometry," is a powerful tool for identifying individual compounds in small samples. Furthermore, the separated components can be split into two streams, one to be submitted to a detection device, the other to be sniffed by a human judge. In spite of the sensitivity and versatility of these methods, however, the complete analysis and identification of the odorous constituents of a mixture as complex as, say, diesel exhaust still elude us.

TECHNOLOGY OF ODOR CONTROL

The control of odors may be regarded as a special case of the general objective of controlling gaseous emission into the atmosphere. The particular characteristics of odors, however, impose some additional requirements and offer some attractive opportunities, as follows.

Because some offensive odorants can be detected at extremely low concentrations, the efficiencies required of control methods are often above 95%. Moreover, because the dilution of an odorant yields a less than proportional reduction in odor intensity, the efficacy of atmospheric dispersion as a control method is similarly reduced.

Another effect related to the low concentrations at which odors can be detected is the fact that minor, episodic, or fugitive emission—such as that associated with occasional spills, imperfect seals or leaky valves, or cleanup operations and the like—may constitute, in the aggregate, a significant cause for community concern, even when the major potential source of odor, such as the gaseous exhaust from a primary process, is fully controlled.

The problems just outlined may also be regarded as constituting an opportunity. If a problem of objectionable odors is indeed caused by emission of small quantities of odorous matter, it is possible that judicious modification of the process, a rigorous program of equipment maintenance to prevent even minor leaks, and a thoughtful regimen of good housekeeping to avoid episodes of spillage and putrefaction will correct the nuisance at modest cost.

If a given air contaminant is objectionable only because of its odor and has no other adverse effects, it is conceivable that the problem can be controlled by modifying olfactory perception, as opposed to cleaning the air.

In accordance with the general principles outlined above, the following methods for controlling odors are available:

- *Modify the process*: Use materials that are less odorous, generate less waste material, operate at more favorable temperatures, apply better maintenance and housekeeping procedures, and so forth.
- *Dilute the odorants in the atmosphere*: Collect them and discharge them through a tall stack, or separate the source from centers of population.
- *Absorb the odorants at ambient temperatures by dissolving them in a suitable liquid*: The process may be greatly aided by a chemical reagent that converts the odorants to products that are more soluble, less odorous, or both. Some of these reactions may occur in the gas phase or in both gas and liquid phases.
- *Adsorb the odorants in a highly porous solid*: The medium of choice is activated carbon. Under some circumstances, the carbon can be used to recover valuable material and return it to the process, leading to a net savings in cost.
- *Oxidize the odorants with air*: This method offers the best opportunity for substantial and irreversible destruction of odorous matter, but it is often the most costly alternative, because high temperatures are required. In many instances, the cost is prohibitive. There are three approaches to improving the economics of oxidation: make an existing boiler do double duty as an incinerator, recover some of the heat of oxidation for reuse, or lower the required oxidation temperature by using a catalyst.

• *Modify the perception of the odor*: This method, sometimes termed "odor masking" or "odor counteraction," is controversial, because it requires the addition of foreign substances to the air (the opposite of air cleaning) and because its effect has yet to be convincingly demonstrated.

Various sections of this report describe the applications and costs of alternative control methods for specific agricultural and industrial processes and for mobile sources. The agricultural applications described include coffee-roasting, canneries, smokehouses, fermentation processes, feedlots, rendering plants, fisheries and fish-processing, and field burning; the industrial applications include sewage treatment, rubber-processing, steel-making, pulp and paper mills, petroleum-refining, and the chemical industry; and the mobile sources include diesel-powered vehicles, gasoline engines, gas turbines, and jet engines.

METHODS OF ASSESSING THE BENEFITS OF ODOR CONTROL

Economic theory and practice suggest techniques for valuing nonmarketed aesthetic phenomena. In particular, property-value studies, substitution studies, and bidding games have shown that many benefits of pollution control traditionally viewed as intangible and hence nonmeasurable may perhaps be assessed and made comparable with economic values as expressed in markets. Although these methods are potentially useful for odor problems, they have not been applied to odors.

LEGAL ASPECTS

The common nuisance law has been used by the states as the primary legal vehicle for responding to complaints about odors. The nuisance approach can be supplemented by more scientific and more comprehensive regulatory approaches to odor control. For example, standards based on an index of odor perception may be established for ambient-air quality and for emission of odorants. Some odors may be controlled through legal economic incentives. Of course, the regulatory aspects of odor control are linked to the problems of measuring odors, to the determination of the effects of odors on people, and to the state of the art of methods of controlling odors.

The establishment of national standards would protect people from malodorous environments. However, because reactions to odor depend heavily on local values and individual aesthetic judgments, national standard-setting will be very difficult.

RECOMMENDATIONS

The establishment of federal ambient-air quality or emission standards for odors would confront various conceptual and technical difficulties. These difficulties may be briefly summarized. First, the adverse effects of odors on people are variable, and our knowledge about the effects is very incomplete. Thus, it will be difficult to define standards that will be widely accepted. Second, although odor perception can be assessed by psychophysical methods and some odorous substances can be measured by modern instrumental methods, the two sets of results are difficult to relate to each other; furthermore, the methods are costly and time-consuming.

If, in spite of these problems, federal ambient-air quality or emission standards for odors were to be established now, the recommended approach would incorporate the following features:

- The standard should be related to a measurement of odor perception.
- The standard should be expressed in terms of the perceived magnitude, or intensity, of the odor. Such odor intensity should be assessed by comparing it with the intensity of a specified concentration of a standard reference odorant.
- The duration of and frequency of exposure to an odor are important determinants of human responses. These factors should therefore be taken into account in the establishment of standards.
- The offensiveness or inoffensiveness of an odor is also an important determinant of its effect on people. Consideration should therefore be given to specifying exemptions or relaxations of the standards when a given odor is known to be inoffensive.
- Exemptions for industries in areas far from population centers or in cases of excessive economic impact of odor abatement should be considered.
- Special types of odor standards for agricultural and mobile sources should be defined.

In recognition of the difficulties that would confront the establishment of federal ambient-air quality or emission standards for odors, various kinds of studies will need to be done.

- There is need for more basic scientific information on the effects of odors on people. Research should include studies on animals when they are relevant.
- Basic research is needed on the mechanism whereby the presence of an odorous airborne contaminant is translated to neural signals that result

in odor perception. If an odor meter is ever to be produced, it must be based on such research.

• Individual sensitivity to odors should be studied to identify differentially sensitive subgroups of the population.

• The modes of dispersion of odorants in the atmosphere should be studied more fully. Such studies should provide an experimental base for improving the currently available mathematical approaches to dispersion models. Attention should be given to durations and magnitudes of human exposures that result from the release of odorous matter to the atmosphere. Differences between point, area, and mobile sources should be taken into account.

1 Introduction

The Clean Air Act of 1970 noted that the growth in the amount and complexity of air pollution brought about by urbanization, industrial development, and the increasing use of motor vehicles had resulted in mounting dangers to the public health and welfare. The identities of various significant air pollutants, such as oxides of sulfur and nitrogen, have long been known. Other substances are classified as atmospheric contaminants of recognized chemical types, such as "oxidants" and "unsaturated hydrocarbons." However, it is estimated that 50% or more of the complaints about air pollution deal with exposure to odors.

In the Clean Air Act Amendments of 1977, the Administrator of the Environmental Protection Agency was directed to study "the effects on public health and welfare of odors or odorous emissions, the sources of such emissions, the technology or other measures available for control of such emissions and the costs of such technology or measures, and the costs and benefits of alternative measures or strategies to abate such emissions."

In September 1977, the National Academy of Sciences entered into a contract with the Environmental Protection Agency whereby the National Research Council would evaluate "the available published information on the effects on public health and welfare; the threshold concentrations for perception of odorous substances and concentrations that have adverse health effects; health effects of combinations of odorous compounds; adaptability to odors; the technology or other measures available for control or abatement of the emissions; and the costs of the strategies for emission control. A part of the study task will be to determine the feasibility and practicality of measuring odors objectively for the purpose of regulatory enforcement."

This report addresses itself to the charge to the Committee on Odors from Stationary and Mobile Sources. Unfortunately, the state of our knowledge about odor is not uniformly advanced in all aspects. At one extreme, highly effective methods are available for removal of odorous contaminants from airstreams. Similarly, powerful, although often costly and time-consuming, instrumental techniques have recently been developed for analysis of complex mixtures of odorants. At the other extreme, the assessment of the adverse aesthetic, social, and health effects of unpleasant odors on people is very imprecise. Furthermore, the problem of relating the sensation of odor to the chemical analysis of odorous substances is complicated by the fact that the physical and chemical determinants of odor have not yet been clearly established. As a result, there is no conceptual basis for the construction of an "odor meter" analogous to a decibel meter for sound or a light meter for light. Nonetheless, methods in all these areas have been proposed or developed. This report describes and critically evaluates these various approaches and makes recommendations for further study.

Attempts have been made to characterize complex odorant mixtures on the basis of the properties of their components. The great advances in methods of chemical analysis and separation have strengthened these efforts. Nonetheless, success has often been elusive. The odor of a mixture may be strongly influenced by barely detectable (or even undetectable) traces of highly odorous constituents. Certainly, it cannot be assumed that the odor of a mixture is that of its major component. Thus, a "phenolic" odor from the curing of a phenolic resin is not the same as the odor of pure phenol, and the pungent odor of burning fat is not the same as that of acrolein, although it is often so characterized.

It is particularly noteworthy that the odors of some nominally pure materials have been shown to be due largely to the presence of impurities. For example, phosphine, whose reported detection threshold ranges from 0.2 to 3.0 ppm, has been shown to be odorless when pure,[3] the reported odors being due to impurities in the form of organic phosphine derivatives.

Because most organic chemists believe that they can identify the functional group (alcohol, amine, ester, etc.) in a compound by smell, it is interesting to determine the degree to which such attempts are successful. Brower and Schafer[1] conducted such a study and found that, for most representative compounds, the functional group was correctly identified in 45% of the cases. The performance was poor for alcohols, ethers, and halides and excellent for amines, sulfur compounds, esters, phenols, and carboxylic acids. When the subjects missed the functional group, they used the labels "alcohol," "ester," and "ketone" twice as often as average. The label "sulfur compound" was misapplied in only 1% of all cases. Bulky hydrocarbon groups near the functional group can weaken or obliterate the odor quality of a compound, but aliphatic amines and sulfur compounds are very resistant to such steric hindrance. In contrast, the odors of amine and sulfur entities are greatly weakened by electron-withdrawing groups. Aliphatic compounds with a multiplicity of methyl groups have the odor of camphor or menthol.

The classification of ambient odorant sources also presents difficulties. The odors produced by emission from a given type of operation (such as rendering of meat byproducts) may range from very intense in the absence of effluent controls to undetectable (if they are treated by effective abatement systems). The mere listing of such a "source" therefore does not predict the seriousness of the problem in any particular instance. However, a compilation of odorant sources can illustrate the variety of industries and operations that have the potential to generate odorous products. One such tabulation has been prepared by Copley International Corporation.[2]

The development of complete lists of odorants or odorant sources becomes extremely difficult when one confronts the task of assessing the *amounts* of odorant produced. Aside from the many problems related to the measurement of odor (see Chapter 4), no national inventory of odorant sources has ever been attempted.

There are no available estimates of the number of people exposed to odors of each source. The only available estimates of the *total* number of people affected by odors in the United States were made in 1969[2] and may be summarized as follows:

- About 25 million residents of the United States would state that air pollution is a problem *and* that odors are a major element.
- About 1.5 million residents would voluntarily state that odors are a disadvantage to living in their areas of their cities.
- Almost 5 million residents would admit that odors bothered them very much.
- More than 5.5 million residents would state that odors are a continuously serious problem.
- About 3 million residents would state that odor pollution has reduced the value of their home property.
- Almost 5 million residents would have seriously considered moving away from their areas because of odor pollution.
- Only 0.5 million residents would have requested some authority or agency to take action concerning *air pollution*. The number of residents

that would have requested action concerning odor problems would be a small fraction of this.

It must be emphasized that, in the years since 1969, any of the following factors could have produced considerable changes in the number of people affected by odors:

- The U.S. population has increased and its geographic distribution has changed.
- People have become more aware of odors, perhaps more anxious about their effects, and more likely to view them as a problem.
- Odor-control technology has improved, and many formerly odorous emissions have been reduced.

The 1969 survey indicated that a large number of residents perceived odors as a problem. Yet only a small percentage of these residents were motivated to seek recourse. The reasons for this apparent apathy could not be inferred from findings of the public-opinion surveys.

REFERENCES

1. Brower, K. R., and R. Schafer. The recognition of chemical types by odor: The effect of steric hindrance at the functional group. J. Chem. Educ. 52:538–540, 1975.

2. Copley International Corporation. A Study of the Social and Economic Impact of Odors. Phase I. National Survey of the Odor Problem. Report to the National Air Pollution Control Administration. Contract No. CPA 22-69-50. Washington, D.C.: U.S. Department of Health, Education, and Welfare, Public Health Service, NAPCA, January 1970. 246 pp.

3. Fluck, E. The odor threshold of phosphine. J. Air Poll. Cont. Assoc. 26:795, 1976.

2 Structure and Function of the Olfactory System

PHYSIOLOGY AND ANATOMY OF THE OLFACTORY SYSTEM

The nose houses an organ—the olfactory organ—that can detect and discriminate among many thousands of different odors and can detect some of them in concentrations lower than those detectable by physical instruments, such as a gas chromatograph. The 10 million or so receptors that constitute the organ approach or reach the theoretical limits of sensitivity: it has been calculated that one molecule of butyl mercaptan can excite a receptor cell. It is with such excitation that the process by which humans perceive odor begins.

But there are two chemosensitive systems in the nose: the olfactory organ, which consists of the olfactory epithelium, lying closest to the brain at the back of the nose; and the receptors of the so-called "common chemical sense," which are the free nerve endings of the trigeminal (fifth) cranial nerve distributed throughout the nasal mucosa. Odorants that cause "irritating," "tickling," or "burning" sensations probably stimulate trigeminal nerve endings; but most, if not all, odorants stimulate both systems in higher concentrations. These two systems connect to different regions of the brain, with different behavioral consequences. A major function of trigeminal reception is to initiate protective reflexes, such as sneezing and interruption of inhalation. Olfaction, however, appears to be less involved in the reflex response to irritating odors.

In addition to these direct consequences of odors on olfactory function,

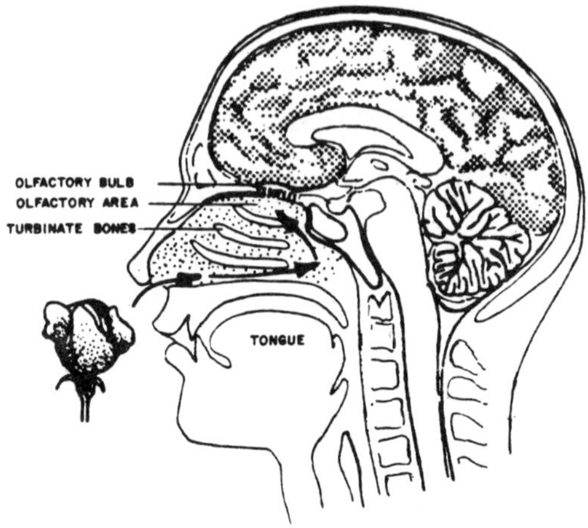

FIGURE 2-1 Cross section showing the three turbinate bones, the olfactory area, and the termination of the endings of nerves from the olfactory bulb in the olfactory epithelium.
Reprinted with permission from Brown.[13 (p. 8-135)]

there may be more indirect influences. The odor receptors of the nose adjoin the respiratory mucosa, whose main function is to humidify and warm incoming air (Figure 2-1). Olfactory function depends, at least partly, on the condition of this mucosa.[84]

Because information based on humans is lacking in many cases, we are often forced to refer to results of animal studies in describing olfactory morphology and function.

NASAL CAVITY

The nasal airways are formed of three paired chambers. Of each pair, the two members are separated from each other by a partition, or septum (Figure 2-2). These airways have a narrow roof (cribriform plate) and a broader floor (hard palate). The first pair of chambers lies just inside the nostrils and is covered with skin; the second lies farther back and is lined with respiratory epithelium (or mucosa); and the third, lying closest to the brain, is lined with olfactory epithelium. Anatomically, the most complex region of each nasal chamber is the lateral wall (Figure 2-2). Protruding downward from this wall like a series of partially folded shelves are three conchae, or turbinates (inferior, middle, and superior). The epithelium covering these bones is so richly supplied with blood vessels that at the back of the nasal cavity there is only a ribbon-like passage between them and between them and the nasal septum. This passage is seldom more than 1 mm and rarely more than 2 mm in diameter. In fact, the inferior turbinates may touch the septum.

The superficial blood vessels in the respiratory mucosa can respond to mucosal irritation by filling the mucosa with blood, in which case a turbinate may change in thickness by as much as 4 mm in a matter of minutes and thus become able to close off much of the passage or markedly reduce odorant access to the receptors. Dilatation of the submucosal tissue, however, is a much slower response to hormonal, allergic, or emotional stimuli and can be chronic.[79] A relatively high degree of either swelling or shrinkage has been found to correlate with high olfactory thresholds for citral in human subjects.[84]

The yellowish olfactory epithelium is an area of about 5 cm^2 at the

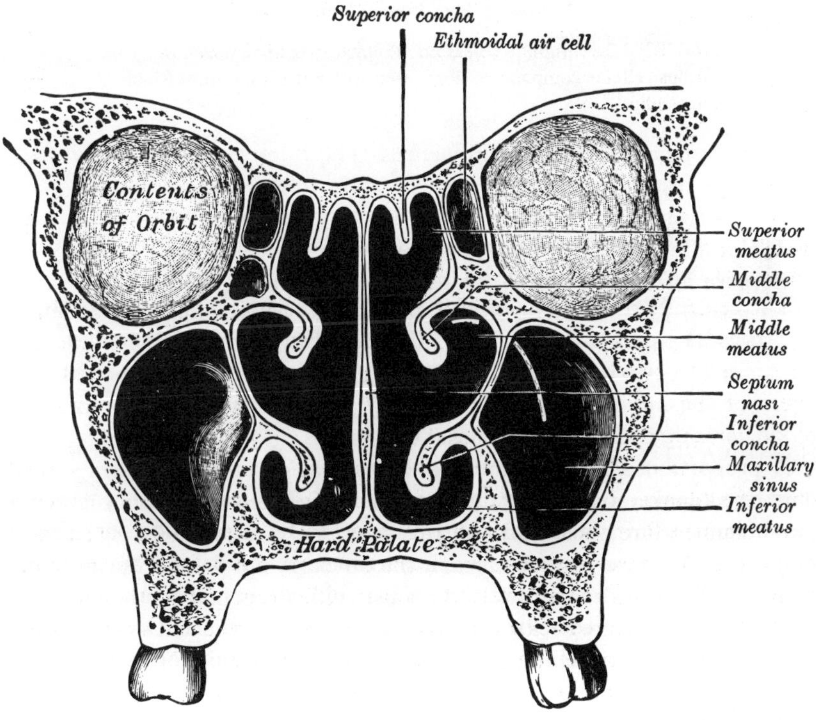

FIGURE 2-2 Coronal section of nasal cavity. Reprinted with permission from Goss.[50 (p. 1118)]

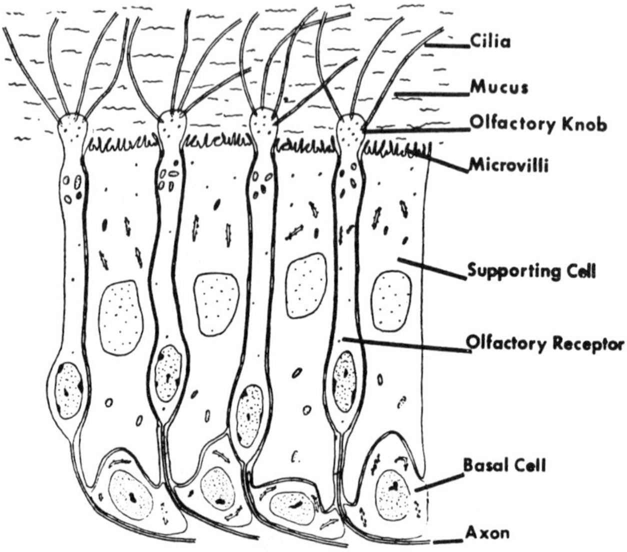

FIGURE 2-3 Simplified diagram of olfactory epithelium, showing the various cellular components. Reprinted with permission from Moulton and Beidler.[76]

highest part of the nasal cavity (10 cm^2 in all), remote from the main respiratory airstream (Figure 2-1).

The centrally directed poles of the olfactory receptors form nerve fibers, or axons (Figures 2-3 and 2-4). The respiratory mucosa, however, is innervated by two branches of the trigeminal nerve. It is also supplied with fibers of the autonomic nervous system, which controls the amount of blood entering and leaving the nasal mucosa.

A proportion of the odorant molecules that enter the nose are sorbed onto the blanket of mucus that covers the ciliated respiratory mucosa (epithelium). Partly because of this loss, only about 3% of odorant molecules that enter the nose reach the olfactory epithelium during quiet breathing.[33] Thus, to smell well, we must sniff. Propelled by the cilia, the mucous layer sweeps backward toward the throat, where it is swallowed with entrapped bacteria, particles, and odorant molecules. The mean flow rate—4.2 mm/min in nonsmokers at 43.6% relative humidity—is sufficient to ensure replacement of the mucus about once every 10 min.[45,89] Because the olfactory surface lacks organized ciliary action, mucus removal appears to depend on traction exerted by mucus moving over neighboring areas of respiratory mucosa—another example of dependence of olfactory function on normal respiratory mucosal function.

TRIGEMINAL SYSTEM IN THE NOSE

Many compounds elicit sensations variously described as "burning," "stinging," and "tickling"—short-chain fatty acids, alcohols, amines, aldehydes, ammonia, etc. They stimulate trigeminal receptors in the nasal mucosa. Some compounds do not have these effects. But electric

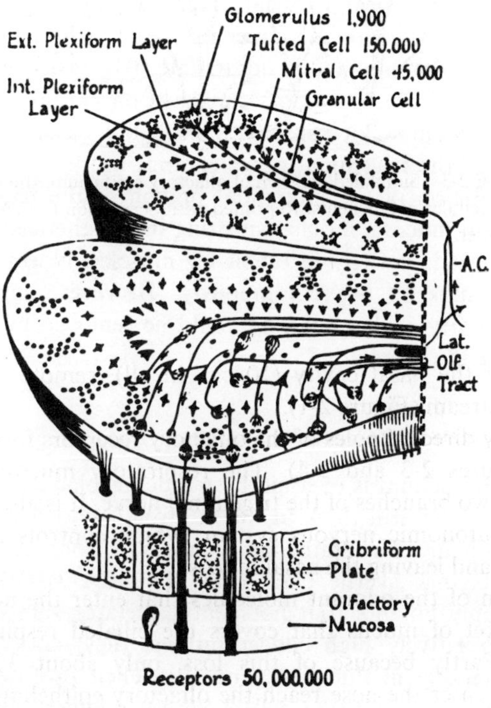

FIGURE 2-4 Structure of the olfactory bulbs and their relations to the nerves and mucosa (modified from Moulton and Tucker,[77] after Gastaut and Lammers). A.C., anterior commissure. The figures are estimates, made by Allison and Warwick,[2] of the numbers of each type of cell in the olfactory bulb and in the olfactory mucosa lining one nasal cavity of the rabbit. Reprinted with permission from Moulton.[74]

recordings of neural activity from a branch of the trigeminal nerve that supplies the nasal mucosa have failed to identify any odorant that does not stimulate this nerve (at least in the rabbit or tortoise). In fact, the trigeminal nerve sometimes responds to lower concentrations than the olfactory nerves, although more commonly the reverse is true, with *n*-pentyl acetate producing the greatest divergence of thresholds between the two systems among the odorants tested. In the tortoise, for example, the thresholds for *n*-pentyl acetate (amyl acetate) are vapor saturations of 10^{-5} at 20°C for the olfactory nerve and 10^{-1} for the trigeminal nerve.[91,92] It may be possible to explain these and other differences in results between behavioral and electrophysiologic studies by assuming that trigeminal receptors can be excited without any perceptual awareness of the response. At any rate, it is not clear to what extent, if any, trigeminal response contributes to the detection and recognition of odors or of particular concentrations of odorants that do not irritate.

The marked influence that odors can have on cardiovascular, respiratory, and even hormonal responses is probably caused largely by the indirect action of the trigeminal nerve endings in the nose. For example, repetitive stimulation of the ethmoidal branch of the trigeminal, which supplies the nasal mucosa, induces sneezing in anesthetized cats.[6] Another case involves electric activity in rabbit leg muscles. Normally, the motor units in these muscles fire continuously. However, nasal stimulation blocked this activity. This inhibition could be removed by sectioning the trigeminal nerve intracranially.[5]

OLFACTORY EPITHELIUM AND OLFACTORY BULBS

The olfactory epithelium is overlaid by a mucous blanket. This blanket is reported to be 10–15 μm thick in the bullfrog[78] —no reliable estimate is available for humans. The epithelium consists of elongated receptors and supporting cells that overlie basal cells (Figure 2-3). When particular odorants—notably butanol (but not pentyl acetate)—are directed at the epithelium in relatively high concentrations, they cause the mucus to increase considerably in thickness, and they elicit profuse and vigorous secretions from the supporting cells.[78] Whether lower concentrations of other odorants would also elicit secretions is not clear, but this response may represent an additional means by which ambient odorants could influence the microenvironment of the receptors.

The olfactory receptors are nerve cells. At one end of each are cilia that project into the mucous blanket (Figure 2-3). Reception sites are assumed to lie in this region of the cell. At the other end is a nerve fiber that connects the receptor directly to the olfactory bulb of the brain without interruption or branching (Figure 2-4). These cells seem to undergo continuous replacement and may live for less than a month in mice—the only mammal on which data are available.[75] Whether their life span is altered by exposure to ambient odors is not known. Nor is it known by what mechanism odorants excite the receptors. The initial events probably involve adsorption on receptor molecules in the outer membrane of the receptor cell. After a conformational change at the receptor site, an ionic exchange may occur across the membrane that initiates electric events that precede the firing of one or more nerve impulses or inhibit the background activity in the receptor.

Olfactory information is processed in the olfactory bulb and then projected to other parts of the brain (Figure 2-5). This information controls or influences complex patterns of behavior, such as feeding and reproduction, as well as emotional responses, such as fear, pleasure, and excitement.

ODOR RECOGNITION

Odor recognition is thought to depend on the presence of receptor molecules—possibly proteins—at various sites on the surface of the cell membrane. Single receptor cells usually respond to most odorants presented, but occasionally cells with higher selectivity are found. There is no evidence, however, that odorants can be classified according to the responses that they elicit. Site types that are more sensitive to a given odorant may be more concentrated in some regions of the epithelium than in others. Both the distribution of metabolic activity and the morphologic changes in cells of the rat olfactory bulb elicited by prolonged stimulation of the mucosa with odorants[63] have shown different patterns for different odorants. In addition, when odorants are streamed through the nasal sac of the bullfrog, some bind more strongly to the mucosal surface than others and thus set up different patterns of activity in the primary neurons. Thus, there are two distinct mechanisms for generating spatiotemporal patterns of excitation in the olfactory epithelium and bulbs. Each may contribute information that assists in odor-quality discrimination.[75]

ODOR PERCEPTION

In the absence of objective means to measure the many varieties of odor pollution, odor perception is the basis for specifying and evaluating offensive odors and for assessing the effect of abatement procedures. This

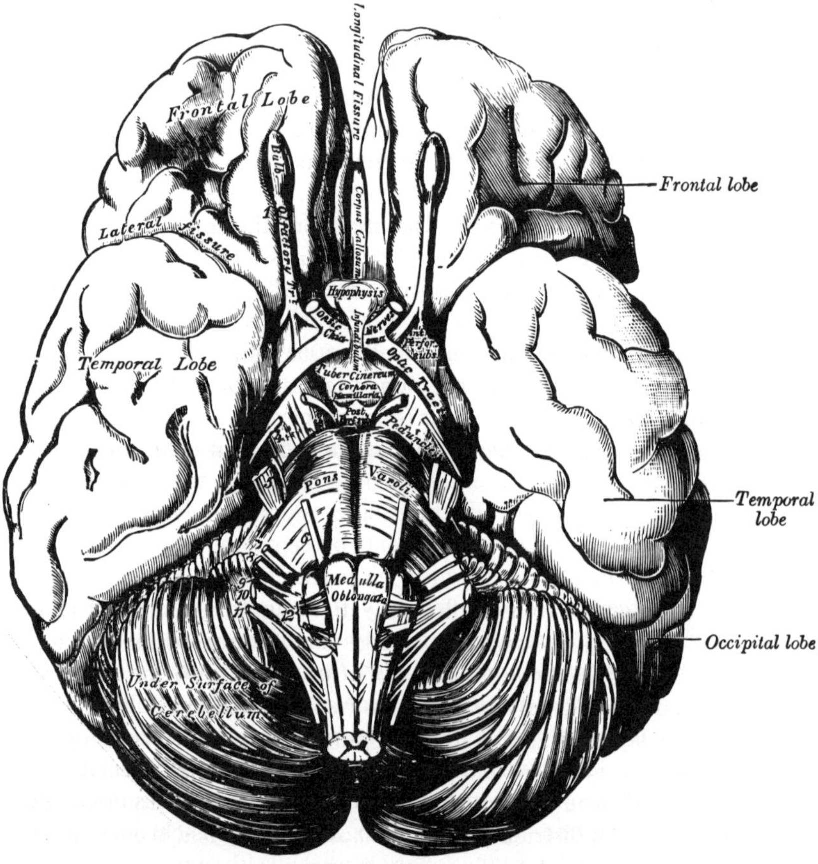

FIGURE 2-5 Ventral view of brain, showing olfactory tract and bulb. Reprinted with permission from Goss.[50 (p. 835)]

section deals with functional principles that characterize both odor perception and its important companion, the common chemical sense. The section highlights issues of particular relevance to odor pollution. Specific topics include the sensitivity of olfaction, the relation between odor intensity and concentration, the role of the common chemical sense, time-dependent processes (adaptation and habituation) in odor perception, the perception of complex stimuli (mixtures), the action of odor modifiers, odor pleasantness, and odor character, or quality.

OLFACTORY SENSITIVITY

Most industrial effluents comprise complex mixtures of odorous and nonodorous constituents. Diesel exhaust, for example, has more than 1,000 constituents. In the face of such a complex olfactory stimulus, a question of primary interest is which constituents contribute most heavily to the resulting odor.

The olfactory system exhibits substantial nonuniformity in its sensitivity to odorants. To illustrate, Table 2-1 displays thresholds for various petrochemicals.[51] The column labeled "absolute" refers to the concentration required for half the judges in a panel (presumably about 10 persons) to just detect odor. The columns labeled 50% and 100% "recognition" refer, respectively, to the concentrations necessary for half or all the panel members to recognize the character (quality) of odor. Even for this limited group of substances, all of which appear in industrial effluents, olfactory sensitivity varies by many orders of magnitude. Hence, the proportions (by mass) of constituents in a mixture offer, by themselves, little or no indication of which ones will predominate perceptually.

Knowledge of the physicochemical properties that enable one substance to stimulate perception more effectively than another remains fragmentary. No single physicochemical property can account for the nonuniformity. Nevertheless, some combinations of properties provide moderately good predictions of relative threshold values. Laffort and colleagues,[56,60,62] for instance, have developed a model that incorporates several properties: molar volume, proton affinity (Brönsted basicity), local polarizability, and ability to donate protons (Brönsted acidity). These properties are assessed from relative retention times on four different gas-chromatographic columns (i.e., four different stationary phases). Although the correlation between predicted and obtained values approaches 0.90, the model holds more theoretical than practical usefulness. Its practical benefits may increase as the amount of data used to develop and assess the model grows. Rather than rely on such a model, a person interested in the threshold for any particular substance will usually consult tabulations of thresholds[46,93] or will measure the threshold under conditions of interest.[35]

Although characterized by keen absolute sensitivity, olfaction has seemed to possess poor differential sensitivity, i.e., poor sensitivity to differences in concentration.[86] Its apparent dullness to small changes in concentration has probably resulted from poor control of the olfactory stimulus. An increase in the precision of control reveals much greater differential sensitivity than previously suspected. Experiments have revealed that persons can reliably resolve differences smaller than 10%.[15,20]

TABLE 2-1 Odor Threshold, Quality, and Hedonic Tone of Odors of Various Petrochemicals[a]

Compound	Absolute, ppm	50% Recognition, ppm	100% Recognition, ppm	Quality	Hedonic Tone
Acetic anhydride	<0.14	0.36	0.36	Sour acid	Neutral to unpleasant
Acetone	20.0	32.5	140	Sweet/Fruity	Pleasant to neutral
Acetophenone	0.30	0.60	0.60	Sweet/Almond	Pleasant
Acrylic acid	0.094	1.04	1.04	Rancid/Sweet	Unpleasant
Amyl acetate, primary (mixed isomers)	0.067	0.15	0.21	Sweet/Ester/Banana	Pleasant
Amyl alcohol	0.12	1.0	1.0	Sweet	Pleasant
1,3-Butadiene	0.45	1.1	1.3	Undefined	Unpleasant to neutral
n-Butanol	0.30	1.0	2.0	Rancid/Sweet	Neutral to unpleasant
2-Butanol	0.12	0.41	0.56	Sweet	Pleasant to neutral
Butyl acetate	0.006	0.037	0.037	Sweet/Ester	Pleasant
n-Butylamine	0.08	0.24	0.24	Sour/Ammoniacal	Unpleasant to pleasant
Butyl Cellosolve	0.10	0.35	0.48	Sweet/Ester	Pleasant
Butyl Cellosolve acetate	0.11	0.20	0.20	Sweet/Ester	Pleasant
n-Butyl chloride	8.82	13.3	16.7	Pungent	Unpleasant
n-Butyl ether	0.07	0.24	0.47	Fruity/Sweet	Pleasant
Butylene oxide	0.07	0.71	0.71	Sweet/Alcohol	Pleasant
Butyraldehyde	<0.0046	0.0092	0.039	Sweet/Rancid	Unpleasant
Carbitol acetate	0.026	0.157	0.263	Sweet	Pleasant to unpleasant
Carbitol solvent	<0.21	1.10	1.10	Sweet/Musty	Neutral
Cellosolve acetate	0.056	0.138	0.250	Sweet/Musty	Pleasant
Cellosolve solvent	0.30	0.55	1.3	Sweet/Musty	Unpleasant to pleasant
Cumene	0.008	0.047	0.047	Sharp	Unpleasant
Cyclohexanone	0.12	0.12	0.24	Sweet/Sharp	Pleasant
Diacetone alcohol	0.28	1.1	1.7	Sweet	Unpleasant to pleasant
Di-*N*-butylamine	0.08	0.27	0.48	Fishy/Amine	Unpleasant to neutral

Dicyclopentadiene	0.011	0.020	0.020	Sweet/Sharp	Unpleasant
Diethylamine	0.02	0.06	0.06	Musty/Fishy/Amine	Unpleasant
Diethyl ethanolamine	0.011	0.04	0.04	Amine	Unpleasant
Diisobutyl carbinol	0.032	0.048	0.160	Sweet/Alcohol	Pleasant
Diisobutyl ketone	<0.11	0.31	0.31	Sweet/Ester	Pleasant
Diisopropylamine	0.13	0.38	0.85	Fishy/Amine/ Ammoniacal	Unpleasant to pleasant
Dimethyl ethanolamine	0.015	0.045	0.045	Amine	Unpleasant
1-4-Dioxane	0.80	1.8	5.7	Sweet/Alcohol	Pleasant
1-3-Dioxolane	16.9	64.0	128.0	Sweet/Musty	Neutral
Di-*N*-propylamine	0.02	0.10	0.10	Ammoniacal/Amine	Unpleasant to neutral
Ethyl acetate	6.3	13.2	13.2	Sweet/Ester	Pleasant
Ethyl acrylate	0.0002	0.00030	0.00036	Sour/Pungent	Unpleasant
Ethylamine (70-72% in water)	0.27	0.83	0.83	Sharp/Ammoniacal	Unpleasant
Ethylene	260	400	700	Olefinic	Unpleasant to neutral
Ethylenediamine	1.0	3.4	11.2	Ammoniacal/Musty	Unpleasant
Ethylene dichloride	6.0	40.0	40.0	Sweet	Unpleasant to neutral
Ethylene oxide	260	500	500	Sweet/Olefinic	Neutral
2-Ethylbutanol	0.07	0.77	0.77	Musty/Sweet	Neutral
2-Ethylhexanol	0.075	0.138	0.138	Musty	Unpleasant to pleasant
Ethylhexyl acetate	0.1	0.21	0.21	Sweet	Pleasant to neutral
2-Ethylhexyl acrylate	0.073	0.18	0.18	Musty/Sharp	Unpleasant
Ethylidene norbornene	0.02	0.073	0.073	Sweet/Aromatic	Unpleasant to pleasant
2-Ethoxy-3,4-dihydro-1,2-pyran	0.020	0.10	0.60	Sweet/Fruity	Pleasant
N-Ethyl morpholine	0.08	0.25	0.25	Ammoniacal	Unpleasant to pleasant
Glycol diacetate	0.093	0.312	0.312	Fruity/Acid	Pleasant
1-Hexanol	0.01	0.09	0.09	Sweet/Alcohol	Pleasant
Isobutanol	0.68	1.80	2.05	Sweet/Musty	Unpleasant to pleasant
Isobutyl acetate	0.35	0.50	0.50	Sweet/Ester	Pleasant
Isobutyl acrylate	0.002	0.009	0.012	Sweet/Musty	Unpleasant to pleasant
Isobutyl cellosolve	0.019	0.114	0.191	Sweet	Pleasant
Isobutyraldehyde	0.047	0.141	0.236	Sweet/Ester	Pleasant to unpleasant

TABLE 2-1 (*Continued*)

Compound	Absolute, ppm	50% Recognition, ppm	100% Recognition, ppm	Quality	Hedonic Tone
Isodecanol	0.020	0.031	0.042	Musty/Alcohol	Unpleasant to pleasant
Isopentanoic acid (mixed isomers)	0.005	0.015	0.026	Goaty	Unpleasant
Isophorone	0.20	0.54	0.54	Sharp	Unpleasant to pleasant
Isopropanol (anhydrous)	3.20	7.50	28.2	Sharp/Musty	Unpleasant
Isopropyl acetate	0.49	0.90	0.97	Sweet/Ester	Pleasant to unpleasant
Isopropylamine	0.21	0.71	0.95	Ammoniacal/Amine	Unpleasant to pleasant
Isopropyl ether	0.017	0.053	0.053	Sweet	Pleasant
Mesityl oxide	0.017	0.051	0.051	Sweet	Pleasant
Methanol	4.26	53.3	53.3	Sour/Sharp	Neutral
Methyl amyl acetate	<0.07	0.23	0.40	Sweet/Ester	Pleasant
Methyl amyl alcohol	0.33	0.52	0.52	Sweet/Alcohol	Unpleasant to pleasant
2-Methyl butanol	0.04	0.23	0.23	Sour/Sharp	Unpleasant to neutral
Methyl Cellosolve	<0.09	0.22	0.40	Sweet/Alcohol	Pleasant
Methyl Cellosolve acetate	0.34	0.64	0.64	Sweet/Ester	Pleasant
Methylethanolamine	1.0	3.4	3.4	Musty/Ammoniacal	Unpleasant
Methyl ethyl ketone	2.0	5.5	6.0	Sweet/Sharp	Neutral to unpleasant
2-Methyl-5-ethyl pyridine	0.006	0.008	0.010	Sour/Pungent	Unpleasant
Methyl isoamyl alcohol	0.07	0.20	0.20	Sweet/Pungent	Pleasant
Methyl isoamyl ketone	0.012	0.049	0.070	Sweet/Sharp	Pleasant
Methyl isobutyl ketone	0.10	0.28	0.28	Sweet/Sharp	Pleasant to unpleasant

Methyl methacrylate	0.05	0.34	0.34	Sweet/Sharp	Unpleasant
2-Methylpentaldehyde	0.09	0.136	0.136	Sweet/Rancid	Unpleasant
2-Methyl-1-pentanol	0.024	0.024	0.082	Sweet/Alcohol	Pleasant
α-Methyl styrene	0.052	0.156	0.156	Sweet/Aromatic	Pleasant
Morpholine	0.01	0.07	0.14	Fishy/Amine	Unpleasant
2,4-Pentanedione	0.01	0.020	0.024	Sour/Rancid	Unpleasant
n-Pentanol	0.21	0.31	0.31	Sweet/Alcohol	Pleasant
2-Picoline	0.014	0.023	0.046	Sweet	Unpleasant
n-Propanol	<0.03	0.08	0.13	Sweet/Alcohol	Pleasant
Propionaldehyde	0.009	0.040	0.080	Sweet/Ester	Pleasant
Propionic acid	0.028	0.034	0.034	Sour	Unpleasant
n-Propyl acetate	0.05	0.15	0.15	Sweet/Ester	Pleasant
Propylene	22.5	67.6	67.6	Aromatic	Neutral to pleasant
Propylenediamine	0.014	0.048	0.067	Sharp/Amine	Unpleasant
Propylene dichloride	0.25	0.50	0.60	Sweet	Pleasant
Propylene oxide	9.9	35.0	35.0	Sweet	Neutral to pleasant
Styrene	0.05	0.15	0.15	Sharp/Sweet	Unpleasant
Styrene oxide	0.063	0.40	0.40	Sweet	Pleasant
Tetraethyl ortho silicate	3.6	5.0	7.2	Sweet/Alcohol	Pleasant
Toluene	0.17	1.74	1.74	Sour/Burnt	Unpleasant to neutral
Triethylamine	<0.09	0.28	0.28	Fishy/Amine	Unpleasant to pleasant
Ucon-11 (trichloromonofluoromethane)	5.00	135.0	209.0	Sweet	Pleasant to unpleasant
Ucon-113 solvent (trichlorotrifluoroethane)	45.0	68.0	135.0	Sweet	Pleasant to unpleasant
Vinyl acetate	0.12	0.40	0.55	Sour/Sharp	Unpleasant
Xylene	0.08	0.27	0.27	Sweet	Neutral to pleasant

[a] Adapted from Hellman and Small.[51] In this table and in most later instances in this chapter, concentrations are in parts per million by volume.

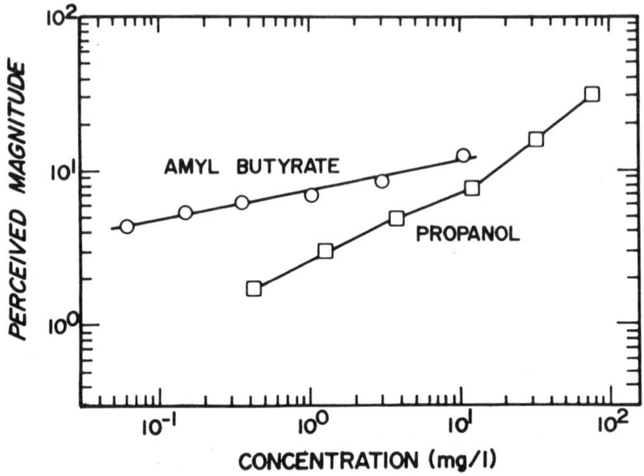

FIGURE 2-6 Odor intensity versus concentration for *n*-amyl butyrate and l-propanol. Reprinted with permission from Cain.[25]

Olfaction, therefore, rivals vision and hearing in its differential sensitivity to stimuli presented successively.

SUPRATHRESHOLD INTENSITY

The nonuniformities seen at just detectable or just recognizable odorant concentrations change with increases in concentration. A given increment in concentration may cause the odor magnitude of one substance to increase markedly and that of another to increase only slightly. Figure 2-6 depicts two psychophysical functions as an example. A 200-fold change in concentration caused a 15-fold change in the perceived magnitude of 1-propanol and only about a twofold change in the perceived magnitude of *n*-amyl butyrate.

Figure 2-7 provides additional examples of differences in the increase in odor intensity with concentration. These various psychophysical functions make it clear that some substances will resist odor abatement more strongly than others.

To a first approximation, odor intensity grows as a power function of concentration, i.e.,

$$\text{odor intensity} = k\,(\text{concentration})^n. \qquad (1)$$

In the logarithmic coordinates of Figures 2-6 and 2-7, such functions are plotted as straight lines, i.e.,

$$\log(\text{odor intensity}) = n \log(\text{concentration}) + \log k, \tag{2}$$

and the constant n (the exponent in Equation 1) equals the slope of the straight line. That constant has characteristically been between 0 and 1.0. Hence, even for compounds that produce relatively steep functions, odor intensity increases as a negatively accelerated function of concentration.[18,52] This rule, that the sense modality compresses the range of physical stimulation into a smaller range of perceived intensity, holds rather commonly throughout the sensory realm.[70]

Laffort and colleagues[61,62] have sought, with modest success, to predict relative exponents (i.e., relative values of n) from the same molecular characteristics that Laffort used to predict threshold. As in the case of the model for threshold, the predictive ability of the model for n will presumably increase as more data accumulate.

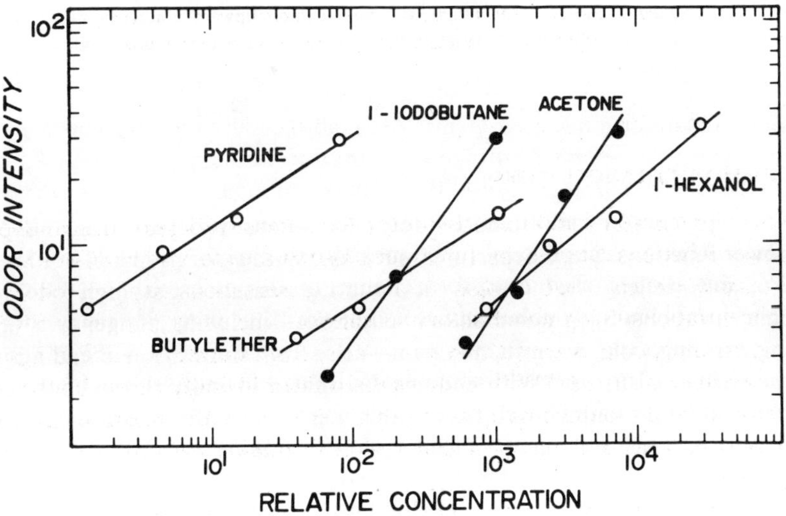

FIGURE 2-7 Psychophysical functions for various odorants. Data from Dravnieks and Laffort,[36] rescaled in accordance with the standardization procedure recommended by Moskowitz *et al.*[72]

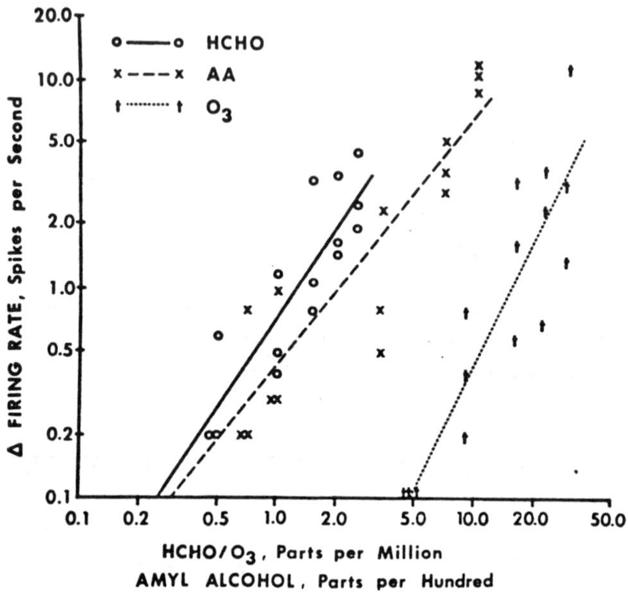

FIGURE 2-8 Response of the nasopalatine branch of the trigeminal nerve in the rat to concentrations of formaldehyde, amyl alcohol, and ozone. The straight lines represent power functions with exponents greater than 1.0. Such functions would exhibit upward concavity in linear coordinates. Reprinted with permission from Kulle and Cooper.[58]

COMMON CHEMICAL SENSE

Some psychophysical functions for odor intensity depart from simple power functions. Some departures, such as that seen for propanol in Figure 2-6, may reflect the addition of irritating sensations at high odorant concentrations. Such nonolfactory sensations—including pungency, stinging, burning, cold, warmth, and pain—arise from stimulation of endings of the trigeminal nerve.[14] With endings distributed liberally throughout both mucosal and nonmucosal tissue, this nerve mediates most cutaneous sensations arising from the facial region, including nasal, oral, and eye irritation.

Figure 2-8 displays the growth of activity in the trigeminal nerve as a function of the concentration of three inhaled substances. Note that the response increases rapidly with concentration—a finding consistent with psychophysical results obtained with various irritants.[54] Figure 2-9 (lower part) shows separate psychophysical functions for the odor and irritation evoked by 1-butanol.[22] Although irritation accounted for only a very small proportion of overall perceived magnitude at low concentrations, the irritating component increased more rapidly than odor with increases in concentration. At the highest test concentrations, irritation equaled odor in perceived magnitude.

Unless specifically instructed to exclude it from consideration, observers will incorporate irritation into their estimates of the perceived magnitude of odorants. Indeed, most odorants seem able to stimulate the trigeminal nerve, at least at high concentrations.[90]

Some of the more potent irritants include low-molecular-weight aldehydes, substances containing halogen atoms, and substances containing unsaturated carbon linkages.[1,71] Table 2-2 displays, in rank order, the

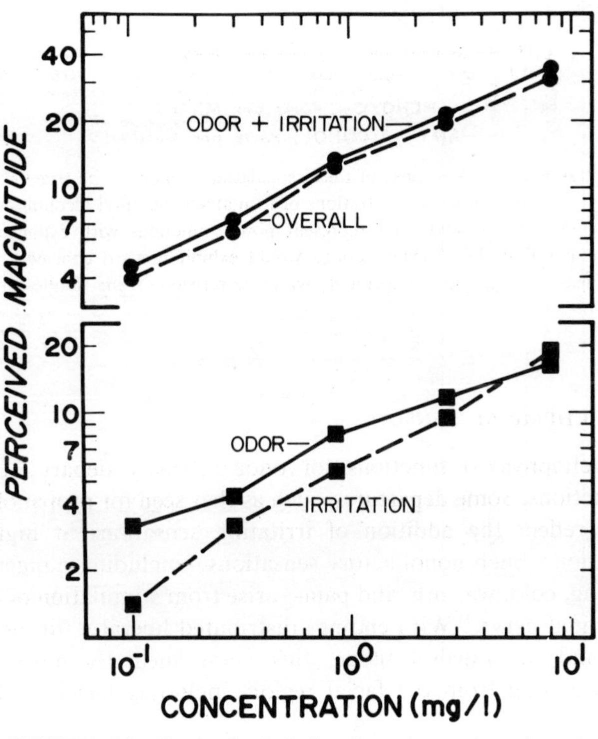

FIGURE 2-9 Psychophysical functions for various intensive attributes of *n*-butyl alcohol. In any given trial, subjects estimated two features of the stimulus: total intensive impact (denoted "overall") and magnitude of odor or magnitude of irritation. Also shown is a function for the algebraic sum of odor plus irritation. Reprinted with permission from Cain.[22]

TABLE 2-2 Mean Intensity Rating Scale Values (± SD) of Anosmic, Trigeminal-Focus, and Normal Experimental Groups[a]

Compound	Anosmic Group		Trigeminal-Focus Group		Normal Group[b]
	Proportion Detecting	Intensity	Proportion Detecting	Intensity	Intensity
1. Decanoic acid	0/15	0.00(0.00)	1/15	0.13(0.05)	4.07(2.14)
2. Vanillin	0/15	0.00(0.00)	0/15	0.00(0.00)	4.20(1.68)
3. Phenyl ethyl alcohol	1/15	0.13(0.50)	4/15	0.80(1.51)	4.40(1.96)
4. Eugenol	1/15	0.13(0.50)	2/15	0.67(1.85)	5.20(1.56)
5. Coumarin	2/15	0.13(0.34)	2/15	0.20(0.54)	4.60(1.36)
6. Nonane	3/15	0.27(0.57)	5/15	1.13(1.71)	4.53(2.25)
7. Octane	3/15	0.27(0.57)	4/15	1.07(2.02)	4.33(1.96)
8. Indole	3/15	0.53(1.20)	5/15	1.13(1.86)	4.60(1.99)
9. α-Terpineol	5/15	0.53(1.02)	7/15	1.20(1.47)	5.60(1.78)
10. Geraniol	2/15	0.60(1.54)	4/15	0.87(1.71)	5.13(1.31)
11. Heptanoic acid	5/15	0.87(1.45)	3/15	0.33(0.70)	4.80(2.01)
12. Limonene	6/15	0.93(1.44)	8/15	1.60(1.96)	5.40(1.86)
13. Hexanoic acid	7/15	0.93(1.39)	4/15	1.07(2.21)	5.33(1.78)
14. Heptane	5/15	1.00(1.86)	4/15	1.13(2.22)	4.67(2.15)
15. Benzyl acetate	7/15	1.40(2.12)	8/15	1.80(2.34)	4.87(2.03)
16. Methyl salicylate	9/15	1.60(1.86)	10/15	2.46(2.25)	6.27(1.88)
17. β-Ionone	9/15	1.93(2.21)	10/15	2.47(2.28)	4.47(2.31)
18. Anethole	8/15	2.73(2.86)	7/15	1.47(2.16)	5.93(1.06)
19. Heptyl alcohol	13/15	2.80(1.80)	9/15	1.93(1.88)	5.13(1.67)
20. Guaiacol	13/15	2.80(1.87)	9/15	2.73(2.77)	5.93(1.34)
21. Citral	12/15	2.87(2.25)	7/15	1.73(2.35)	5.53(1.75)
22. Camphor	14/15	3.53(2.09)	12/15	3.87(2.90)	6.00(1.51)

23. 4-Methyl valeric acid	9/15	3.93(3.68)	6/15	1.07(1.84)	6.20(2.43)
24. Linalool	13/15	4.00(2.37)	9/15	2.53(2.47)	6.00(1.82)
25. *n*-Butyl ether	13/15	4.00(2.10)	12/15	3.73(2.70)	6.53(1.41)
26. Valeric acid	15/15	5.00(2.16)	14/15	3.80(2.66)	6.00(2.22)
27. 2,4-Pentanedione	15/15	5.57(1.29)	14/15	5.27(2.65)	7.13(1.20)
28. Furfural	15/15	6.07(1.24)	14/15	5.33(2.55)	6.00(1.93)
29. Menthol	15/15	6.14(0.92)	14/15	5.80(2.20)	6.60(1.41)
30. iso-Amyl acetate	15/15	6.67(1.19)	13/15	5.73(3.02)	6.67(1.81)
31. *n*-Butyl alcohol	15/15	6.67(1.30)	14/15	5.87(3.01)	6.13(1.54)
32. Acetaldoxime	15/15	6.71(0.80)	14/15	5.93(2.32)	7.00(1.41)
33. 2-Heptanone	15/15	6.73(1.00)	15/15	6.80(2.34)	7.53(1.02)
34. iso-Valeric acid	15/15	6.73(1.24)	14/15	6.27(2.32)	7.47(1.26)
35. Ethyl benzene	15/15	6.87(2.00)	14/15	6.60(3.34)	6.73(1.24)
36. *n*-Butyl acetate	15/15	7.33(1.08)	13/15	5.93(2.93)	6.93(1.48)
37. Ethyl acetate	15/15	7.53(1.02)	15/15	7.40(1.93)	7.60(0.95)
38. Methanol	15/15	7.67(1.14)	15/15	6.80(2.23)	6.93(1.29)
39. Benzaldehyde	15/15	7.73(0.93)	15/15	7.87(1.36)	7.33(1.08)
40. Cyclohexanone	15/15	7.80(1.38)	14/15	6.27(2.54)	7.40(1.25)
41. Toluene	15/15	7.87(1.09)	14/15	7.13(2.60)	6.80(1.51)
42. Butyric acid	15/15	7.87(0.96)	15/15	7.00(2.42)	7.93(1.34)
43. Acetal	15/15	8.13(1.15)	14/15	7.87(2.28)	7.93(1.12)
44. Ethyl methyl ketone	15/15	8.40(0.61)	14/15	7.33(2.09)	8.40(0.71)
45. Pyridine	15/15	8.47(0.72)	15/15	8.13(2.00)	8.13(1.31)
46. Acetone	15/15	8.53(0.88)	15/15	8.13(1.41)	7.93(1.73)
47. Propionic acid	15/15	8.73(0.57)	15/15	8.27(1.73)	8.47(0.88)

[a] Reprinted with permission from Doty *et al.*[34] Means based on data from all observers, with zero values given to responses of persons who reported no stimulus detection. Compounds listed in order of increasing perceived trigeminal intensity for the anosmic observers.

[b] All 15 normals reported detection of all stimulants.

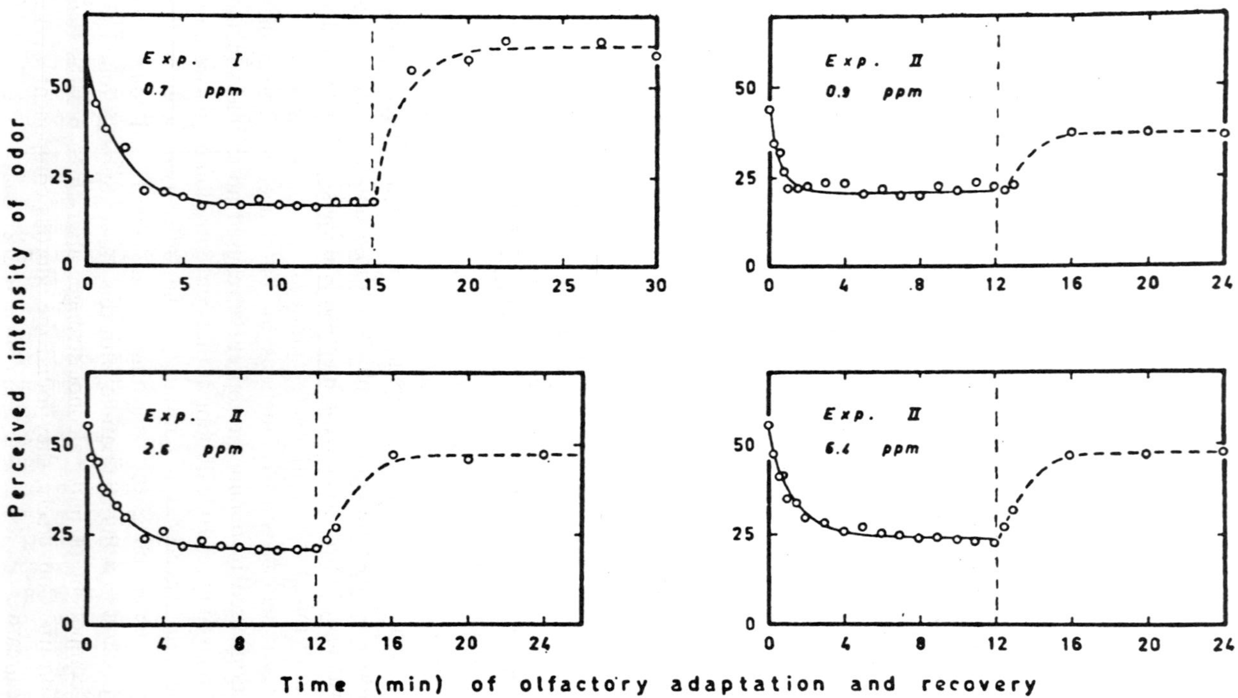

FIGURE 2-10 Adaptation to hydrogen sulfide at various concentrations. Continuous lines represent exponential decay. Dashed lines were drawn to indicate general trend of recovery process. Reprinted with permission from Ekman *et al.*[37]

trigeminal effectiveness of 47 odorous substances presented for evaluation at full strength to both normal subjects ("trigeminal focus group") and subjects with complete absence of olfaction ("anosmic group").[34] The final column of the table represents judgments obtained from a second group of normal subjects who were asked merely to judge overall intensity. For the more potent irritants in the list, the judgments of overall intensity seem to reflect primarily the magnitude of irritation. Precedence for this effect can be found in Katz and Talbert's observation:[54] "The odor of some irritants in higher concentrations is lost entirely in the pain of irritation in the nose."

With mixtures containing an irritant (carbon dioxide at concentrations above 10%) and a benign odorant (amyl butyrate at low concentrations), W. S. Cain (unpublished data) has verified mutual inhibitory interaction between odor and irritation. That is, an increase in perceived magnitude of odor will be associated with a decrease in irritation and vice versa. The finding will presumably generalize beyond the two stimuli in these mixtures, inasmuch as the inhibitory interaction occurs strongly even when the odorant is delivered to one nostril and the irritant to the other (dichorhinic mixtures). An implication of these results for environmental health is that the presence of high odor intensity may inhibit the perception of normally irritating and possibly harmful vapors, and the presence of moderate or high irritation may inhibit the perception of possibly toxic odorants or odorous warning agents.

TIME DEPENDENCE OF ODOR PERCEPTION

Perceived odor magnitude wanes rapidly during continuous stimulation. Figure 2-10 depicts the course of adaptation to hydrogen sulfide at various concentrations[57] These functions, typical of those obtained for various other substances,[23] imply that odor magnitude decays to approximately one-third of its initial value within a few minutes.

Figure 2-11 depicts similar functions for butyl acetate. The highest concentration, used to obtain the top curve, caused noticeable stinging, and the perceived magnitude of odor at that concentration decreased more slowly than the perceived magnitude at the other two. This finding suggests that the action of the trigeminal system may retard adaptation. Figure 2-12 shows results compatible with this view.[22] Subjects found the irritating component of butyl alcohol stronger after three breaths than after one breath. Conversely, they found the odorous component weaker after three breaths.

Olfactory adaptation operates on a short time scale. Note, for instance,

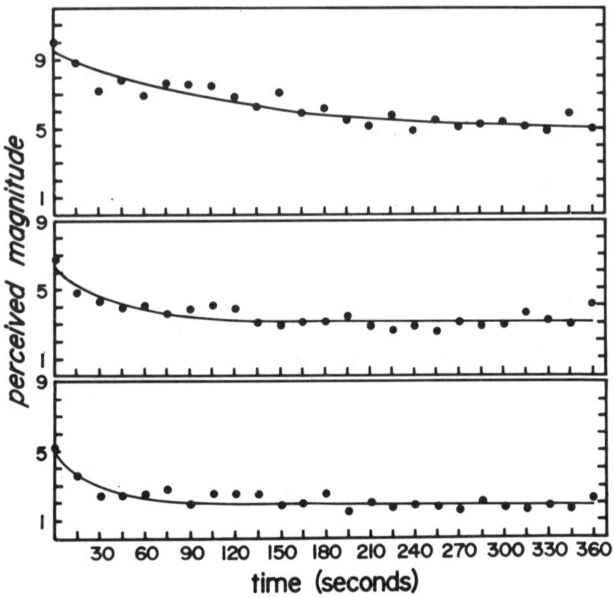

FIGURE 2-11 Course of adaptation to butyl acetate at three concentrations: 0.8 mg/L (bottom), 2.7 mg/L (middle), and 18.6 mg/L (top). From Cain.[23] Reprinted from *ASHRAE Transactions* by permission of the American Society of Heating, Refrigerating and Air-Conditioning Engineers, Inc.

that cessation of exposure to the stimulus leads to rapid recovery of sensitivity (Figure 2-10). Another time-dependent phenomenon, habituation, operates on a longer time scale. This phenomenon, of uncertain physiologic origin, involves a change in the effective impact of odorous vapors. The phenomenon reveals itself in industrial situations, where workers report that an initially repulsive odor eventually seems less repulsive. Habituation has been found in laboratory experiments with animals,[29] human infants,[41,42] and human adults.[26] It occurs with both pleasant and unpleasant odors, but much about its magnitude, limits, and course remains unknown. Hence, anecdotal reports, such as the following (from an interview with a mechanic in a rendering plant), dominate its description:[88 (p. 111)]

The odor was terrible, but I got used to it. It was less annoying when you stayed right in it. When you left for a week or so, a vacation, you had to come back and get used to the thing all over again. I've had people that say, "How do you stand it?" I say it's like anything else. I don't say you get exactly used to it, but it does get less annoying in time.

ODOR MIXTURES

Regarding odor mixtures, Zwaardemaker[98 (p. 220)] observed that "purely olfactive stimuli are liable to mutual weakening." Little has ever contradicted this observation that an odor mixture smells less intense than would be expected on the basis of the sum of its unmixed components, i.e., displays hypoadditivity.[11,12,47,53] Only very rare exceptions have occurred.[57] Hypoadditivity seems to be due to more than physical or neural interactions at the olfactory mucosa. Dichorhinic mixtures, in which one component is inhaled through one nostril and the other component through the other nostril, also exhibit hypoadditivity.[19,97] Figure 2-13 offers

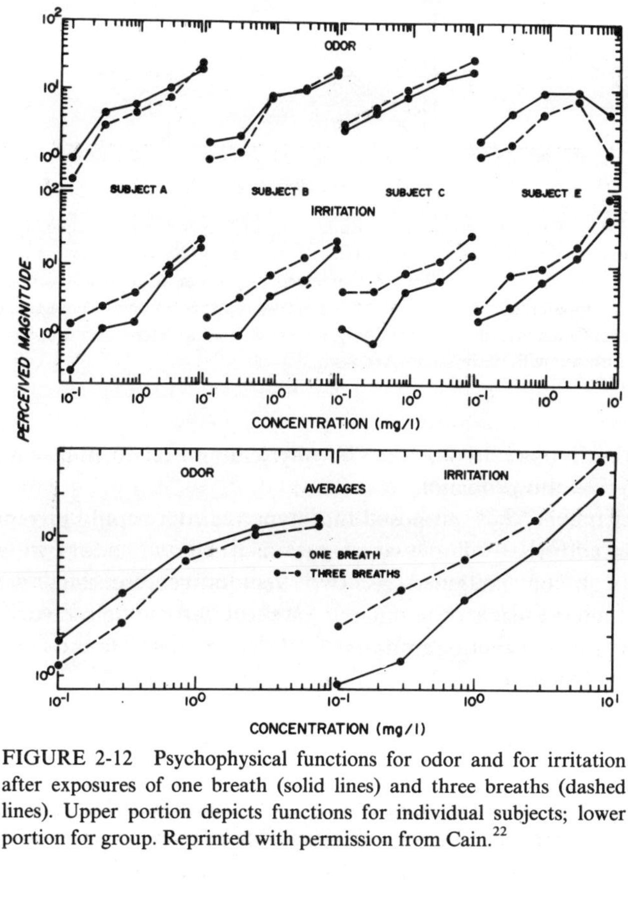

FIGURE 2-12 Psychophysical functions for odor and for irritation after exposures of one breath (solid lines) and three breaths (dashed lines). Upper portion depicts functions for individual subjects; lower portion for group. Reprinted with permission from Cain.[22]

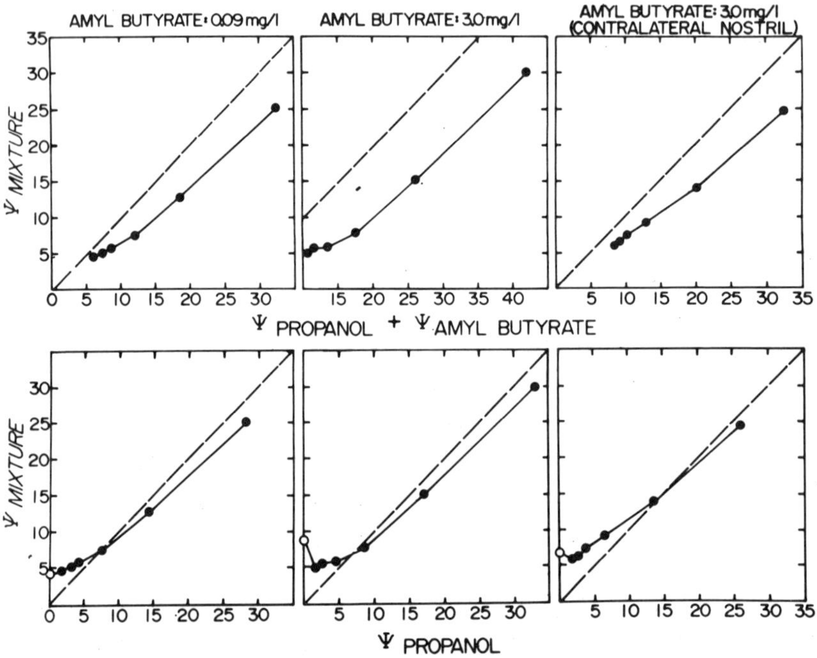

FIGURE 2-13 Top, perceived magnitude (ψ) of mixtures of propanol at various concentrations and amyl butyrate at a fixed concentration versus the sum of the perceived magnitudes of the components. Bottom, perceived magnitude of mixtures versus perceived magnitude of propanol smelled alone. Unfilled circles represent perceived magnitude of amyl butyrate alone. Concentrations of amyl butyrate shown at top refer to both upper and lower portions. Reprinted with permission from Cain.[19]

examples of hypoadditivity for both physical and dichorhinic mixtures of amyl butyrate and propanol.

Zwaardemaker[96 (p. 502)] proposed that hypoadditivity could be represented by vector addition: "The two sensations [i.e., two odors in a binary mixture] can be imagined as two vectors representing two forces counterbalancing each other in our intellect." Although Zwaardemaker never used this concept quantitatively, Berglund *et al.*[12] found considerable merit in the formula,

$$\psi_{ab} = (\psi_a{}^2 + \psi_b{}^2 + 2\,\psi_a \psi_b \cos \alpha)^{1/2}, \tag{3}$$

where ψ_{ab} is the perceived magnitude of the mixture and ψ_a and ψ_b are the perceived magnitudes of the unmixed components. The constant α, commonly estimated from a portion of the data, represents the angle between the vectors ψ_a and ψ_b. Figure 2-14 shows how well the vector formula described the perceived magnitude of mixtures of amyl butyrate and propanol. The various symbols represent different experiments, such as the three experiments represented in the three pairs of curves in Figure 2-13. Berglund[10] has found that the formula also accounts moderately well for the magnitude of mixtures of more than two components (Figure 2-15). Generally, however, the formula tends to overestimate the magnitude of complex mixtures.

ODOR MASKING AND COUNTERACTION

The study of binary mixtures has revealed that a mixture may smell less intense than its stronger component smells alone. The reduction in intensity, although hardly dramatic in the cases reported so far, lends

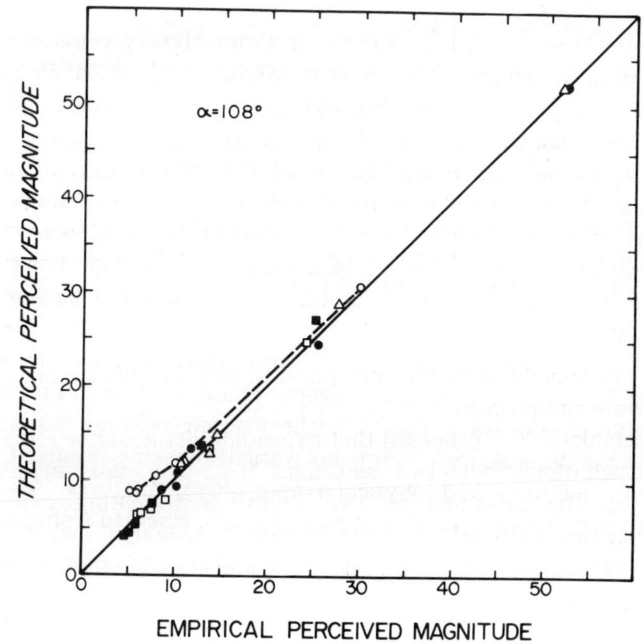

FIGURE 2-14 Theoretical perceived magnitude of both physical and dichorhinic mixtures of propanol and amyl butyrate versus empirical perceived magnitude. Theoretical values were derived from a vector summation model with an angle of 108° between vectors. Reprinted with permission from Cain.[19]

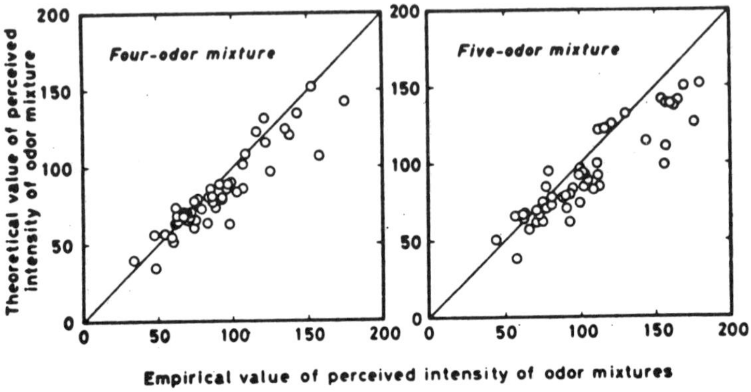

FIGURE 2-15 Theoretical perceived magnitude of mixtures of hydrogen sulfide, dimethyl disulfide, pyridine, and dimethyl sulfide (left) and these four substances plus methyl mercaptan (right) versus empirical perceived magnitude. Reprinted with permission from Berglund.[10]

credence to claims of odor counteraction. Commercially available counteractants usually comprise unreactive mixtures of essential oils and fragrance chemicals, but sometimes contain in addition a patented "active ingredient" of more or less unproven worth. The formulation normally has a pleasant odor (e.g., pine or floral) or a "sanitary" odor. The addition of the proper amount of the counteractant to malodorous air can take advantage of the hypoadditivity of odor mixtures and can thereby lead to some reduction in overall odor magnitude. In addition, however, the counteractant adds its own acceptable odor quality and may mask the presence of the malodor.

Figure 2-16 gives an example of how amyl butyrate masked the perceived "propanol component" in vapor-phase mixtures of propanol and amyl butyrate. For each function, the amount of amyl butyrate was constant and the amount of propanol varied. Masking occurred in both dichorhinic mixtures and physical (vapor-phase) mixtures. Aside from such a finding, no general rules of masking have emerged from laboratory investigations.

ODOR PLEASANTNESS

Up to this point, the discussion of odor perception has focused on the quantification of odor intensity under various conditions of stimulation. Odor pleasantness and quality also permit a degree of quantification. For example, Figure 2-17 shows how both perceived intensity and perceived

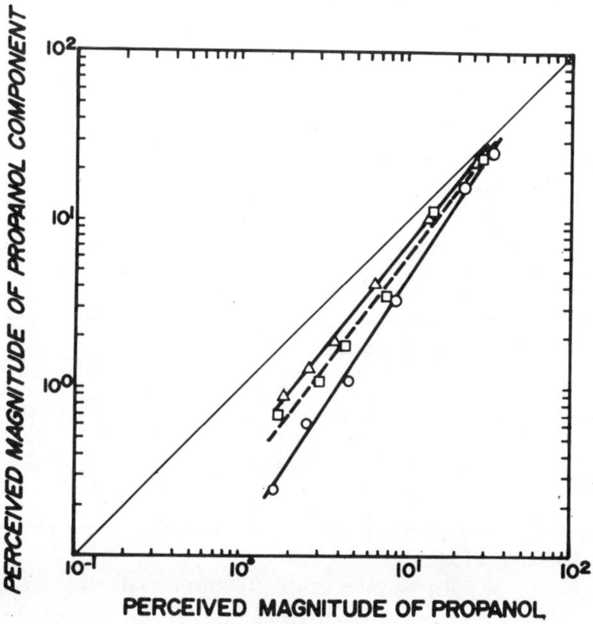

FIGURE 2-16 Perceived magnitude of apparent propanol component in mixtures of propanol and amyl butyrate versus perceived magnitude (overall intensity) of propanol smelled alone. Circles and squares represent results obtained with physical mixtures; triangles represent results obtained with dichorhinic mixtures; see Figure 2-13. Reprinted with permission from Cain.[19]

pleasantness of 31 odorants vary with concentration.[73] With few exceptions, pleasantness decreased as intensity increased. An equation relating pleasantness, P, to concentration, C, took the form:

$$P = k_1 + k_2 C^{k_3}, \tag{4}$$

The various determinants of odor pleasantness are only poorly specified. Steiner[85] has argued that facial expressions reveal clear positive and negative affective responses to odorants, even in the newborn. This view implies inherent, biologic determinants. Engen[40] has observed, however, that young children (e.g., 3-yr-olds) are indifferent to most odors, but exhibit an increasing range of pleasantness-unpleasantness with age. This view suggests, although it does not prove, that learning and culture may participate in the development of olfactory hedonics. Nevertheless, the affective habituation that workers may experience with relation to

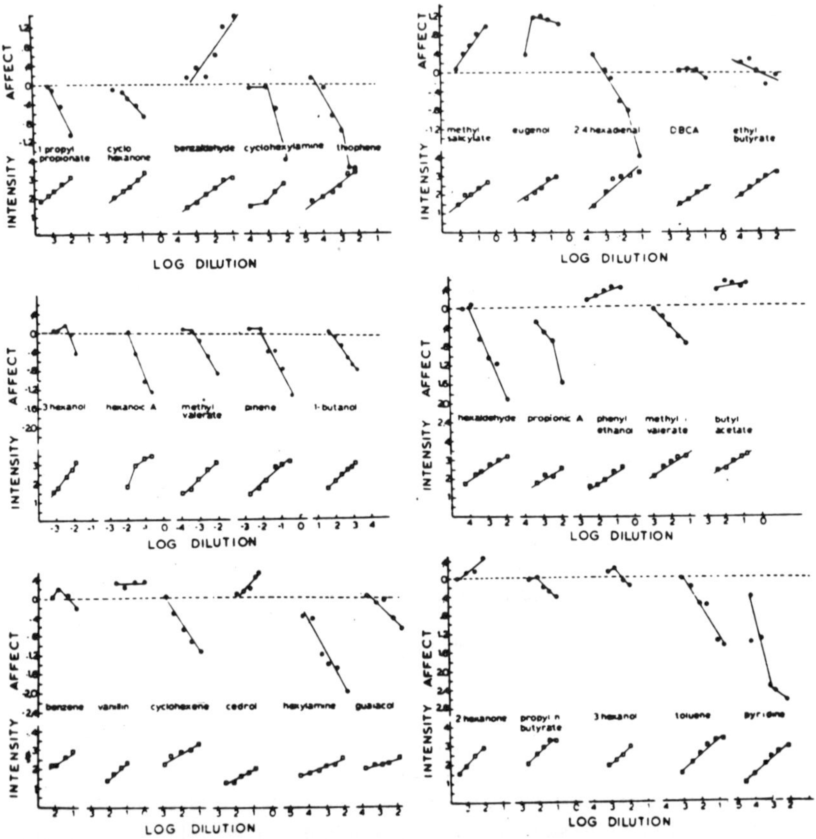

FIGURE 2-17 Relation between odor pleasantness and concentration (dilution) and between odor intensity and concentration (dilution) of 31 odorants. Dashed lines represent hedonic neutrality. Reprinted with permission from Moskowitz *et al.*[73]

unpleasant job-related odors implies at least some plasticity in affective reactions to olfactory stimulation.

ODOR QUALITY

Odor quality (character) can be quantified through procedures of multidimensional scaling applied to psychophysical judgments (e.g., numerical ratings) of qualitative similarity.[82] Often a two- or three-dimensional solution will account for the perceived differences in quality among a large battery of odorants. Figure 2-18 shows a two-dimensional

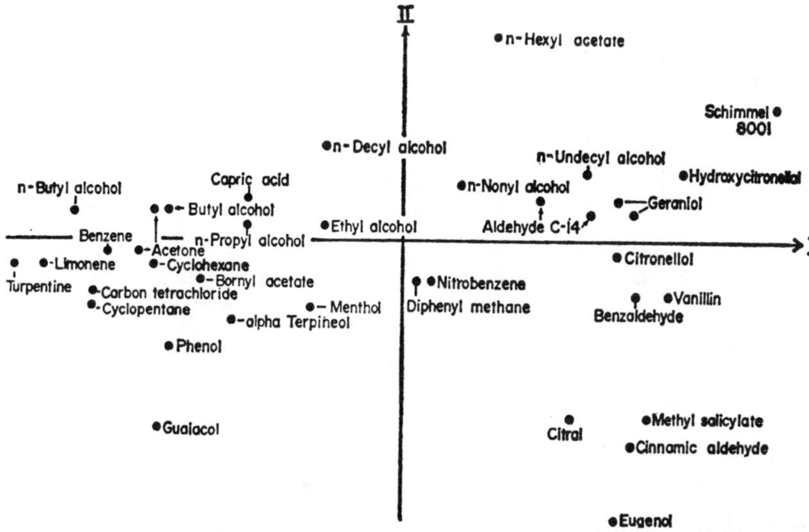

FIGURE 2-18 Two-dimensional representation of the psychologic distances among the qualities of various odorants. The multidimensional solution does not require the experimenter to name the dimensions. In this solution, dimension I seemed related to the pleasantness of the odorants. Dimension II admitted to a less certain interpretation. Reprinted with permission from Schiffman.[81] Copyright 1974 by the American Association for the Advancement of Science.

"psychologic map" for 50 odorants.[81] Figure 2-19, which replaces the names of the substances in Figure 2-18 with their molecular formulas, shows one step in a search for physicochemical correlates of quality. Construction of three-dimensional models of the molecules can represent another step—a step designed to search for stereochemical correlates. The use of weighted combinations of physicochemical variables (e.g., molecular weight, number of double bonds, and presence of particular functional groups or nuclei) provides yet another step and, when successful, can permit a reconstruction of the psychologic space from physicochemical variables. As more data accumulate, this approach may offer some insight into the properties that endow a molecule with a particular odor quality.

Multidimensional scaling offers only one of many approaches to the study of odor quality. Specialists in structure-activity relationships in olfaction often perform painstaking experiments on how subtle changes in molecular structure alter odor quality or, perhaps, maintain quality and alter intensity. These experiments may require the investigators to synthesize new molecules or to purify samples rigorously. Only recently have techniques to assess purity made it possible to decide with certainty

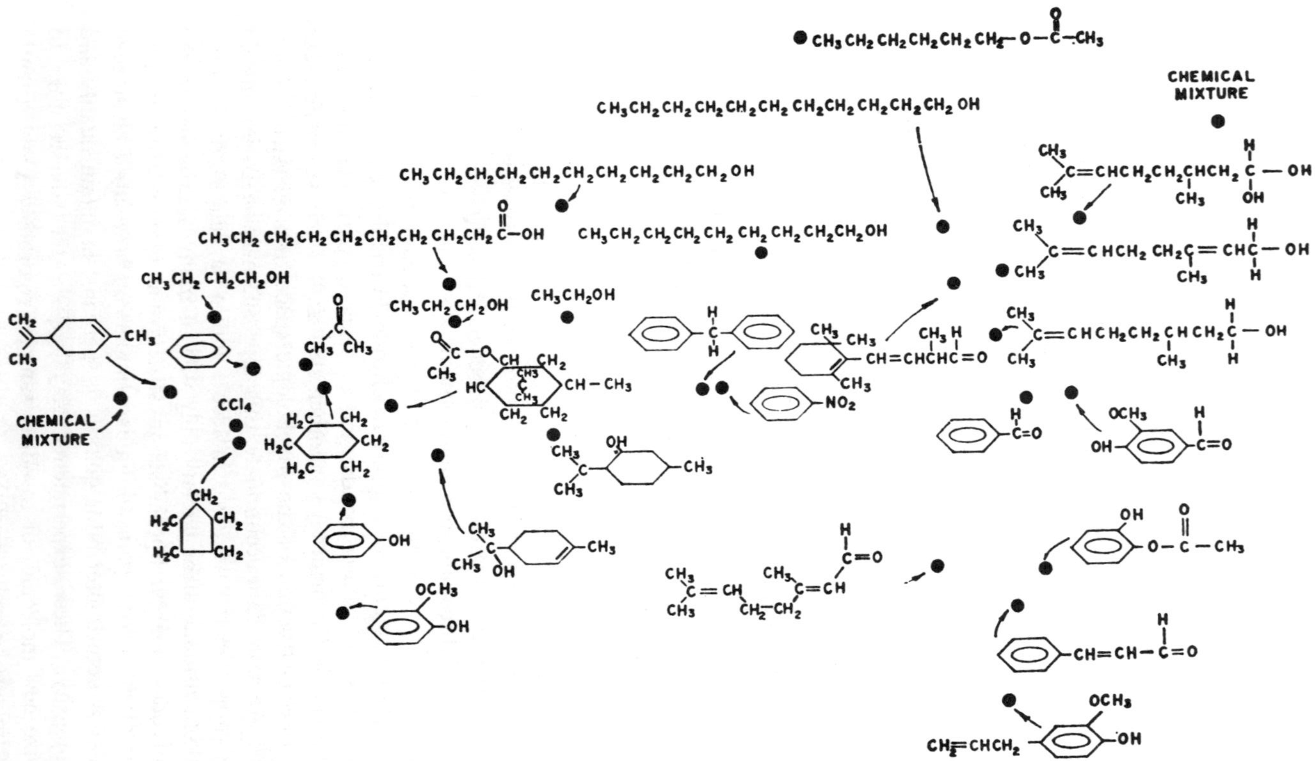

FIGURE 2-19 Molecular formulas associated with the odorants shown in Figure 2-18. Reprinted with permission from Schiffman.[8] Copyright 1974 by the American Association for the Advancement of Science.

whether some pairs of optical isomers (e.g., *d*- and *l*-carvone) produce the same or different qualities.[48,66,80] The finding that highly purified *d*-carvone smells like caraway and highly purified *l*-carvone like spearmint implies some chiral specificity in receptor sites and seems to rule out some theories of odor quality. Nevertheless, not all enantiomeric pairs produce different qualities—a finding that complicates the quest to discover which properties endow a molecule with a given quality (see Beets[7]).

Although subtle differences in structure may sometimes cause large alterations in quality, reasonably large differences may leave odor quality unaltered.[8] To cite one of many examples, nitrobenzene and benzaldehyde both evoke the odor of burnt almonds. There is no certainty that such structurally different molecules interact with the same receptor sites.[9] Because of convergence and divergence of neurons at the various levels of the olfactory pathways, different patterns of activity at the most peripheral level (i.e., in first-order neurons) may give rise to the same sensation.[68] Hence, the search for the molecular correlates of odor quality must stretch beyond the study of mere perceived similarity. One notable extension involves investigation of specific anosmia, a putatively congenital insensitivity to one or more odorants.[4,95] The investigations generally chart the range and the magnitude of the insensitivity. In theory, the outcome can uncover pivotal structural similarities.[3] Similar reasoning provides an incentive for investigations of cross-adaptation, a temporary desensitization to one odorant or a range of odorants after exposure to some adapting odorant.[17,56] Only the application of a variety of techniques, including some not highlighted in this brief account, will provide the converging operations necessary to develop a full, predictive theory of odor quality. The recent emphasis on collection of data, rather than the common theory-spinning of the past, is a sign of maturity in odor science.

ORGANISMIC VARIABLES

The factors of age, sex, and cigarette-smoking exert some influence on olfactory sensitivity, but the effects are surprisingly small.[39,94] Hence, although age eventually takes its toll on olfactory sensitivity, the decline shows up only statistically and, even then, markedly only in the elderly.[83]

Females (young adults) have displayed slightly higher sensitivity than males in some, but not all, studies.[55] Females may possess an advantage over males for only some odorants. Evidence of an odorant-specific phenomenon has come from studies of how sensitivity varies through the menstrual cycle, reaching a maximum at approximately the time of ovulation and a minimum during the menses.[49] Whereas previous experiments had produced mixed results regarding the reality of this variation, Mair *et al.*[69] have implicated the tendency of molecules to diffuse through the mucus barrier as a determining feature: sensitivity oscillates for odorants that diffuse slowly, but not for odorants that diffuse rapidly. The physicochemical characteristics, thickness, and differential filtering of mucus might account similarly for why cigarette-smokers may appear slightly less sensitive in some studies but not in others.[39] These various investigations, although hardly of uniform merit methodologically, teach a valuable lesson collectively—viz., charting olfactory sensitivity with merely one odorant chosen by convenience or availability may be misleading with regard to the reality, specificity, and magnitude of the dependence of sensitivity on one or another organismic variable.

COGNITIVE FUNCTIONING

Because the sense of smell serves as a channel of information, it is relevant to consider how human beings store, process, and retrieve this information. Research on this matter, only now gaining momentum, has uncovered a very consistent and easily summarized pattern:

- Persons can recognize, but not necessarily name, previously smelled odorants over very long intervals.[44,64] That is, recognition memory decays very slowly—more slowly than for most other sensory stimuli (see Figure 2-20). In fact, Engen and Ross[44] found little difference in odor recognition between intervals of 1 day and 1 yr. This seems to verify the anecdotal observation, relatively common in literature, that an adult will recognize an aroma not experienced since childhood. Indeed, such writers as Proust, Huysmans, and Nabokov have noted that nothing can revive the past so completely as a seemingly forgotten aroma.
- Persons (young adults) learn to name odors only very slowly (Figure 2-21).[24,30,31] Once learned, however, the association between an odor and a name or label resists easy replacement.[65] In the language of learning theory, the association exhibits little sensitivity to retroactive interference by a second label.
- Persons can usually identify (i.e., name) the odors of only about half of commonly smelled substances, such as popcorn, beer, chocolate, lemon, and rubber.[27,32,43,87] When faced with the task of odor identification, subjects often report themselves in a "tip-of-the-tongue" state. That is, they recognize the odor as familiar, but cannot retrieve its name.[65] When prompted out of this state (i.e., reminded of the correct label), they can identify many substances very accurately.[27,32]

The data on cognitive processing of olfactory information imply a weak

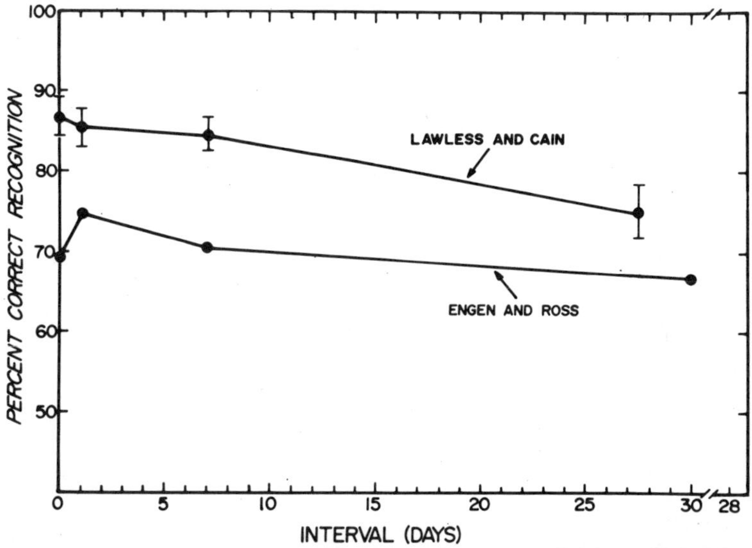

FIGURE 2-20 Percent correct recognition in a two-alternative forced-choice task versus interval after a single exposure to various test odorants. Lower function represents results obtained by Engen and Ross;[44] upper function, results obtained by Lawless and Cain.[64] The conditions in the two studies were similar, but not identical. The main point of interest is that both functions decline very slowly. The points plotted at day *zero* were obtained from recognition tests performed only a few minutes after initial exposure. Reprinted with permission from Cain.[24]

connection between odors and language.[38] That is, as mentioned above, it takes a long time to learn a verbal label for an odor and a long time to dislodge an old label with a new one. Even a familiar label may fail to get past the "tip-of-the-tongue" state.[38] This situation may arise from what could be termed the low evident intrinsic structure of odors. To illustrate, the odor of coffee offers little information regarding the many chemical constituents of the stimulus. By comparison, a picture of an object has high evident intrinsic structure (lines, angles, size, shape, color, and shading). In contrast with most visual or auditory stimuli, olfactory stimuli must be encoded cognitively as units (i.e., wholes), rather than as a collection or pattern of features. Such holistic encoding may explain why olfaction has never given rise to its own glossary. That is, there are no verbal descriptors of odors *per se*. Odor descriptors are derived from the stimulus objects for smell (e.g., musky and fruity) or are borrowed from other senses (e.g., sweet).

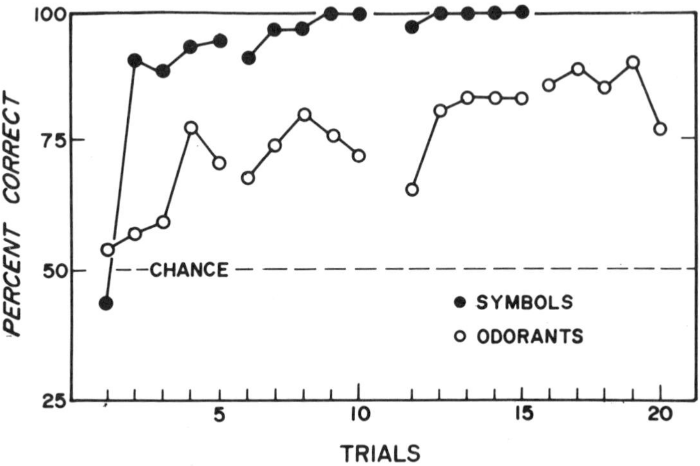

FIGURE 2-21 Percent correct versus number of trials in a paired-associate learning task where the stimuli were four schematic symbols (i.e., visual patterns) or four odorants and the responses were the numerals 1 through 4. A rest interval of 5 min intervened between trials 5 and 6 and between trials 15 and 16; 24 h intervened between trials 10 and 11. Note the clear superiority of performance with the visual stimuli. Reprinted with permission from Cain;[24] modified from Davis.[31]

CONCLUSIONS

Olfactory psychophysics has specified various quantitative rules that characterize the perception of odors. The laboratory data are relevant to the issue of odor pollution insofar as they set general expectations about odor perception in field situations. For instance, knowledge of the immense sensitivity of olfaction for some odorants, but not others, makes it possible to understand why some effluents seem perceptible at much greater distances than others. Knowledge that odor intensity grows more slowly with concentration for some odorants than for others makes it possible to understand why a relatively mild-smelling effluent may show less attenuation with distance than some relatively strong-smelling effluents. Knowledge that most odorants can appeal to the trigeminal nerve makes it possible to understand why an effluent with few if any corrosive constituents can still evoke complaints of nasal, eye, and throat irritation. Knowledge that common chemical sensations exhibit slower adaptation, but that odor and irritation display inhibitory interaction, makes it possible to understand why discomfort may remain even after the odor quality of an effluent has faded. Knowledge that mixtures display hypoadditivity makes it possible to understand why elimination of some fraction of the constituents in an effluent may cause no discernible reduction in odor intensity. The principle of hypoadditivity also makes it possible to understand why the addition of a "counteractant" may cause no discernible increase in odor intensity, whereas the counteractant may, through masking, provide a more acceptable odor quality.

No matter how good the laboratory data, however, their predictive power remains limited, and that limitation gives rise to the need for field-based psychophysics.[67] But the same holds true in other realms. For example, the results from the psychoacoustics laboratory, although more highly refined than those from the odor laboratory, have offered only rather imprecise predictive power regarding the impact of nondeafening environmental noise. There is a similar lack of predictive precision in the study of photochemical smog and other pollutants.[28] Nevertheless, odor pollution presents the added complication that even barely perceptible and barely measurable amounts of some unpleasant-smelling odorants will give rise to concern. An unpleasant-smelling effluent may arouse mere uneasiness as a person drives through an industrial park, but may arouse alarm at the dinner table. Hence, the ever-present affective impact of odors depends heavily, but in a poorly predictable way, on context. Even when the context mutes affective impact, however, persons may view unpleasant-smelling air as potentially hazardous. With this in mind, ventilation and air-conditioning engineers have typically set ventilation at rates that will combat virtually all indoor malodors.[16] The prevailing rule in buildings equipped with mechanical ventilation has been that the only acceptable odor is no odor at all.[21]

REFERENCES

1. Alarie, Y. Sensory irritation of the upper airways by airborne chemicals. Toxicol. App. Pharmacol. 24:279–297, 1973.

2. Allison, A. C., and R. T. T. Warwick. Quantitative observations on the olfactory system of the rabbit. Brain 72:186–197, 1949.

3. Amoore, J. E. Evidence for the chemical olfactory code in man. Ann. N.Y. Acad. Sci. 237:137–143, 1974.

4. Amoore, J. E. Molecular basis of odor. Springfield, Ill.: Charles C Thomas, Publisher, 1970.

5. Andersen, P. Inhibitory reflexes elicited from the trigeminal and olfactory nerves in rabbits. Acta Physiol. Scand. 30:137–148, 1954.

6. Batsel, H. L., and A. J. Lines. Neural mechanisms of sneeze. Amer. J. Physiol. 229:770–776, 1975.

7. Beets, M. G. J. Pharmacological aspects of olfaction, pp. 445–472. In D. G. Moulton, A. Turk, and J. W. Johnston, Jr., Eds. Methods in Olfactory Research. London: Academic Press, 1975.

8. Beets, M. G. J. Structure-response relationships in chemoreception, pp. 225–295. In C. J. Cavallito, Ed. Section 5. Structure–Activity Relationships. Vol. 1. International Encyclopedia of Pharmacology and Therapeutics. Oxford: Pergamon Press, 1973.

9. Beets, M. G. J., and E. T. Theimer. Odour similarity between structurally unrelated odorants, pp. 313–321. In G. E. W. Wolstenholme and J. Knight, Eds. Taste and Smell in Vertebrates. London: J. & A. Churchill, 1970.

10. Berglund, B. Quantitative and qualitative analysis of industrial odors with human observers. Ann. N.Y. Acad. Sci. 237:35–51, 1974.

11. Berglund, B., U. Berglund, and T. Lindvall. On the principle of odor interaction. Acta Psychol. 35:255–268, 1971.

12. Berglund, B., U. Berglund, T. Lindvall, and L. T. Svensson. A quantitative principle of perceived intensity summation in odor mixtures. J. Exp. Psychol. 100:29–38, 1973.

13. Brown, J. L., Ed. Sensory systems, pp. 8-1—8-165. In J. R. Brobeck, Ed. Best & Taylor's Physiological Basis of Medical Practice. Baltimore: The Williams & Wilkins Company, 1973.

14. Cain, W. S. Contribution of the trigeminal nerve to perceived odor magnitude. Ann. N.Y. Acad. Sci. 237:28–34, 1974.

15. Cain, W. S. Differential sensitivity for smell: "Noise" at the nose. Science 195:796–798, 1977.

16. Cain, W. S. Minimum rates of air intake into buildings: Role of odors. In J. A. J. Stolwijk, Ed. Energy Conservation Strategies in Buildings: Comfort, Acceptability, and Health. New Haven: Yale University Press. (in press)

17. Cain, W. S. Odor intensity after self-adaptation and cross-adaptation. Percept. Psychophys. 7:271–275, 1970.

18. Cain, W. S. Odor intensity: Differences in the exponent of the psychophysical function. Percept. Psychophys. 6:349–354, 1969.

19. Cain, W. S. Odor intensity: Mixtures and masking. Chem. Sens. Flav. 1:339–352, 1975.

20. Cain, W. S. Odor magnitude: Coarse vs. fine grain. Percept. Psychophys. 22:545–549, 1977.

21. Cain, W. S. Odors and ventilation requirements. In P. E. McNall and M. Reppert, Eds. Ventilation vs. Energy Conservation in Buildings. Proceedings of the Engineering Foundation Conference, New England College, Henniker, New Hampshire, July 25–29, 1977. N.B.S. Special Publication. Washington, D.C.: National Bureau of Standards. (in press)

22. Cain, W. S. Olfaction and the common chemical sense: Some psychophysical contrasts. Sens. Proc. 1:57–67, 1976.

23. Cain, W. S. Perception of odor intensity and the time-course of olfactory adaptation. Trans. Amer. Soc. Heat. Refrig. Air-Cond. Eng. 80:53–75, 1974.

24. Cain, W. S. Physical and cognitive limitations on olfactory processing in human beings, pp. 287–302. In D. Muller-Schwarze and M. H. Mozell, Eds. Chemical Signals in Vertebrates. New York: Plenum Press, 1977.

25. Cain, W. S. The odoriferous environment and the application of olfactory research. Chapter 10 in E. C. Carterette and M. P. Friedman, Eds. Handbook of Perception, Vol. 6A, Tasting and Smelling. New York: Academic Press, 1978.

26. Cain, W. S., and F. Johnson, Jr. Lability of odor pleasantness: Influence of mere exposure. Perception 7:459–465, 1978.

27. Cain, W. S., and R. J. Krause. Olfactory testing: Rules for odor identification. Neurolog. Res. (in press)

28. Copley International Corp. Survey of eye irritation and lacrymation in relation to air pollution, by K. Wilson. Final report to Coordinating Research Council, Inc. Contract No. CAPM-17-71 (1-73), 1974.

29. Cornwell, C. A. Golden hamster pups adapt to complex rearing odors. Behav. Biol. 14:175–188, 1975.

30. Davis, R. G. Acquisition and retention of verbal associations to olfactory and abstract visual stimuli of varying similarity. J. Exp. Psychol.: Hum. Learn. Mem. 3(1):37–51, 1977.

31. Davis, R. G. Acquisition of verbal associations to olfactory stimuli of varying familiarity and to abstract visual stimuli. J. Exp. Psychol.: Hum. Learn. Mem. 104:134–142, 1975.

32. Desor, J. A., and G. K. Beauchamp. The human capacity to transmit olfactory information. Percept. Psychophys. 16:551–556, 1974.

33. de Vries, H., and M. Stuiver. The absolute sensitivity of the human sense of smell, pp. 159–167. In W. A. Rosenblith, Ed. Sensory Communication: Contributions to the Symposium on Principles of Sensory Communication, July 19–August 1, 1959, Endicott House, M.I.T. Cambridge, Mass.: MIT Press, 1961.

34. Doty, R. L., W. E. Brugger, P. C. Jurs, M. A. Orndorff, P. J. Snyder, and L. D. Lowry. Intranasal trigeminal stimulation from odorous volatiles: Psychometric responses from anosmic and normal humans. Physiol. Behav. 20:175–185, 1978.

35. Dravnieks, A. Threshold of smell and measurement, pp. 27–44. In P. N. Cheremisinoff and R. A. Young, Eds. Industrial Odor Technology Assessment. Ann Arbor: Ann Arbor Science Publishers, Inc., 1975.

36. Dravnieks, A., and P. Laffort. Physico-chemical basis of quantitative and qualitative odor discrimination in humans, pp. 142–148. In D. Schneider, Ed. Olfaction and Taste. IV. Stuttgart: Wissenschaftliche Verlagsgesellschaft MBH, 1972.

37. Ekman, G., B. Berglund, U. Berglund, and T. Lindvall. Perceived intensity of odor as a function of time of adaptation. Scand. J. Psychol. 8:177–186, 1967.

38. Engen, T. Odor memory. Perfum. Flav. 2:8–11, 1977.

39. Engen, T. Taste and smell, pp. 554–561. In J. E. Birren and K. W. Schaie, Eds. Handbook of the Psychology of Aging. New York: Van Nostrand Reinhold Co., 1977.

40. Engen, T. The potential usefulness of sensations of odor and taste in keeping children away from harmful substances. Ann. N.Y. Acad. Sci. 237:224–228, 1974.

41. Engen, T., and L. P. Lipsitt. Decrement and recovery of responses to olfactory stimuli in the human neonate. J. Comp. Physiol. Psychol. 59:312–316, 1965.

42. Engen, T., L. P. Lipsitt, and H. Kaye. Olfactory responses and adaptation in the human neonate. J. Comp. Physiol. Psychol. 56:73–77, 1963.

43. Engen, T., and C. Pfaffmann. Absolute judgments of odor quality. J. Exp. Psychol. 59:214–219, 1960.

44. Engen, T., and B. M. Ross. Long-term memory of odors with and without verbal descriptions. J. Exp. Psychol. 100:221–227, 1973.

45. Ewert, G. On the mucus flow rate in the human nose. Acta Otolaryngol. [suppl.](Stockholm) 200:1–62, 1965.

46. Fazzalari, F. A., Ed. Compilation of Odor and Taste Threshold Values Data. Sponsored by ASTM Committee E-18 on Sensory Evaluation of Materials and Products. ASTM Data Series DS#48A. Philadelphia: American Society for Testing and Materials, 1978. 497 pp.

47. Foster, D. Odors in series and parallel, pp. 1–6. Proceedings of the Scientific Section, Toilet Goods Association, No. 39, May 1963.

48. Friedman, L., and J. G. Miller. Odor incongruity and chirality. Science 172:1044–1046, 1971.

49. Good, P. R., N. Geary, and T. Engen. The effect of estrogen on odor detection. Chem. Sens. Flav. 2:45–50, 1976.

50. Goss, C. M., Ed. Gray's Anatomy. 29th ed. Philadelphia: Lea & Febiger, 1973. 1,466 pp.

51. Hellman, T. M., and F. H. Small. Characterization of the odor properties of 101 petrochemicals using sensory methods. J. Air Poll. Cont. Assoc. 24:979–982, 1974.

52. Hyman, A. M. Factors influencing the psychophysical function for odor intensity. Sens. Proc. 1:273–291, 1977.

53. Jones, F. N., and M. H. Woskow. On the intensity of odor mixtures. Ann. N.Y. Acad. Sci. 116:484–494, 1964.

54. Katz, S. H., and E. J. Talbert. Intensities of odors and irritating effects of warning agents for inflammable and poisonous gases. Technical paper 480, Bureau of Mines, U.S. Department of Commerce. Washington, D.C.: U.S. Government Printing Office, 1930. 37 pp.

55. Koelega, H. S., and E. P. Köster. Some experiments on sex differences in odor perception. Ann. N.Y. Acad. Sci. 237:234–246, 1974.

56. Köster, E. P. Adaptation and cross-adaptation in olfaction. Rotterdam: Bronder–Offset N.V., 1971.

57. Köster, E. P. Intensity in mixtures of odorous substances, pp. 142–149. In C. Pfaffmann, Ed. Olfaction and Taste. III. New York: Rockefeller Univ. Press, 1969.

58. Kulle, T. J., and G. P. Cooper. Effects of formaldehyde and ozone on the trigeminal nasal sensory system. Arch. Environ. Health 30:237–243, 1975.

59. Laffort, P. A linear relationship between olfactory effectiveness and identified molecular characteristics, extended to fifty pure substances, pp. 150–157. In C. Pfaffmann, Ed. Olfaction and Taste. III. New York: Rockefeller Univ. Press, 1969.

60. Laffort, P. Mise en évidence de relations linéaires entre l'activité odorante des molécules et certaines de leurs caractéristiques physicochimiques. Comp. Rend. 256:5618–5621, 1963.

61. Laffort, P., and A. Dravnieks. An approach to a physico-chemical model of olfactory stimulation in vertebrates by single compounds. J. Theoret. Biol. 38:335–345, 1973.

62. Laffort, P., F. Patte, and M. Etcheto. Olfactory coding on the basis of physicochemical properties. Ann. N.Y. Acad. Sci. 237:193–208, 1974.

63. Laing, D., and H. Panhuber. Neural and behavioral changes in rats following continuous exposure to an odor. J. Comp. Physiol. 124:259–265, 1978.

64. Lawless, H. T., and W. S. Cain. Recognition memory for odors. Chem. Sens. Flav. 1:331–337, 1975.

65. Lawless, H., and T. Engen. Associations to odors: Interference, mnemonics, and verbal labeling. J. Exp. Psychol.: Hum. Learn. Mem. 3:52–59, 1977.

66. Leitereg, T. J., D. G. Guadagni, J. Harris, T. R. Mon, and R. Teranishi. Evidence for the difference between the odours of the optical isomers (+)- and (−)-carvone. Nature 230:455–456, 1971.

67. Lindvall, T. On sensory evaluation of odorous air pollutant intensities. Nordisk Hygienisk Tidskrift, Supplementum 2, 1970.

68. MacLeod, P. Structure and function of higher olfactory centers, pp. 182–204. In L. M. Beidler, Ed. Handbook of Sensory Physiology, Vol. IV, Chemical Senses, Part 1, Olfaction. Berlin: Springer Verlag, 1971.

69. Mair, R. G., J. A. Bouffard, T. Engen, and T. H. Morton. Olfactory sensitivity during the menstrual cycle. Sens. Proc. 2:90–98, 1978.

70. Marks, L. E. Sensory Processes: The New Psychophysics. New York: Academic Press, 1974. 334 pp.

71. Moncrieff, R. W. The Chemical Senses. 3rd ed. Cleveland: CRC Press, 1967. 760 pp.

72. Moskowitz, H. R., A. Dravnieks, W. S. Cain, and A. Turk. Standardized procedure for expressing odor intensity. Chem. Sens. Flav. 1:235–237, 1974.

73. Moskowitz, H. R., A. Dravnieks, and L. A. Klarman. Odor intensity and pleasantness for a diverse set of odorants. Percept. Psychophys. 19:122–128, 1976.

74. Moulton, D. G. Olfaction in mammals. Amer. Zool. 7:421–429, 1967.

75. Moulton, D. G. Spatial patterning of response to odors in the peripheral olfactory system. Physiol. Rev. 56:578–593, 1976.

76. Moulton, D. G., and L. M. Beidler. Structure and function in the peripheral olfactory system. Physiol. Rev. 47:1–52, 1967.

77. Moulton, D. G., and D. Tucker. Electrophysiology of the olfactory system. Ann. N.Y. Acad. Sci. 116:380–428, 1964.

78. Okano, M., and S. F. Takagi. Secretion and electrogenesis of the supporting cell in the olfactory epithelium. J. Physiol. (London) 242:353–370, 1974.

79. Proctor, D. F. Physiology of the upper airway, pp. 309–345. In W. O. Fenn and H. Rahn, Eds. Handbook of Physiology. Section 3: Respiration. Vol. 1. Washington, D.C.: American Physiological Society, 1964.

80. Russell, G. F., and J. I. Hills. Odor differences between enantiomeric isomers. Science 172:1043–1044, 1971.

81. Schiffman, S. S. Physicochemical correlates of olfactory quality. Science 185:112–117, 1974.

82. Schiffman, S. S., and C. Dackis. Multidimensional scaling of musks. Physiol. Behav. 17:823–829, 1976.

83. Schiffman, S. S., J. Moss, and R. P. Erickson. Thresholds of food odors in the elderly. Exp. Aging Res. 2:389–398, 1976.

84. Schneider, R. A., and S. Wolf. Relation of olfactory acuity to nasal membrane function. J. Appl. Physiol. 15:914–920, 1960.

85. Steiner, J. E. Discussion paper: Innate, discriminative human facial expressions to taste and smell stimulation. Ann. N.Y. Acad. Sci. 237:229–233, 1974.

86. Stone, H. Behavioral aspects of absolute and differential olfactory sensitivity. Ann. N.Y. Acad. Sci. 116:527–534, 1964.

87. Sumner, D. On testing the sense of smell. Lancet 2:895–897, 1962.

88. Terkel, L. Working, p. 111. New York: Pantheon Books, 1974.

89. Tremble, G. E. The dynamics of nasal mucus. Laryngoscope 59:639–652, 1949.

90. Tucker, D. Nonolfactory responses from the nasal cavity: Jacobson's organ and the trigeminal system, pp. 151–181. In L. M. Beidler, Ed. Handbook of Sensory Physiology, Vol. IV, Chemical Senses, Part 1, Olfaction. Berlin: Springer-Verlag, 1971.

91. Tucker, D. Olfactory vomeronasal and trigeminal receptor responses to odorants, pp. 45–69. In Y. Zotterman, Ed. Olfaction and Taste: Proceedings of the First International Symposium, Stockholm, Sep. 1962. Oxford: Pergamon Press, 1963.

92. Tucker, D. Physical variables in the olfactory stimulation process. J. Gen. Physiol. 46:453–489, 1963.

93. van Gemert, L. J., and A. N. Nettenbreier, Eds. Compilation of odour threshold values in air and water. TNO, Zeist, Netherlands: Central Institute for Nutrition and Food Research, 1971. 79 pp.

94. Venstrom, D., and J. E. Amoore. Olfactory threshold in relation to age, sex or smoking. J. Food Sci. 33:264–265, 1968.

95. Whissell-Buechy, D., and J. E. Amoore. Odour-blindness to musk: Simple recessive inheritance. Nature 242:271–273, 1973.

96. Zwaardemaker, H. An intellectual history of a physiologist with psychological aspirations, pp. 491–516. In C. Murchison, Ed. A History of Psychology in Autobiography. Vol. 1. Worcester, Mass.: Clark Univ. Press, 1930.

97. Zwaardemaker, H. L'Odorat. Paris: Librairie Octave Doin, 1925. 305 pp.

98. Zwaardemaker, H. C. Smell compensation. (Trans. of "Die compensation von Geruchsempfindungen," originally published in Arch. Physiol., Leipzig.) Perfum. Essent. Oil Rec. 50:217–221, 1959.

3 Effects on Health and Welfare

This chapter focuses on the various consequences of odor pollution. Its coverage includes health, social, and behavioral effects of odors. Although it is generally substantive, the text becomes discursive and even methodologic at points. The discussion is related mainly to attempts to use meager data on a small number of odorous materials to determine the effects of all odors as a class.

When a document intended to be mainly retrospective uncovers a pressing need for more information on a matter of public concern, it must become in part prospective. A small amount of methodologic material in Appendix B is intended to give a glimpse of some techniques that, if applied, might permit a more definitive assessment of the social impact of odors.

ODORS AND WELL-BEING: A HISTORICAL SKETCH

Pleasantness and unpleasantness form perhaps the most salient dimension of olfactory experience. As Aristotle noted, some odors seem pleasant in relation to physiologic needs. Regarding the odor of food and drink, he remarked that "their pleasantness and unpleasantness belong to them contingently. . . . These smells are pleasant when we are hungry, but when we are sated and not required to eat, they are not pleasant. . . ." [40 (p. 75)]

Some odors, like the aromas of flowers, seem pleasant without respect to physiologic needs. Aristotle suspected that such aromas preserve health.

This suspicion and its corollary—that unpleasant-smelling aromas are injurious—formed the basis for man's behavior toward the olfactory environment throughout history.

The famous physician Galen (ca. 180 A.D.), whose influence stretched over 14 centuries, concluded that the olfactory receptors lay within the brain, specifically in the ventricles. Even as late as the seventeenth century, physicians thought that nasal mucus represented cerebrospinal fluid that escaped from pores leading directly from the ventricles to the nasal cavities. Such views reinforced the notion that odors could exert direct poisonous action on the brain. At the time, no distinction was made between odor (the sensation) and odorant (the stimulus for sensation).

Some odors (e.g., that of a freshly extinguished oil lamp) were thought to cause abortion. Others were thought to lead to chronic disease, convulsions, and even death. Putrid odors seemed particularly dangerous and were even considered the cause of the plague. Regarding a thirteenth-century outbreak of the plague, Winslow noted that no commentator at the time had even hinted at the possibility of a *contagium animatum* as the element by which the disease was transmitted: "It was a chemical property of the air which all of these commentators visualized—as was to be the case for many centuries to come."[50 (p. 103)] Indeed, Walter Charleton, commenting on an outbreak of the plague four centuries later, explained: "Nay, scarce an author who hath written of the plague and its causes, but abounds in relations of those accursed miscreants, who have kindled most mortal infections, by certain veneficious practices, and compositions of putrid and noysom odors."[10 (p. 236)]

If bad odors caused disease, it seemed to follow logically that good odors might combat or prevent it. Accordingly, apothecaries concocted formulations of "therapeutic" perfumes and aromatics. Salmon's *Dispensatory* of 1696 contained such aromatic formulations as Apoplectick Balsam of Horstius and Balsam for the Loss of Memory.[34] In eighteenth-century France, persons stuffed their ears and nostrils with sweet-smelling thyme, rue, and pennyroyal during epidemics, still unaware that fleas from rats generally transmitted the plague. As Haggard explained:

> The physicians of those days protected themselves against the disease by means of suits of leather with leather gauntlets and masks with glass coverings for the eyes and a long snout filled with fumigants for the nose. . . . They lit fires on which were burned aromatic substances to purify the air; and for the same purpose sprinkled perfumed water in the rooms and on their clothing. Eau de Cologne is a survival of one of these plague waters or essences.[25 (p. 210)]

Not until the middle of the nineteenth century did physicians realize that microscopic organisms transmitted mainly by direct contact, rather than chemicals transmitted through the air, were the prime mediators of infection and contagious disease. Even Lister, the father of antisepsis, thought it necessary to treat the air in hospitals in order to keep wound infection under control. He sprayed carbolic acid (phenol) into the air during surgery until his colleagues, who experienced great discomfort from the vapors, persuaded him to cease.

The Industrial Revolution forced confrontation with the possible link between odors and disease. Factories exposed both workers and neighbors to many noxious vapors. Although some vapors were clearly toxic, some disagreeable vapors had no evident deleterious effects. In commenting on the paper "Report of the Physical and Mathematical Class of the Institute, upon the Question, Are those Manufactures which emit a disagreeable Smell prejudicial to Health?" the editors of the *Edinburgh Medical and Surgical Journal* (1806) said: "A disagreeable smell is by no means a certain criterion of an unwholesome atmosphere. And, on the other hand, the air is often pestilential, when, to our senses, it seems uncontaminated. . . . "[21 (p. 299)] The editors still suspected that putrid-smelling air held danger.

The perceived connection between odors and disease, although rooted to some extent in idiosyncratic superstitions, arose primarily from an actual association between disease and poor hygiene. When hygiene is poor, both odors and the incidence of disease are high. Throughout the last century, hygienic conditions have improved, but odors from industrial sources have generally increased. Insofar as these odors are unpleasant they have been viewed with the same suspicion as the unpleasant odors of hundreds of years ago. The average citizen believes firmly and justifiably that his senses serve in part to warn of danger. In the absence of explicit information to the contrary, the citizen will draw the reasonable conclusion that, if something smells bad, it is likely to be bad. In the case of ambient malodors, today's citizen has no greater reason to reject this conclusion than his ancestors.

MORPHOLOGIC AND PHYSIOLOGIC EFFECTS OF ODORANTS

Odors may affect well-being by eliciting unpleasant sensations, by triggering reflexes and other possibly harmful physiologic reactions, and by modifying olfactory function.

Unfavorable responses include nausea, vomiting, and headache; shallow breathing and coughing; disturbances of sleep, stomach, and appetite; irritation of eyes, nose, and throat; and decreases in the sense of well-being and in the enjoyment of food, home, and external environment.[43] Exposure to some odorous substances may also decrease heart rate, constrict blood vessels of the skin and muscles, increase epinephrine concentration in the blood, and even bring about alterations in the size and condition of cells in the olfactory bulbs of the brain.[31] However, the relationships between the intensity or duration of the exposure to odor and the magnitudes of these symptoms have not been established.

These statements of possible ill effects must be interpreted with caution, inasmuch as some odorous substances have a primary toxic action that is exerted as a systemic poison, and a dose-response relation has not been established for odorous substances that lack overt toxic properties.

We consider here the physiologic and morphologic changes that can follow exposure to odorous substances. Much of the evidence comes, of necessity, from experiments involving animals. Extrapolation of the results to humans requires considerable caution, because of the concentrations used, duration of exposure, and manner of odor presentation. But the evidence alerts us to the possibility that similar changes may occur in humans.

NASAL MUCOUS-MEMBRANE FUNCTION

The condition of the respiratory epithelium, which borders the olfactory epithelium, is important in olfactory function.[41] If odorants disturb its condition, they may alter the perceived intensity or character of other odorants. Such disturbances may take several forms. For example, the continuous movement of the secretions overlying the olfactory epithelium seems to depend, at least partly, on the pull of mucus flowing over the adjoining respiratory epithelium. This in turn depends on the rhythmic beating of cilia. Therefore, compounds that arrest or slow ciliary action may, by this route, alter the microenvironment of the olfactory receptors. The ciliostatic action of some airborne pollutants, in higher concentrations at least, is well established.[14,15] Some of the effects may result in part from alteration of the pH of mucus. Compounds that lower the normal pH of nasal mucus—about 6.7–6.9—are thought to transform the mucus from a sol to a gel.[8,52] Other substances that have a detrimental action on ciliary activity include volatile oil vapors and menthol (1% concentration). In the case of prolonged exposure to odorous irritants, damage is severe.[14] Because mucus acts as a protective interface between infectious particles in the ambient air and the respiratory mucosa, anything that slows or stops ciliary motion may reduce or destroy the effectiveness of mucus as a barrier. Further work must establish the mechanism by which these changes can occur.

In addition to the response of the respiratory epithelium to odorants, the olfactory epithelium can be stimulated to secrete an excess flow of mucus and thus change the efficiency of olfactory function.[38]

TRANSPORT TO THE BRAIN IN ANIMALS

The olfactory receptor can act as a bridge between the odorous environment and the brain. Cilia of olfactory receptor cells terminate distally in the nasal mucus covering the olfactory epithelium. The opposite end of the olfactory cell is in an olfactory bulb of the brain. Because a virus is a living organism, it may not offer a close analogy to an inert substance, such as an odorant. However, De Lorenzo[19] offered electron microscopic data that colloidal gold particles placed on the surface of the olfactory epithelium can enter the olfactory receptor and travel toward the brain. Radioactively labeled leucine placed in the nasal cavity can enter the olfactory receptor, be incorporated into proteins manufactured in the cell body, and be transported in the axoplasm of the olfactory axon toward the olfactory bulb at about 400 mm/day. Because both large proteins and inorganic material can move toward the brain, it may be possible for odorous substances continuously present in the environment to move in a similar manner and enter the brain. This is of particular concern for persons who work in an environment of odorous vapors or particles. The little cited evidence of such transport comes from animal studies, and the implication for man is yet to be established. Additional research is urgently needed to investigate the extent of such transport and the possible accumulation of these foreign materials in the brain.

MORPHOLOGIC CHANGES IN OLFACTORY-BULB CELLS AFTER PROLONGED STIMULATION WITH ODORANTS

Little is known about the effects of prolonged exposure to environmental odorants on the function of the olfactory system itself. It is clear, however, that atrophic changes occur in the morphology of some cell types in the olfactory bulbs of rats in association with exposure to a constant flow of a single odorant.[31,39] Mitral cells shrink, sometimes to about half their original size, and the nucleus and cytoplasm stain darkly. The odorants tested were chosen irrespective of their pleasant or unpleasant odors and covered a wide range of chemical structure (e.g., cyclopentanone, cyclohexanol, benzyl mercaptan, limonene, naphthylene, pyridine, and 1-menthol). They were presented in relatively low concentrations, although concentration was measured accurately only by Laing and

Panhuber,[31] who exposed rats to acetophenone at 0.2 ppm and cyclohexanone at 4 ppm. The receptor axons were not affected. Higher centers of the brain and the olfactory epithelium were not examined. The phenomenon has been called "selective degeneration." However, there is no evidence of cell death, and electrophysiologic recordings have shown that the affected mitral cells still respond to odor, if less effectively than normal cells (B. Oakley, personal communication). Rats exposed to odorants and to deodorized air for 2 months and then left in room air for 5 months showed signs of restoration of mitral cells to nearly normal appearance. This suggests that the effects are ultimately reversible.[31]

Rats exposed to odorants and then tested in a behavioral-test apparatus showed no alteration in sensitivity to the exposure odorant, but had lowered sensitivity to a similar but novel odorant. Rats exposed to acetophenone had a significantly lower sensitivity to cyclohexanone than rats exposed to either room air or cyclohexanone.

The results of both behavioral and histologic studies imply that changes in the morphology of mitral cells may occur because the cells are understimulated, owing to the rats' being deprived of a sufficiently rich odor environment. (In the extreme case, when animals are exposed to deodorized air alone, the changes are severe and nonselective.) They also suggest that any odorant in low concentration may be effective in inducing the alterations, regardless of whether the odorant is pleasant or unpleasant, whether exposure is for a few weeks or a few months, and whether the diluent is deodorized air or room air.[20] It is yet to be determined whether the effects are more pronounced for concentrations or for odorants different from those tested, in infants as opposed to adults, or in higher levels of the central nervous system than the olfactory bulb.

If the effects occur in human beings—and it has not yet been shown—then pleasant-smelling substances could pose the more serious hazard, inasmuch as a person will generally avoid continuous exposure to odorants that produce unpleasant odors more readily than those with pleasant odors. Workers exposed to the same industrial effluent on the job and in nearby residences could be at risk, as could infants reared in an area of heavy odor pollution.

Aside from a modicum of research on workers, there appear to have been no controlled studies of olfactory sensitivity in human populations exposed to odorants. Consequently, we do not know whether any subgroups of the population have a higher incidence of altered olfactory function of a kind attributable to morphologic changes in mitral cells. However, a study by Nauš[36] has relevance to this question. Workers in a factory that produced menthol candies could not smell five of 14 test substances (not related to menthol). A control group of workers not exposed to the menthol odor could smell all 14 substances. Nauš claimed that occupational hyposmia (reduced sensitivity to odors) is not rare. If this is so, it may interfere with detection of warning odors.

INFLUENCE OF ODORANTS ON AUTONOMIC AND HORMONAL FUNCTIONS

Odorant stimulation of receptors in the nasal mucosa can elicit marked respiratory and cardiovascular responses. The reported effects, documented in various species, include reduction in breathing, sneezing, bronchodilatation, lowering of heart rate (bradycardia), increase in arterial blood pressure, reduction in cardiac output, and vasoconstriction in skin, muscle, splanchnic, and renal vascular beds.[3,6,27] For example, Alarie[1,2] exposed mice to 51 airborne chemicals, many of them odorants, and found a characteristic decrease in respiratory rate.

In the rabbit, nasal stimulation with ether vapor from a Woulff bottle for 2 s elicited an increase in vascular resistance owing to vasoconstriction in the nose; the effects were abolished by local anesthesia of the nasal mucosa. The effects occurred in two phases: an early phase, seen within 10 s of stimulus application, and a less marked late phase occurring about 50 s after stimulation. The early phase was apparently mediated by efferent fibers of the sympathetic nervous system, whereas the late phase was mediated by adrenal hormones.[4]

The autonomic nervous system consists of two parts: the sympathetic and the parasympathetic. The various lower senses—i.e., olfaction, taste, vestibular sense, and somesthesis (warmth, cold, touch, irritation, pain)—all seem to bear an intimate, often reflex, connection to autonomic functioning. Not surprisingly, therefore, stimulation of these senses often carries strong affective impact (pleasantness–unpleasantness). The affective reactions generally drive the organism toward a condition of homeostasis (a tendency to stability in the normal body states of an organism).

EFFECTS OF ODOROUS EMISSION ON HUMAN BEINGS

Virtually all information regarding adverse reactions to environmental odors arises from self-reports, either spontaneous or solicited, rather than from objective measures of physiologic responses.

For example, 495 residents exposed to the malodors of a pulp mill in Lewiston, Idaho, signed a petition that stated that "this contamination of our air and its odor affects us from headaches, watery eyes, runny noses, and breathing difficulties. . . . "[37] It is likely that these symptoms resulted in part from odorant stimulation of trigeminal receptors—the same receptors that mediated the respiratory and cardiovascular reactions in the laboratory experiments cited above.

Systematic surveys of odor pollution caused by pulp mills have uncovered a constellation of symptoms similar to those enumerated in the Lewiston petition. In Swedish and American surveys described by Jonsson,[28] 5–10% of respondents experienced shortness of breath, nasal irritation, and runny nose, as well as coughing (which could be elicited by stimulation of vagal receptors in the throat); 10% experienced eye irritation and headache; and 15% experienced nausea and sinus congestion. Nausea is the one symptom that does not readily fit the constellation of trigeminally mediated symptoms.

The nasal branch of the trigeminal nerve adds the perceptual attribute of pungency (stinging, burning, sharpness, etc.) to inhaled vapors. The olfactory nerve adds most other perceptual attributes of odor character (e.g., putrid, musky, fruity, goaty, and floral). Some odors (e.g., the putrid smell of rotting flesh and the smells of vomitus and raw sewage) cause most persons to withdraw vigorously from these odorant sources. These odors seem unpleasant without respect to any trigeminal impact. If forced to endure them for more than a few seconds, people will often report adverse physiologic reactions, such as headache, dizziness, and nausea.

The relationship between olfactory stimulation by such odorants as food and alimentary functioning makes itself apparent in everyday life, commonly evoking salivary and other digestive enzyme secretions, but unpleasant smells depress appetite—a phenomenon confirmed readily in the laboratory by Winslow and colleagues[49,51]—or they may even induce vomiting. In recent years, experiments have revealed that mammals will rapidly learn an association between an odor and subsequent nausea, even if the odorant does not actually cause the nausea. In the experimental case, an animal ingests a flavored liquid and later receives a nausea-inducing injection of lithium chloride. Although the nausea may not begin for an hour or more, the animal will develop a specific aversion to the flavored liquid. Aversions to odors or tastes will develop readily under these circumstances, whereas aversions to sensory stimuli not normally associated with eating will fail to develop. Aversions to sounds will develop, however, if followed by electric shock. This phenomenon has given rise to the notion of inherent "preparedness" to form particular associations (odor-nausea, sound-shock).

Irrespective of the exact physiologic mechanism of action, persons who live in malodorous environments report adverse somatic symptoms. For instance, Winneke and Kastka[48] found that the majority of persons living within 1 km of a tar-oil plant in Duisburg and an insulation plant in Cologne experienced occasional-to-frequent periods of "odor-induced" nausea and headache. Anyone who wishes to argue against the conclusion that these adverse reactions result directly from the odor sensation must account for the immediacy of the reactions. One or two inhalations of the malodorous air surrounding many industrial operations will often induce nausea, just as one or two inhalations of rancid leftovers in the home can cause nausea.

Self-reports of adverse reactions to odorous pollutants should not automatically be categorized simply as "annoyance." Undoubtedly, many complaints regarding environmental odors fall neatly into this category, which has been defined as "an effect which may not be demonstrably pathogenic but which involves a negative factor for an individual's comfort and well-being."[32] Odors deemed neutral or pleasant within a proper context (e.g., bakery odor) may indeed give rise to annoyance when present at unwanted times. Odors that cause annoyance are analogous to noise, commonly defined as unwanted sound. Even odors deemed somewhat unpleasant, irrespective of context, may generate nothing more serious than annoyance. Nevertheless, to categorize all negative reactions to community odors as annoyance reactions trivializes the problem of odor pollution. Some industrial odorants produce malodors that are simply too revolting and sickening to warrant the designation of sources of annoyance.

The above discussions of mucous-membrane function and morphologic changes reveal that the label "demonstrably pathogenic" can depend on where in the body and over what duration a physician chooses to look for pathologic change. Even aside from these considerations, the symptoms of nausea, headache, and dizziness seem to transcend annoyance. Unfortunately, these three symptoms lend themselves poorly to objective verification and measurement. This leaves the victim of odor pollution, particularly the victim exposed only briefly, medically ignored. The transitory victim or the victim of episodic odor pollution may never have any symptom other than unconditioned nausea. The subjective magnitude of this symptom may exceed that of any other immediate reaction to pollution. But its status as a health effect may await an objective validation currently unavailable.

PUBLIC-HEALTH ASPECTS*

This section discusses various kinds of responses to odorants and examines the public-health implications that flow from each. By definition, chemicals hazardous to human health are considered to be toxic; hence, they are subject to control in accordance with existing laws or regulations, e.g., the Toxic Substances Act and the Clean Air Act. Toxic odorous substances in the atmosphere are automatically subject to standard-setting under the Clean Air Act of 1970 and its amendments, and reduction of their presence to below toxic thresholds is mandatory.

There is very little available information on the toxicity or hazardous effects of odorous substances in man. It is known from studies on experimental animals that some odorous substances may damage tissue, but there is no direct evidence of the same phenomena in man. This lack of evidence may be due to a lack of research; as new information is found, the number of odorous substances known to be toxic may increase.

In some cases, a large fraction of the population are affected by odorous substances described as ill-smelling, rank, foul, fetid, or stinking. There are many odors to which the term "foul" would be applied universally. Those commonly encountered include odors from poorly designed and badly operated sewage-treatment facilities, rendering and fishmeal plants, cattle feedlots, farms with garbage-fed pigs, and a variety of rubber, petroleum, and chemical manufacturing operations, including wood-pulping. These responses include not only mild discomfort, but other symptoms that may be associated with stimulation of the autonomic nervous system, such as those in the Swedish and American surveys described by Jonsson.[28] Distressing symptoms reported to result from such odors include "nausea, headache, loss of sleep, loss of appetite, impaired breathing, and even allergic reactions,"[5] vomiting, general malaise, and sleep disturbance.

Note that in all these cases, although there may be an absence of disease and infirmity, there is not a state of complete mental, social, and physical well-being. Many people show distaste (disgust) and annoyance in response to foul odors of moderate or even low intensity. The scientific basis for this response is still unknown, but may depend on some physiologic changes, as mentioned above, or may have a more vague origin in the psychology of the individual.

In addition to malodors, the population can respond adversely to

*Dr. Melvin W. First, a member of the Committee, holds some views different from those of the Committee as a whole, with respect to the effects of odors on public health. He has developed a categorization of odorous substances based on their toxicity and their effects on public health. That categorization is discussed in some depth in Appendix A, prepared by Dr. First.

pleasant odors. A great variety of manufacturing and agricultural odors are almost universally considered to be pleasant. These include the characteristic smells of baking and cooking, perfume-blending, candy-making, and haying. In spite of the pleasant feelings usually associated with these products and activities, air-pollution control agencies receive complaints about emission of odors from restaurant kitchens, bakeries, and similar establishments that make products that we take pleasure in eating, wearing, or using. Thus, even odors that are universally thought of as pleasant may become unacceptable when they are present in the air at an unusually high intensity and for a long period.

There are many types of odorants commonly encountered on which there is no consensus regarding effects on human beings; and there are odorants that produce odors that a small fraction of the population considers unpleasant, but that are acceptable to most when they are present in commonly encountered concentrations. Exposures to odorants that have unpleasant odors are often associated with a small number of chronic and persistent complainers, a vast majority of indifferent neighbors, and a bemused and troubled air-pollution control officer who experiences great difficulty in trying to decide whether a situation merits official control action in the face of an odor exposure that is thought to be trivial. Normally, the intensity, duration, character, and degree of unpleasantness of an odor in the mind of the public are important for judging when exposures to odors constitute a matter of public concern.

Among the attempts to develop procedures for abating foul odors have been a number of examinations of the dose-response relationships of odors in the atmospheric environment.[17,27,30,35] These investigations have used polling techniques and questionnaires in areas with persistent odor problems. Unfavorable responses were found to be most numerous when perceived odor intensity was highest; single-point sources of large discharge volume and high odor intensity (as in uncontrolled kraft pulp mills) disturbed residents 20 km away; and there were sex-, health-, and age-related differences in annoyance reactions to odors.[23] The correlation between degree of chemical exposure and degree of annoyance reported is often weak. These studies have many inherent methodologic complications (e.g., reactions of individuals cannot be measured objectively).

To overcome the methodologic deficiencies, a number of performance measurements have been proposed as indexes of annoyance. The assumption is that, as the degree of annoyance increases, the performance of tasks that require skill and close attention (such as problem-solving) is affected adversely. Because these tests measure reactions resulting from annoyance, rather than the annoyance itself, it has been hypothesized that some somatic responses mediated by the central nervous system are more closely related to the perceived annoyance. Measurements have been proposed to monitor central-nervous-system activity. But the great need in using physiologic indicators of annoyance is to establish their validity.[32] The development of unifying principles would go a long way toward resolving some seeming inconsistencies in odor-control efforts. In short, what is needed is a synthesizing and integrating effort in science and technology, to assure the public that adequate knowledge will become available to safeguard its health and comfort and to keep the environment in a condition close to the natural pristine state.

From the standpoint of public health, it is highly desirable that the effects of atmospheric odors on humans be analyzed in conventional epidemiologic terms—i.e.;

- Identification of the physical and emotional symptoms that odorants cause.
- Development of firm dose-effect and dose-response relationships that approach zero in the absence of stimulus but are unlikely to reach 100%, even under the strongest provocation (i.e., where there is some degree of immunity).
- Identification of susceptible populations, with respect to age, sex, occupation, geographic distribution, etc., and measurement, in acceptable statistical terms, of the nature, incidence, and prevalence of unfavorable responses among identified susceptible groups.
- Development of a theoretical rationale that makes it possible to use the foregoing information for preventive purposes—the ultimate public-health goal.

SOCIAL AND BEHAVIORAL EFFECTS OF ODOR POLLUTION

Community annoyance due to odors has been studied primarily in Sweden, the United States, and the Federal Republic of Germany. A series of Swedish studies, beginning in the early 1960s,[9,23,28] have refined the techniques for the measurement of annoyance and were the first to suggest the use of such measurement to establish legal standards. The results also pointed to the risk of relying on voluntary complaints for enforcement purposes.

In the United States, seven major studies of community odor problems have been completed since 1969. Four were conducted in California—two in Eureka, one in Anderson, and one in Carson.[18,24,29] These four studies focused on both annoyance and health aspects and indicated the existence of dose-response relationships. The remaining studies were conducted as a series to determine the social and economic impact of odors and to develop procedures for the identification and assessment of community odor problems.[11–13] The second in this series, carried out in Los Angeles, included an attempt to determine the effects of odor pollution on property value.

The social and behavioral effects described here include annoyance and interference with such activities as use of home and yard, working, attending school, shopping in desired locations, use of recreational activities (parks, libraries, etc.), and maintaining comfort in confined situations (hospitals, institutional homes, etc.). For present purposes, the term "social effects" excludes consequences describable directly in monetary terms.

Recent studies of the social and economic impact of odors[12,13] indicated that feelings of annoyance were the most frequently cited problem. In fact, whereas annoyance sometimes occurred without other undesirable effects, the reverse never held true. This evidence seems consistent with a conclusion voiced at the Third Karolinska Institute Symposium on Environmental Health: "In environmental health the most important dimension of an odor is probably its acceptability, e.g., what percentage of the population is annoyed by the smell."[45 (p. 16)]

Interference with activities has not been well documented in the United States. Social surveys have been concerned chiefly with annoyance and symptoms of illness. A few surveys have included open-ended followup questions, such as "How have odors bothered you?" and "How have odors reduced the value of your home?" The answers have included statements of interference with activities, but the frequency of responses to these questions has been too low and too sporadic to permit firm conclusions or generalizations.[11,12]

A series of laboratory and field studies recently conducted by Winneke and Kastka[48] in the Federal Republic of Germany have sought to develop concepts and strategies for odor-control legislation. The first study began in 1974 with a face-to-face survey of 704 inhabitants selected at random from one known odor-problem area and two odor-free areas in Düsseldorf. Although the questionnaire was not included in the publication cited here, it is apparent that the inhabitants were asked questions about the interference of odors with their activities. The authors subjected the responses to factor analysis. Three factors—F1, F2, and F3—were found to account for more than 50% of the total variance. The analysis produced the structure of response components shown in Table 3-1. F1 consists of items that stress sensory aspects of annoyance and is therefore termed the "sensory experience" dimension of annoyance. F2 consists mainly of social and behavioral effects of odors beyond annoyance and includes most of the items that denote interference with activities; it is called the "social-

TABLE 3-1 Response Components of Annoyance Factors Derived from a Survey of 704 Inhabitants of Düsseldorf[a]

F1, Sensory Experience
Distinct perception of odors in neighborhood
Intensity of perceived odors
Frequency of odor perception
Odor annoyance of people "does exist"
Degree of annoyance due to malodors
Degree of disturbance due to malodors
Windows often shut because of odors
Discomfort due to environmental odors
Foul-smelling air
Ventilation of apartment difficult
"Stinking" air in neighborhood
Sleeping with windows shut
Environmental odors unbearable
Reduced pleasure in taking a walk
Strange smell in apartment
F2, Social-Emotional Disturbance
Reduced social contacts
No pleasure in coming home
Odor leads to tensions within family
Odor disturbs communication
Odor spoils appetite
Odor interferes with comfort of living
Odor interferes with outdoor activities
Odor induces anger
F3, Somatic Disturbance
Odor interferes with falling asleep
Odor disturbs sleep
Odor induces headache
Odor induces nausea
Odor induces fits of coughing
Odor interferes with reading and thinking
Odor induces bad mood
Odor interferes with recreation
Odor induces vomiting
Odor interferes with normal breathing

[a] Data from Winneke and Kastka.[48]

emotional disturbance" dimension. F3 contains mainly somatic reactions and is called the "somatic disturbance" dimension. This series of studies is discussed more fully later. The point to be made here is that items of interference with activities have been successfully measured by at least one group of investigators. The authors have shown that these measures, as well as measures of somatic disturbance, are indicative of more severe degrees of annoyance than are mere measures of perception and reports of bother.

There is a large body of evidence in the United States and other countries that shows annoyance to be a common reaction of inhabitants in communities where unpleasant odors are encountered. There is far less evidence of the relative severity of annoyance caused by various odors among various social groups and of social effects other than annoyance. To increase the data base so that decisions can be made on when to control unpleasant odors and what degree of control to exert, there must be many additional studies in many different locales. The choice of methods used in performing those studies is important, if the results are to be helpful.

Two methods have been used: one relies on spontaneous complaints, the other on social surveys. Both have assisted state and local air-pollution control agencies and other public authorities in gathering evidence on community odor problems. The method that involves spontaneous complaints has evolved as a routine part of agency activities and is by far the more widely used.

SPONTANEOUS COMPLAINTS

Odor complaints are almost always initiated spontaneously by citizens in the community. Solicitation of complaints by local authorities is permitted (but seldom attempted) in some jurisdictions and strictly avoided in others, depending on local policy or interpretation of state law. Almost all grievances are received by telephone and handled by inspectors or clerical personnel. Most agencies maintain records of odor complaints on some type of form, examples of which are presented in Figure 3-1. When a complaint is received by the Los Angeles County Air Pollution Control District, for example, a summary of the information given by the complainant is recorded on a Radiophone Message Log sheet (Figure 3-1, top) and then dispatched to an inspector in the field. If the complaint involves odors, code 963 is indicated. The inspector is required to drive to the home of the complainant to verify the presence and, if possible, the source of the odors. A Nuisance Complaint Form (Figure 3-1, bottom) is completed by the inspector at that time, in accordance with the complainant's responses to the questions on the form.

TIME RECEIVED	Air Pollution Control District—County of Los Angeles Enforcement Division RADIOPHONE MESSAGE LOG	TIME DISPATCHED
REC'D. FROM:	ASSIGNED TO:	OPR.
930 | | 931	SOURCE:	
DO NOT CONTACT	ADDRESS:	
961-SMOKE	COMPLAINANT:	
962-OPEN FIRE	ADDRESS:	
963-ODORS	ADDITIONAL INFORMATION:	
964-DUST		
967-BREAKDOWN		
968-REPORT TO COURT		
969-REPORT TO H. BD.		
990-CAR SERVICE		
TEN 19-COME TO HDQRS.		
TEN 21-TELE. HDQRS.	TEN 98 - COMPLAINT NO:	TIME-TEN 98
TEN 88-TELE. NUMBER?	SOURCE:	
TEN 91-ROLL CALL CHG.	ADDRESS:	
CODE:	DISPOSITION:	

10-40D129A

AIR POLLUTION CONTROL DISTRICT - COUNTY OF LOS ANGELES 40D261
434 South San Pedro Street, Los Angeles, California 90013 R. 2-69

NUISANCE COMPLAINT FORM

Date________

1. Source complained of: Name ________ (Print or Type)
   Address ________
2. Complainant: Name ________ (Print or Type) Phone No. ________
3. Home address ________ (Print or Type) How long there ________
4. Business address ________ (Print or Type) How long there ________
5. Approximate distance from source ________
6. Wind direction when nuisance occurred, circle. N E S W
7. Dates and hours nuisance occurred ________
8. Nature of nuisance complained of, circle. Smoke, charred paper, dust, soot, grime, carbon, acid, gas, odors. (other ________)
9. Have you observed the above-described emissions coming from the source complained of? ________
10. Have you or any member of your household become ill from the nuisance? ________
    If "yes" who? ________
11. Describe nature of illness ________
12. How is your home address zoned ________ (Residential, Commercial or Manufacturing)
13. Has the property at this address depreciated in value due to the nuisance? ________
14. State any damage done to your property, furniture, automobile, clothing, etc. ________
15. Will you appear in court to testify? Yes ( ) No ( )
16. Signature of complainant ________ (Married woman should use her own first name)
17. Complainant's remarks ________

INSPECTOR ________ BADGE NO. ________ NOTICE NO. ________

FIGURE 3-1.

Traditionally, odor-complaint data have been used for the following purposes.[13]

- To alert local authorities that odors were detected at specific times and locations and that possible odor problems exist.
- To help determine whether local law governing odors has been violated.
- To describe the conditions under which odors were detected and to enumerate the effects of odors experienced by the persons who complained.
- To help identify the offending odor sources, so that steps can be taken to eliminate the odor problems.

Enforcement agencies have relied on odor complaints as a measure of community reaction for a variety of reasons:[12,13]

- In some jurisdictions, local law requires that a given number of complaints be received before initiation of enforcement procedures.
- Odor complaints provide an inexpensive means to gather information.
- Use of complaints offers a simple, straightforward approach for dealing with odor problems—viz., those responsible for the sources of offending odors can be pressured until they apply whatever process controls are necessary to satisfy the complainants.
- Other methods of measuring community reaction are considered by agency officials to be too elaborate and expensive for purposes of enforcement.

In some jurisdictions, odor complaints are so numerous that it is sometimes difficult for enforcement agencies to attend to them in a timely manner. In recent years, the number of complaints about disagreeable odors has approached the total of all other air-pollution complaints. Data from California illustrate this fact. A national survey of odor problems, conducted in 1969, ranked the Los Angeles and San Francisco metropolitan areas among the 10 greatest potential odor-problem areas in the United States.[11] The relative importance of odor complaints, compared with other air-pollution complaints, in these areas, as well as in Humboldt County, is shown in Table 3-2.[22] The survey found the national average to be 29 odor complaints per 100,000 persons, which is exactly that shown for the San Francisco metropolitan area (Table 3-3). Note that the per-capita figure for Humboldt County is more than 25 times as great as the national average. Virtually all the complaints were directed against reduced-sulfur odors produced by two sulfate pulping plants operating near Eureka. The

TABLE 3-2 Comparison of Numbers (and Percentages) of Odor, Smoke, and Other Air-Pollution Complaints Received by Los Angeles County, Bay Area, and Humboldt County Air Pollution Control Districts During 1971[a]

	No. (%) Complaints		
Reason for Complaint	Los Angeles County APCD	Bay Area APCD	Humboldt County APCD
Odor	3,812 (60%)	1,202 (33%)	746 (61%)
Smoke	1,423 (23%)	1,418 (39%)	388 (32%)
Other forms of air pollution	1,074 (17%)	985 (27%)	88 (7%)
TOTAL	6,309	3,605	1,222

[a] Data from Flesh and Turk.[22] Sources: R. E. George, Los Angeles County APCD; T. F. Brennan, Bay Area APCD; and J. L. Caufield, Humboldt County APCD.

per-capita figures for Los Angeles County and the Bay Area are much lower, although still at or above the national average. The majority of complaints in these two locations were also aimed at reduced-sulfur odors. However, these odors were produced by numerous oil refineries and chemical plants.

When complaints are received in large numbers, one might assume that the complainants are anxious for public authorities to hear their concerns.

TABLE 3-3 Complaints Received per 100,000 Persons Residing in Los Angeles County, Bay Area, and Humboldt County Air Pollution Control Districts[a]

District	District Population (1970)	Estimated No. Odor Complaints per 100,000 Persons
Los Angeles County	7,036,463	54
Bay Area[b]	4,174,233	29
Humboldt County	99,692	748

[a] Complaint data from Table 3-2; population data from U.S. Bureau of the Census.[47]

[b] Includes Alameda, Contra Costa, Marin, San Francisco, San Mateo, and Santa Clara Counties only.

One might also assume that the information volunteered can be used to measure community reaction to odors accurately, without need for more elaborate methods. Unfortunately, most data from studies of spontaneous complaints suggest that such information offers a poor means of gauging community reaction, because the number of complaints and the quality of the information volunteered are often due to an interaction of many factors and cannot be expected to represent true feelings of the complainants. For example, the following statement regarding volume of complaints appeared in the report of Lindvall and Radford:[32 (p. 15)]

> In general, very few people will register a formal complaint with the authorities about any environmental problem. In Swedish, British and American annoyance surveys less than 10% of the population reported making any formal complaints by writing letters, telephoning or making personal visits to officials.[7,33,44] In British and American surveys only 20–23% of those who felt they had a serious local problem ever felt like calling or writing to an official. In contrast, reports of annoyance in these surveys showed that only a small fraction of those who actually report annoyance complaints take action in any spontaneous way.[26] A study of annoyance to aircraft noise showed that the main characteristics of individuals discriminating complainants from noncomplainants were those of education, value of their home, and membership in organizations. Thus the volume of complaints received by officials may reflect not so much the amount of discomfort experienced by the exposed population as its social class composition and level of community organization.[42,46]

Although not specifically stated in the paragraph, the correlation between the volume of complaints and the social class composition or level of social organization in a community is known to be positive. That is, the higher the social class or the more organized the social structure, the larger the volume of complaints one might expect to receive as a result of the presence of odors.

Of equal importance, studies have found that persons who volunteer their opinions tend to exaggerate or overstate their concern. An example occurred in Sweden:[9] 1,200 persons signed a petition demanding that health authorities eliminate the nuisance of offensive flue gases caused by a sulfate cellulose factory. In response to a later survey sponsored by the health authorities, almost 600 of these persons stated that they had not experienced annoyance during the period specified in the petition. The contradiction could not be explained by a diminution in exposure after the petition was submitted. The petition was therefore rejected as an unreliable index of the extent of concern. This is not to say that all petitions are unreliable, but rather that community-initiated petitions are likely to contain unmeasurable bias. Social surveys, however, when conducted by properly trained investigators, can be used to estimate the true feelings of the average citizen in the community by including controls for bias.

SOCIAL SURVEYS

For the reader who is unfamiliar with the techniques of social surveys, basic elements are described in Appendix B. Also included there are suggested uses of survey methods as they are applicable to odor cases. The remaining paragraphs of this chapter return to the social surveys conducted by Winneke and Kastka.

After the face-to-face survey conducted in 1974 (described earlier), Winneke and Kastka[48] performed followup studies on small subsamples of the original 704 inhabitants of the test area and two control areas. One purpose of the followup studies was to examine the stability of annoyance over time. Stability coefficients were generated from the application of a "test-retest" procedure. The authors reported that "these stability-coefficients are sufficiently high to support the conclusion that in dealing with odor annoyance we are obviously dealing with a markedly stable attitude, amenable to scientific study."[48 (p. 474)]

Winneke and Kastka then studied changes of annoyance with distance from the source. Surveys were conducted in communities near a tar-oil plant in Duisburg and an insulation-material plant in Cologne. The effluent from the tar-oil plant (an area source of odors) consisted mainly of hydrocarbons and sulfur compounds, and that from the insulation plant (a point source), mainly of phenolic compounds. As illustrated in Figure 3-2, the change of annoyance with distance was found to be completely different for these plants. By closely examining the patterns of annoyance near the insulation plant (broken lines), one can see that there are marked decreases in the degree of annoyance between about 100 and 300 m from the plant boundary for factors F2 ("social-emotional disturbance") and F3 ("somatic disturbance"). Statistical comparisons (Scheffé method) between these points were highly significant for both factors, but not for F1 ("sensory experience"). The authors concluded that "this supports our hypothesis that social-emotional and somatic aspects of annoyance are in fact indicative of more severe degrees of disturbance than are sensory or stimulus-centered odor experiences. This, furthermore, illustrates the superiority of a multi-dimensional approach to odor-annoyance as opposed to one-dimensional odor-indexes."[48 (p. 477)]

The difference in the patterns of the multidimensional annoyance reaction at increasing distances from the tar-oil and insulation plants led the authors to attempt to relate these patterns to patterns of odor exposure. A program was developed to measure ambient-odor concentrations in the vicinity of the two sources. The program included acquisition of a mobile laboratory equipped with olfactometers and an initial phase of operation to cover 1 yr. Odor concentrations were determined according

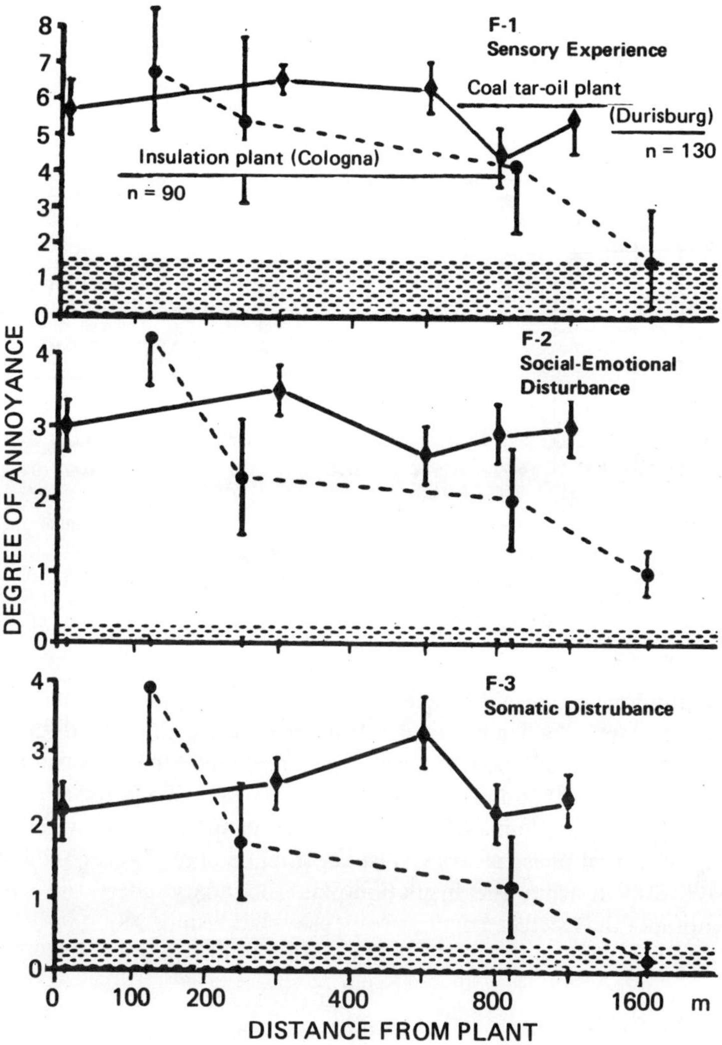

FIGURE 3-2 Change of annoyance with distance from plant boundary for three annoyance dimensions and two different plants. Width of shaded areas represents tentative background annoyance, as estimated from control areas surveyed in Düsseldorf. Reprinted with permission from Winneke and Kastka.[48]

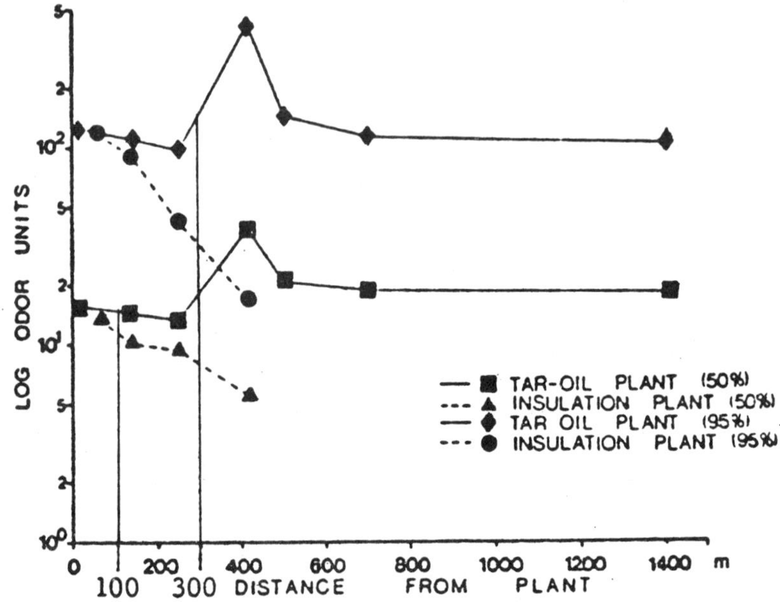

FIGURE 3-3 Median (50%) and extreme (95%) values of odor concentration, expressed as odor units (i.e., the ratio of the volume of odor-free air to the volume of the ambient-air sample at odor threshold). Reprinted with permission from Winneke and Kastka.[48]

to the dilution-to-threshold method. The measurements were made in the lee of the plants. Two subjects at a time were asked to make dilution-to-threshold determinations, with each determination taking approximately 2 min. Median (50%) and extreme (95%) values were computed at different distances from the plants. The values shown in Figure 3-3 were obtained during the initial phase of work. Visual comparison of Figures 3-2 and 3-3 for 100–300 m from the insulation-plant boundary offers the clearest indication of the existence of dose-response relationships.

CONCLUSIONS

There is extensive evidence from laboratory studies with animals that some odorants can induce marked physiologic and morphologic changes—notably cardiovascular and respiratory. These may be distinct from toxic effects. Corresponding studies with humans are lacking and, in some cases, impossible. Nevertheless, reported symptoms of those exposed to air pollution show close parallels with such responses. This should alert us to the strong possibility that comparable physiologic changes underlie human responses to odors associated with expressions of irritation or annoyance. Moreover, some changes may also occur without being perceived by the person who undergoes them. Their impact on health, if any, is yet to be established and urgently requires study.

Regardless of their underlying cause, many of the unpleasant sensations associated with exposure to odorants pose a serious threat to the sense of well-being: nausea, depression, headache, and even the belief itself that malodors are threatening health are conditions that must be considered as potentially, or in fact, damaging to health.

Groups that are particularly susceptible include asthmatics and others with preexisting respiratory and cardiovascular disease and allergic persons.

The use of information obtained by means of spontaneous complaints about odors is a poor method of measuring community reaction.

The social-survey method of measuring community reaction to odors is a more accurate method than relying on odor complaints. Properly and expertly developed and applied, this method can yield valid and significant results related to annoyance in the community.

RECOMMENDATIONS

Two sets of criteria should be considered in establishing standards for odor emission. The first depends on complaints that the emission has an objectionable odor or causes objectionable symptoms. The second concerns evidence that exposure to odors results in some measurable departure from a normal or control condition—be it anatomic, physiologic, biochemical, or behavioral. Those exposed need not be aware of the change (although they may be); the odors may be pleasant, unpleasant, or neutral; and the receptors that mediate the change may be anywhere in the nasal cavity. The second set of criteria offers some possibility of defining emission standards for a range of substances in terms of specific concentrations or ranges of concentrations, because it depends on objective measures of selected indicators.

There are established methods for measuring most if not all of the effects, and research with animal models should be undertaken to determine whether representative odors or emissions elicit changes that constitute substantial departures from the normal. Heart rate, respiratory rate, and electric activity in the olfactory system can be monitored with chronically implanted electrodes in freely moving rodents in an odor exposure chamber such as that described by Laing and Panhuber.[31] Indicators of nasal mucosal function include rates of ciliary activity and mucus flow, pH, mucus thickness, and changes in nasal resistance.[14,16] Techniques for investigating altered morphology in mitral cells have been described by Laing and Panhuber.[31] If odorants have adverse effects on body functions, they may not be drastic. Thus, a battery of techniques, each with its own resolving power, may be needed to establish the nature and importance of any phenomena that emerge.

Despite the difficulty of such investigations, populations exposed to both pleasant and offensive odors should be studied to determine whether they contain higher proportions of people who suffer from alterations of the senses of smell and taste, including reduced sensitivity to odors, than do control populations. If increased proportions are found, an attempt should be made to establish whether those people have a higher incidence of work-related accidents associated with failure to detect odors or have reduced enjoyment of food and poor nutrition, which might be associated with altered appreciation of flavor. Particular attention might be paid to workers who live near an odorant source to which they are also exposed during working hours; to the very young and very old; and to those with cardiovascular or respiratory disease.

The regulation of substances that are released to the atmosphere and are of public concern for no other reason than their characteristic foul odor is a necessary function of air-pollution control agencies and should continue to be pursued with vigor. This follows from the recognition that prolonged exposure to foul odors usually evokes undesirable reactions in people that can range from unease, discomfort, irritation, and anger to violent physiologic manifestations, including circulatory and respiratory effects, nausea, vomiting, and headache severe enough to lead to prostration. Psychophysical and other studies are needed to make it possible to predict, before a potential odor-emission source is constructed, the degree of odor control that will be needed at the source to avoid community displeasure. It will be especially important for planning and enforcement purposes to learn what intensities of odor above a recognition threshold are tolerable for what periods and exposure frequencies. This research effort should inform the setting of additional regulations for control of malodors.

The social-survey method of measuring community reaction to odors should be refined and tested for use by enforcement agencies and others as a substitute for using odor-complaint information. Consideration should be given to the relative usefulness of the different dimensions of annoyance in performing this work.

Further attempts should be made to develop dose-response information for odorous emission from pulp and paper mills, oil refineries, chemical plants, feedlots and stockyards, and other sources of widespread emission of odorous substances.

REFERENCES

1. Alarie, Y. Irritating properties of airborne materials to the upper respiratory tract. Arch. Environ. Health 13:433–449, 1966.

2. Alarie, Y. Sensory irritation of the upper airways by airborne chemicals. Toxicol. Appl. Pharmacol. 24:279–297, 1973.

3. Allison, D. J., and D. A. Powis. Adrenal catecholamine secretion during stimulation of the nasal mucous membrane in the rabbit. J. Physiol. 217:327–339, 1971.

4. Allison, D. J., and D. A. Powis. Early and late hind-limb vascular reponses to stimulation of receptors in the nose of the rabbit. J. Physiol. 262:301–317, 1976.

5. American Medical Association. The Physician's Guide to Odor Pollution. Prepared by J. A. Bell. Chicago: American Medical Association, 1973.

6. Batsel, H. L., and A. J. Lines. Neural mechanisms of sneeze. Amer. J. Physiol. 229:770–776, 1975.

7. Borsky, P. N. Effects of Noise in the Community. Paper presented at 138th meeting of the American Association for the Advancement of Science, Philadelphia, December 26–31, 1971.

8. Breuninger, H. Uber das physikalisch-chemische verhalten des Nasenschleims. Archiv fuer Ohren-Nasen-und Kehlkopfheilkunde 184(2):133–138, 1964.

9. Cederlöf, R., L. Friberg, E. Jonsson, L. Kaij, and T. Lindvall. Studies of annoyance connected with offensive smell from a sulphate cellulose factory. Nord. Hyg. Tidskr. 45:39–48, 1964.

10. Charleton, W. (1654) Physiologia Epicuro-Gassendo-Charltoniana: Or a fabrick of science natural, upon the hypothesis of atoms. New York: Johnson Reprint, 1966.

11. Copley International Corporation. A Study of the Social and Economic Impact of Odors. Phase I. National Survey of the Odor Problem. Report to the National Air Pollution Control Administration. Contract No. CPA 22-69-50. Washington, D.C.: U.S. Department of Health, Education, and Welfare, Public Health Service, NAPCA, January 1970. 246 pp.

12. Copley International Corporation. A Study of the Social and Economic Impact of Odors. Phase II. Report to the U.S. Environmental Protection Agency. Contract No. CPA 70-116. La Jolla, Calif.: Copley International Corporation, November 1971. 308 pp.

13. Copley International Corporation. A Study of the Social and Economic Impact of Odors. Phase III. Development and Evaluation of a Model Odor Control Ordinance. Report to the U.S. Environmental Protection Agency. Publ. No. EPA-650/5-73-001. Contract No. 68-02-0095. Washington, D.C.: U.S. EPA, February 1973. 203 pp.

14. Dalhamn, T. Mucous flow and ciliary activity in the trachea of healthy rats and rats exposed to respiratory irritant gases. Acta Physiol. Scand. 36:Suppl. 123, 1956. 161 pp.

15. Dalhamn, T., and J. Rhodin. Mucous flow and ciliary activity in the trachea of rats exposed to pulmonary irritant gas. Brit. J. Indust. Med. 13(2):110–113, 1956.

16. Davis, T. R. A., S. P. Battista, and C. J. Kensler. Mechanism of respiratory effects during exposure of guinea pigs to irritants. Arch. Environ. Health 15:412–419, 1967.

17. Deane, M., and J. R. Goldsmith. Health Effects of Pulp Mill Odor in Anderson, California. APTIC No. 39242. Presented at the Conference on the Dose–Response Relationships Affecting Human Reactions to Odorous Compounds, Cambridge, Mass., April 26–30, 1971. Preprint, U.S. Environmental Protection Agency, Washington, D.C., 1971. 22 pp.

18. Deane, M., G. Sanders, and E. Jonsson. Trends and community annoyance reactions to odors from pulp mills Eureka, California 1969–1971. Environ. Res. 14:232–244, 1977.

19. de Lorenzo, A. J. D. The olfactory neuron and the blood–brain barrier, pp. 151–176. In G. E. W. Wolstenholme and J. Knight, Eds. Taste and Smell in Vertebrates. London: J. and A. Churchill, 1970.

20. Doving, K. B., and A. J. Pinching. Selective degeneration of neurones in the olfactory bulb following prolonged odour exposure. Brain Res. 52:115–129, 1973.

21. Editors, Edinburgh Medical and Surgical Journal. Observations by the Editors, pp. 297–300. In Chap. V. Report of the Physical and Mathematical Class of the Institute, upon the Question, Are those Manufactures which emit a disagreeable Smell prejudicial to Health? In Pt. 1, Original Communications. The Edinburgh Medical and Surgical Journal, July 1, 1806. Vol. 2.

22. Flesh, R. D., and A. Turk. Social and economic effects of odors, pp. 57–74. In P. N. Cheremisinoff and R. A. Young, Eds. Industrial Odor Technology Assessment. Ann Arbor: Ann Arbor Science Publishers, Inc., 1975.

23. Friberg, L., E. Jonsson, and R. Cederlof. Studies on sanitary nuisance of smoke from a sulphate cellulose factory (I). An interview study. Nord. Hyg. Tidskr. 41:41–43, 45–50, 1960. (in Swedish, summary in English)

24. Goldsmith, J. R. Health and Annoyance Impact of Odor Pollution. Publ. No. EPA-650/1-75-001. Contract No. 68–02–0083. Research Triangle Park, N.C.: U.S. Environmental Protection Agency, Human Studies Laboratory, National Environmental Research Center, 1973.

25. Haggard, H. W. Devils, Drugs, and Doctors. New York: Harper & Bros. Publishers, 1929. 405 pp.

26. Heathrow II. Second survey of aircraft noise annoyance around London (Heathrow) Airport. Report prepared by MIL Research Ltd. for the Office of Population Censuses and Surveys, Social Survey Division. London: Her Majesty's Stationery Office, 1970.

27. James, J. E. A., and M. de B. Daly. Nasal reflexes. Proc. Roy. Soc. Med. 62:1287–1293, 1969.

28. Jonsson, E. Annoyance reactions to environmental odors, pp. 329–333. In A. Turk, J. W. Johnston, Jr., and D. G. Moulton, Eds. Human Responses to Environmental Odors. New York: Academic Press, 1974.

29. Jonsson, E., M. Deane, and G. Sanders. Community reactions to odors from pulp mills: A pilot study in Eureka, California. Environ. Res. 10:249–270, 1975.

30. Kendall, D. A., and T. Lindvall, Eds. Evaluation of Community Odor Exposure. Cambridge, Mass.: Arthur D. Little, 1971. 80 pp.

31. Laing, D. G., and H. Panhuber. Neural and behavioral changes in rats following continuous exposure to an odor. J. Comp. Physiol. 124:259–265, 1978.

32. Lindvall, T., and E. P. Radford, Eds. Measurement of annoyance due to exposure to environmental factors. A report from the Fourth Karolinska Institute Symposium on Environmental Health. Environ. Res. 6:1–36, 1973.

33. McKennell, A. C. Aircraft noise annoyance around London (Heathrow) Airport: An inquiry carried out for the Office of the Minister for Science at the request of the Wilson Committee for Inquiry on the Problem of Noise. Publication No. S.S.337. London: Social Survey Division, Office of Population Censuses and Surveys, 1963.

34. McKenzie, D. Aromatics and the Soul: A Study of Smells. London: William Heinemann (Medical Books) Ltd., 1923. 164 pp.

35. Medalia, N. Z. Community perception of air quality: An opinion survey in Clarkston, Washington. U.S. Public Health Service Publ. No. 999-AP-10. Cincinnati, Ohio: U.S. Department of Health, Education, and Welfare, Public Health Service, Robert A. Taft Sanitary Engineering Center, 1965.

36. Nauš, A. Alterations of the smell acuity caused by menthol. J. Laryngol. Otol. 82:1009–1011, 1968.

37. Nelson, B. Air pollution: The "Feds" move to abate Idaho pulp mill stench. Science 157:1018–1021, 1967.

38. Okano, M., and S. F. Takagi. Secretion and electrogenesis of the supporting cell in the olfactory epithelium. J. Physiol. (London) 242:353–370, 1974.

39. Pinching, A. J., and K. B. Doving. Selective degeneration in the rat olfactory bulb following exposure to different odours. Brain Res. 82:195–204, 1974.

40. Ross, G. R. T. Aristotle: De sensu and De memoria. Cambridge: Cambridge Univ. Press, 1906. 303 pp.

41. Schneider, R. A., and S. Wolf. Relation of olfactory acuity to nasal membrane function. J. Appl. Physiol. 15:914–920, 1960.

42. Smith, W. S., J. J. Schueneman, and L. D. Zeidberg. Public reaction to air pollution in Nashville, Tennessee. J. Air Poll. Cont. Assoc. 14:418–423, 1964.

43. Sullivan, R. J. Preliminary Air Pollution Survey of Odorous Compounds. A Literature Review. Prepared for the Department of Health, Education, and Welfare. National Air Pollution Control Administration Pub. No. APTD 66-42. Bethesda, Md.: Litton Systems, Inc., 1969. 244 pp.

44. Swedish Institute for Building Research. Traffic Noise in Dwelling Areas: An investigation by the National Swedish Institute for Building Research and the National Swedish Institute of Public Health. Report No. 36E. Stockholm: Statens institut for byggnadsforskning, 1968.

45. The Third Karolinska Institute Symposium on Environmental Health. Methods for Measuring and Evaluating Odorous Air Pollutants at the Source and in the Ambient Air. Nord. Hyg. Tidskr. 51(2):1–77, 1970.

46. Tracor, Inc. Community Reaction to Airport Noise. Vol. 1: Contractor Report No. 1761, NASA Contract NASW-1549. July 1971. 97 pp. Vol. 2: NASA No. CR-111-316. Sept. 4, 1970. 245 pp. Washington, D.C.: National Aeronautics and Space Administration.

47. U.S. Bureau of the Census. County and City Data Book, 1972 (A Statistical Abstract Supplement). Washington, D.C.: U.S. Government Printing Office, 1973.

48. Winneke, G., and J. Kastka. Odor pollution and odor annoyance reactions in industrial areas of the Rhine–Ruhr region, pp. 471–479. In J. Magnen and P. MacLeod, Eds. Olfaction and Taste VI. London: Information Retrieval, Ltd., 1977.

49. Winslow, C.-E. A. Fresh Air and Ventilation, pp. 29–31. New York: E. P. Dutton and Company, 1926.

50. Winslow, C.-E. A. The Conquest of Epidemic Disease. Princeton: Princeton University Press, 1943. 411 pp.

51. Winslow, C.-E. A., and L. P. Herrington. The influence of odor upon appetite. Amer. J. Hyg. 23:143–156, 1936.

52. Zippel, R., and R. Ebel. Vergleichende pH-Messungen an Schleimhäuten der Nase und des Mundrachens bei Gesunden und Laryngektomierten. Arch. Orh.-, Nas.-d, und Kehlk., Heilk. 201:49–56, 1972.

4 Measurement Methods

Odorants carried with emission into the atmosphere disperse and sometimes react in the atmosphere and produce odors in the ambient air that are perceived by people in communities. The ambient odors constitute a dose to which the affected population may respond in a variety of ways, as shown in Figure 4-1. Thus, the measurement of odors can be considered at three levels: odors can be measured at their source, they can be measured in the ambient air, and their impact on the community can be measured if appropriate scientifically valid methods are available. This chapter deals with analytical and sensory measurement methods that may be applicable to the evaluation of the impact of air-pollution odors in emission and in ambient air. Techniques for the measurement of the social and economic impact of air pollution odors on the community are dealt with in Appendix B.

The intent of this review is to reflect the present state of the art of odor measurement. It is based on information obtained from the open literature, from working papers and reports of various societies and industrial organizations, and from direct correspondence and interviews with groups currently engaged in odor measurement or development of methods for odor measurement. A considerable fraction of the information resulted from various current and past activities of the American Society for Testing and Materials (ASTM) E-18 Sensory Evaluation Committee (including cooperative exercises, e.g., with the TT-4 Odor Committee of the Air Pollution Control Association).

The following principal topics are discussed:

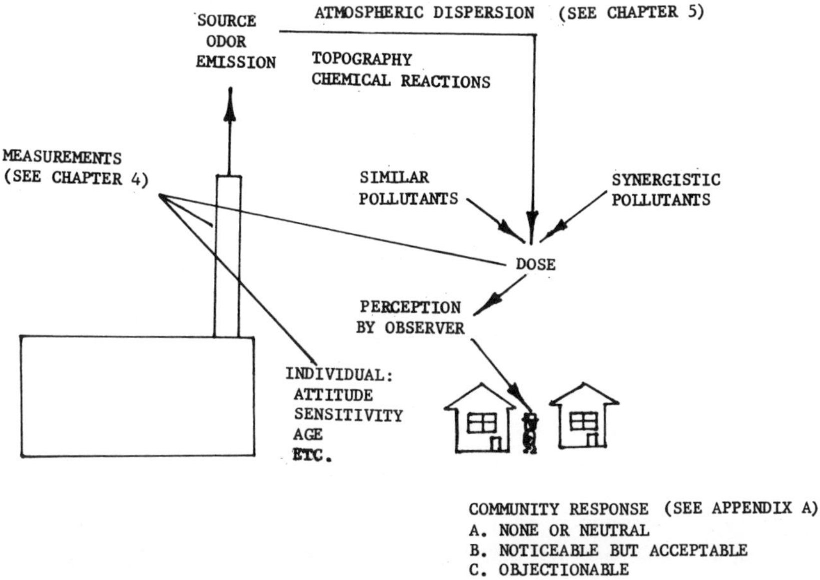

FIGURE 4-1 Factors affecting odor dose response.

- Analytical measurements
- Sensory measurements of odor detectability, odor threshold, odor intensity, the dose-response function (change in intensity with dilution), odor character, and odor pleasantness and unpleasantness
- Panelist selection for odor evaluation
- Sampling for analytical and sensory measurements

Each of these topics is very complex, so only the principal features of the state of the art can be highlighted here; for further details, the original references must be consulted. (However, if the availability of information allows, comparisons of methods of measuring the same property of odors are discussed.) The references have been selected to illustrate the main points of the discussion and reflect the prevailing variety of opinions, rather than to produce an exhaustive but unwieldy bibliography.

The art of odor measurement is quite mature; the goal now is to refine techniques so that methods that are both scientifically sound and reasonably practical can be selected. The discussion section of this chapter deals with such selection. However, after the selection, large gaps will remain in relating the measurement data to the impact of an odor on a community. Only after such gaps are filled can useful criteria for odor control be developed. Possible approaches to this problem are also outlined in the discussion.

An odorous emission or odorous ambient air usually contains large numbers of different odorous substances. The resulting odor sensation is a composite effect of such a mixture and depends on the concentration and odorous properties of the component compounds. Accordingly, an odorous gaseous sample can be characterized by two approaches, serving somewhat different needs: in *analytical terms*, by determining the components and their concentrations, and in *sensory terms*, by measuring human responses to the odor of a sample.

Analytical data are more precise than sensory data and may provide information on the origin of the odorous components. This is useful in modifying a manufacturing process or raw materials and in selecting emission treatment methods to reduce the odor. In the case of ambient-air odors, analytical data may assist in pinpointing the odor source and in plotting the dispersion of odorous emission in a complex meteorology and topography. It is not yet possible to describe the odor of a composite sample from its analytical composition, because the available knowledge on odors of mixtures of odorants is meager.

Sensory data provide direct information on the effect of an odorous sample on the human sense of smell. The principal sensory properties of odors are intensity, change in intensity with dilution, threshold, quality or character, and hedonic tone (pleasantness-unpleasantness).[11,80] The annoyance potential of odors in the context of odorous pollution depends on all these properties.

ANALYTICAL MEASUREMENTS

An odor in air is usually a result of the presence of many odorants; only rarely is a single substance responsible for an air-pollution odor. There are some single-substance odors in the manufacture of specific chemicals, such as ammonia and hydrogen sulfide. In such cases, a suitable analytical method can be applied.[63]

ANALYTICAL SENSITIVITY NEEDED

The sensitivity of the analytical method must exceed the sensitivity of the human sense of smell to the particular odorous substance for which a sample is analyzed. Because the sensitivity of the human sense of smell for different odorants may cover a broad range of concentration, the required instrumental sensitivity may be quite different for different odorants.

As a generalization, analytical sensitivity down to 0.1 ppb (vol/vol) will be adequate for any odorant, inasmuch as this is the lowest known average odor detection threshold for humans. Most thresholds are above this benchmark; some people, however, may be more sensitive than the population mean by a factor of 10.[14,130,133] For most single odorants, less sensitivity may suffice. For example, the analytical sensitivity need not exceed 10 ppm for ammonia or 1 ppb for hydrogen sulfide.[133]

In a few cases, odor-control regulations specify the maximal permissible concentrations of specific odorants in emission and the analytical methods for measuring them. The concentrations that are specified are based on odor detection thresholds and dispersion equations that permit the ambient concentrations of odorants to be estimated from the emission concentrations.[56] Thus, the Bay Area Air Pollution Control District[56] regulates the emission of trimethylamine, phenol, methyl mercaptan, dimethyl sulfide, and ammonia; there are adequate specific analytical methods for these odorants. In West Germany, selected odorants are classified in three groups on the basis of their potential odorosity, and different values of permissible emission concentration are specified for each group; the method of analysis is gas chromatography (GC).[129]

MIXTURES OF ODORANTS

In most cases of odorous air pollution, many odorants are present; and in many of these cases, many odorants are present at concentrations significantly above their odor thresholds. There are indications that subthreshold concentrations of several odorants together may result in an odor above the detection threshold.[16,83] Usually, however, a few odorants are responsible for the characteristic odor of a complex mixture, such as diesel-engine exhaust,[15,101] rendering-plant emission,[117] and tobacco smoke.[47,53]

Food scientists have been the leaders in developing methods for analysis of complex mixtures of odorants and flavors, and many of their methods are now in use in research on air-pollution odors. Such methods deal with sample acquisition, gas-chromatographic separation of the sample components (e.g., with capillary columns), and mass-spectrometric (MS) identification of the components sequentially eluted from the GC column. In the GC methods, chemical compounds are separated on the basis of their vapor pressures and solubilities in the GC materials. In the MS methods, these compounds are identified by their ionized molecular fragmentation patterns. With proper selection of GC column material and GC temperature programing procedures, methods can be adapted for analyzing any odorant or many odorants simultaneously.

These methods have several weak points. They are expensive and therefore are suitable principally for exploratory research, rather than for routine monitoring. Their main use is in discovering the usual components of some odorous emission type. They do not indicate which sample components are relevant to the odor of the complete sample—that is, have occurred at significantly odorous concentrations and have exhibited characteristic odors. This information cannot be deduced from concentration data, unless the odor-concentration relationship is known for all the components measured. In the present state of the art, it is common for many components identified in an odorous emission—alcohols, many noncyclic hydrocarbons, etc.—to have little to do with the odor of the sample. It is also common for many identifications to remain ambiguous or questionable.

"ODOROGRAM" METHOD

In a variant method using GC, a sensory evaluation is conducted on the GC column effluent. In this "odorogram" method, the column effluent is split; one part is delivered to a hydrogen-flame ionization detector, the other part to a sniffing port.[15,39,46,53,60,62,101,117] The detector is sensitive to almost all organic compounds (some exceptions are formaldehyde, carbon tetrachloride, compounds with no CH group, and compounds with a substantial content of oxygen atoms in a small molecule) and responds approximately in proportion to their mass flow rate.[77] A recorder monitors the detector's response for each component eluted from the GC column, forming only one peak for each component. However, peaks for some substances can overlap significantly or coincide. The analyst evaluates the odors of the components as their vapors emerge from the sniffing port simultaneously with the appearance of the peaks on the recorder chart and marks the odor notes by the peaks. A given component produces a GC peak at the same characteristic elution time in different samples, subject to minor fluctuations caused by instrumental factors. The area transcribed under the peak is proportional to the amount of the component introduced into the chromatograph. An example of an odorogram is shown in Figure 4-2.

Odors have no relation to the relative peak sizes for different components. Some components may exhibit very strong odors even if their peaks are small. Others, such as aliphatic hydrocarbons may produce large peaks without exhibiting an odor. If the sample size is sufficient, significant information on odor can be derived from the chromatogram: components that have strong and characteristic odors can be indexed in terms of their odors and GC elution times. Thus, a catalog of the GC elution times and of the odors of the sample components may be compiled and serve as a "signature" of the sample. If the chemical identity of some components is of interest, the mass-spectrometric identification can concentrate on these components. As an example, the odorogram technique led to a conclusion that, although diesel exhaust contains many hundreds of compounds, only a few dozen were reasonably relevant to odors.[15,48,101]

In odor-control efficiency studies, odorograms can be compared without resorting to MS. Comparisons of the odorograms of untreated odorous emission and of emission treated by some odor-control process may indicate which odorants have decreased in their concentration and to what extent. For example, odorograms can indicate (as sometimes happens during incomplete oxidation) whether new odorous components were generated by the control process. Emission odorograms can be compared with those from raw materials and materials at different stages of an industrial process to pinpoint the origin of emitted odorants. In the case of odors in ambient air, odorograms of ambient-air samples can be compared with the odorograms (sometimes named "chemical signatures") of odorous emission of several possible sources.

The sampling of odorous gases for analytical measurement is discussed later in this chapter.

METHODS BASED ON CORRELATIONS OF ANALYTICAL MEASUREMENTS WITH SENSORY MEASUREMENTS

Presumably, if the analytical composition of enough samples of a specific odorous-pollution type is known and the odor characteristics of the samples are measured by sensory means, statistical methods can be used to develop predictive equations relating, for example, odor intensity to the composition. The statistical methods are usually forms of multiple stepwise regression analysis. Much work in this direction has been done in the food and beverage industry,[15,34,52,107] and similar methods have been applied to tobacco-smoke odor[14] and rendering odor.[31]

Short-cut methods may be based on the premise that some chemical families of compounds are primarily responsible for odors of specific types of emission. Composite analytical indexes can be used to characterize the content of such groups of compounds.

Diesel Odors

The diesel-odor analytical system (DOAS) is based on the content of aromatic and oxygenated organic compounds in diesel exhaust.[15] A large sample of diesel exhaust (500 L) is collected by adsorption in a polymeric adsorbent and extracted with cyclohexane for analysis in a liquid

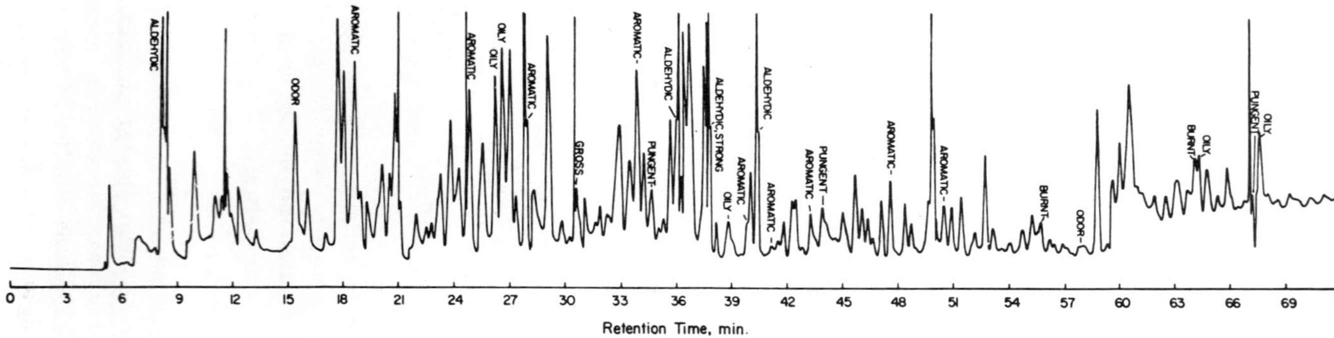

FIGURE 4-2 Typical odorogram of diesel-engine exhaust.[101] Ordinate, flame ionization-detector response. Column, 200-ft 0.020-in.-i.d., support-coated, open tubular GC column with Apiezon *L* stationary phase, programed at 2°C/min from 40°C. From A. Dravnieks (personal communication.)

chromatographic system with an ultraviolet detector to produce a response for pooled aromatics and for oxygenates. Instrumental and procedural improvements in the original method have been reported. The correlations between perceived odor intensity and odor index data based on DOAS were found to be poor;[30] the unburned hydrocarbon content correlated with the odor intensity better than this index. Exhausts from different engines did not show the same correlation trends.

Paper- and Pulp-Mill Odors

The odors from paper and pulp mills usually are caused by hydrogen sulfide and lighter mercaptans and organic sulfides. Instruments that measure overall constituents, such as total sulfur or organic reducible sulfur, may produce data that correlate reasonably with the analytical data on specific contents of hydrogen sulfide (H_2S), methyl mercaptan (CH_3SH), methyl sulfide (CH_3SCH_3), and methyl disulfide (CH_3SSCH_3), because, as long as the industrial process is not significantly changed, similar mechanisms lead to the production of these odorants (Cederlof *et al.*;[26] R. Blosser, personal communication). Multiple correlation coefficients with statistical significance as strong as $p < 0.001$ (confidence level larger than 99.9%) have been obtained for some emissions of this type.

Rendering Odors

In an exploratory study on rendering-plant emission,[132] infrared absorption due to organic acid groups and methyl disulfide was measured with a MIRAN long-path infrared spectrophotometer and total sulfur content was measured with a Meloy sulfur-hydrocarbon analyzer. With a limited number of samples, indications of possible correlations between odor dilution thresholds and the above analytical data were obtained. Electrochemical determination of concentrations of amines and sulfides has recently been explored.[100] These substances are significant contributors to rendering odors.[31,117]

SENSORY MEASUREMENTS

Sensory measurements of odors are based on human responses to samples of odorous emission or odorous ambient air. (Sampling methods are discussed later.) These responses are supplied by evaluators commonly called panelists (or sensory judges or subjects). The procedures for their selection and their training are discussed later in this chapter.

SENSORY PROPERTIES OF ODOR

The types of human responses sought depend on the particular sensory dimension that is measured. These include odor intensity, detectability, character, and hedonic tone (pleasantness and unpleasantness). The combined effect of these properties is related to the annoyance that may be caused by an odor.

Odor intensity is the strength of the perceived odor sensation and depends in a complex way on the odorant concentration, with which it should not be confused. The intensity of an odor is perceived directly, without knowledge of the odorant concentration or of the degree of dilution of the odorous sample needed to eliminate odor.

When an odorous sample is progressively diluted, the concentration of odorants decreases, and the odor intensity weakens, but not in direct proportion to the extent of the dilution. With further dilution, the intensity eventually becomes so low that detection or recognition of the odor is very difficult. At some statistically defined point of dilution, the *detection threshold* is reached. Somewhat less dilution (i.e., somewhat higher odorant concentration) is needed to recognize the odor; this dilution is termed the *recognition threshold*.

In the context of odorous air pollution, the two properties, intensity and threshold, have different functional significance. In locations where an odor is above the detection threshold, the perceived odor intensity decreases at a rate much lower than would be in proportion to dilution. Thus, the effect of a reduction in the odorous emission on the perceived intensity of the odor may not be obvious. However, even a partial reduction in the odorous emission rate will reduce the geographic area in which the odor will be noticeable.

Odors of equal intensity may differ in *character*. An odor that has a distinct, readily recognizable character tends to be annoying when encountered as an air-pollution odor.

The *hedonic tone* of an odor—its place on a scale of pleasantness and unpleasantness—is closely linked to its character. However, an otherwise pleasant odor may be considered objectionable by the exposed population in the context of industrial emission.

Annoyance experienced by a population exposed to an air-pollution odor is a combined result of the intensity, character, and hedonic tone of the odor, as well as of the frequency and duration of the exposure. Measurement of annoyance is one of the most difficult tasks in odor-pollution work.

ODOR INTENSITY

The most common devices for measuring the perceived intensity of odors are category scales, scales based on magnitude estimates, and reference-sample scales. Odorant concentration, as related to the odor-detection or recognition threshold concentration, is not a satisfactory measure of perceived intensity. Odor-intensity measurements may be used for both emission and ambient-air odors.

Category Scales

A category scale consists of a series of numbers with or without a definition of their meaning. One of the most widely used is that by Katz and Talbert:[81]

0 no odor
1 very faint odor
2 faint odor
3 easily noticeable odor
4 strong odor
5 very strong odor

Other scales may be mentioned to illustrate the diversity of category scales in use. Some have fewer categories, and others have more. For example, there is a nine-category scale:[90]

1 no odor
3 slight odor
5 moderate odor
7 large odor
9 extreme odor

In that scale, the intermediate numbers have intermediate meanings. The following scale evolved from an initial 0–3 scale:[121]

)(threshold odor
1/2 very slight odor
1 slight odor
1 1/2 slight to moderate odor
2 moderate odor
2 1/2 moderate to strong odor
3 strong odor

There is no evidence that larger scales, especially those with more than nine categories, improve the reproducibility of the intensity ratings. Standard deviation tends to increase in proportion to the number of categories, and the 0–5 scale is both simple and probably adequate.

The advantage of the category scales is in their simplicity. But category scales also have disadvantages. First, the numbers are not proportional to the perceived intensities of the odors (rather, they are very approximately proportional to the logarithms of the perceived intensities, except at the ends of the scale). An odor of category 3 is felt to be much more than 50% stronger than an odor of category 2; in fact, on a 0–5 scale, the perceived intensity of an odor increases by a factor of approximately 4 per category;[40] thus, an odor of category 3 will be considered, in a direct comparison, 4 times stronger than an odor of category 2. Second, people differ in their use of a given category scale; this necessitates specialized training of panelists. Third, there tends to be a compression of judgments at the upper end of the scale; although the scale ends with a given category, odors in that category may be quite different in intensity.

Scales Derived from Magnitude Estimates

In its simplest form, a magnitude-estimate scale is developed from a free assignment of a number—any number—thought by a panelist to represent the magnitude of an odor sensation experienced.[24] This permits comparison of perceived intensities of odors of several samples. If odor A has been assigned 15 by a particular panelist, and odor B, 30, odor B is twice as strong as odor A for that panelist. Ratios of numbers in such a scale are related as the ratios of the perceived intensities. Different panelists may use different numbers, but the ratios of the numbers for given samples tend to be similar.

In a more standardized form of the magnitude scale, the panelists are instructed to use one odorous sample as an anchor or reference point, with a definite intensity magnitude number—say, 20—assigned to this sample. Each panelist then compares other samples with this reference sample and assigns numbers to them in proportion to their perceived intensities.

The advantage of the magnitude-estimate scale, compared with the category scale, is that relative intensities are immediately evident. The method is particularly suitable for evaluating changes in odor intensity with dilution.

The disadvantages of this kind of scale may vary. In a comparison of only two different odors, difficulty may arise as to which odor property was actually judged; it may be intensity, as requested, or some other form of impact, such as relative unpleasantness or pungency. If panelists have been using a category scale, it may be difficult for them to change to a magnitude scale. Also, people differ in their perception of odor intensity, as well as in their use of numbers; therefore, many (e.g., over 20) panelists would have to be used to calculate statistically stable mean magnitudes.

One way of producing an anchor or reference point is to ask panelists to give an intensity estimate of an odor that is, for example, imagined to be "moderately strong." Magnitude estimates of different panelists can then be normalized by calculations. "Moderately strong" is permanently assigned a magnitude number—say, 10—and the individual ratings are normalized against this reference number. Thus, in a free choice of numbers, if a panelist rated the intensity of an odor 40 and had assigned a rating of 50 to the "moderately strong" intensity, the normalized intensity of the new odor would be $40/50 \times 10 = 8$. Such a process results in a numerically anchored, essentially free-ended scale in which the magnitudes of numbers are related as the perceived odor intensities, but the individual differences in the use of numbers are smoothed when the mean magnitude is calculated from the normalized values. This process does not account for interpanelist differences in the range of numbers used. Many other techniques for analyzing magnitude-estimate data are in a rapid state of development in industry.

In a graphic form of the magnitude-estimate scale method, panelists are asked to draw lines, or mark off line lengths, in proportion to perceived odor intensities. A line that is twice as long as another indicates an odor that is twice as strong as another odor.

Some researchers have experimented with other types of magnitude responses, e.g., finger span[86] and hand-grip strength (Lindvall;[86] Stevens;[33] A. Dravnieks, personal communication). Apart from scientific interest in and confirmation of laws of different types of psychophysical responses,[24,120] these do not appear to have practical advantages over other methods discussed here.

The measurement of responses by monitoring such physiologic reactions as galvanic skin response and breathing rate is a little-researched field, but it is believed that these responses are influenced by many other factors and are difficult to use for odor measurement.

Reference-Sample Scales

In the reference-sample approach, a series of different concentrations of a selected reference odorant is provided, and each odorous sample is compared with this series to locate the position in the series that most closely matches the odor intensity of the sample. The method has the advantages of avoiding semantic definitions and of not needing any but rudimentary training or data normalization.

Two typical examples of such scales are the Turk *et al.* diesel-exhaust odor-intensity kit[125] and the ASTM E-544 scale based on 1-butanol.[6] MThe Turk *et al.* kit uses a mixture of odorants, imitating diesel-exhaust odor, at 12 concentrations, with an increase in odorant concentration by a factor of 2 per step. The ASTM butanol scale (Figure 4-3) uses 1-butanol vapor as the odorant at eight concentrations in the range between aproximately 10 and 2,000 ppm (vol/vol) in air. These represent eight different odor intensities. The concentration increase is by a factor of 2 per step. The vapor samples are supplied in a steady stream from elliptic 20 $\times$ 35-mm glass sniffing nozzles at the rate of 0.16 L/min. Panelists use these eight intensities for matching to the odorous sample, but they can also select between-intensities positions and positions below the weakest or above the strongest intensity.

In France and some other countries, a pyridine odor reference scale is in use in the household gas industry and has recently been used in measurements of diesel-exhaust odors.[30]

Although some practitioners favor the use of specific reference odor scales for specific odors, that is not essential, inasmuch as different odors can be compared with the same scale, without much loss in reproducibility. Consistency has been observed in matching intensities of different odors to a heptanal scale, compared with the direct matching of the same odors.[85] In developing the ASTM butanol scale, an experiment was conducted, in which 22 meeting attendants matched three odors against the butanol scale.[9] Before the test, some doubted the validity of matching different odors. The standard deviation for butanol itself versus the butanol scale was 0.9 scale step; whereas for hexanal it was 1.1 and for *n*-butylacetate it was 1.2 scale steps. These values were thus only slightly worse than for 1-butanol matched against itself. Judgments by the doubters were not out of line with the judgment means.

In the butanol scale method, odor intensities are recorded merely in terms of the concentration, in parts per million (vol/vol), of 1-butanol vapor that has the odor intensity that matches the intensity of the odorous sample. The value of the odor intensity of 200-ppm butanol is not perceived as twice as large as that of 100-ppm butanol. Thus, butanol parts-per-million values do not provide information on the relative perceived intensities of odors. However, such information can be derived from the parts-per-million values with an equation:[96]

$$S = 0.261\ (C)^{0.66}$$

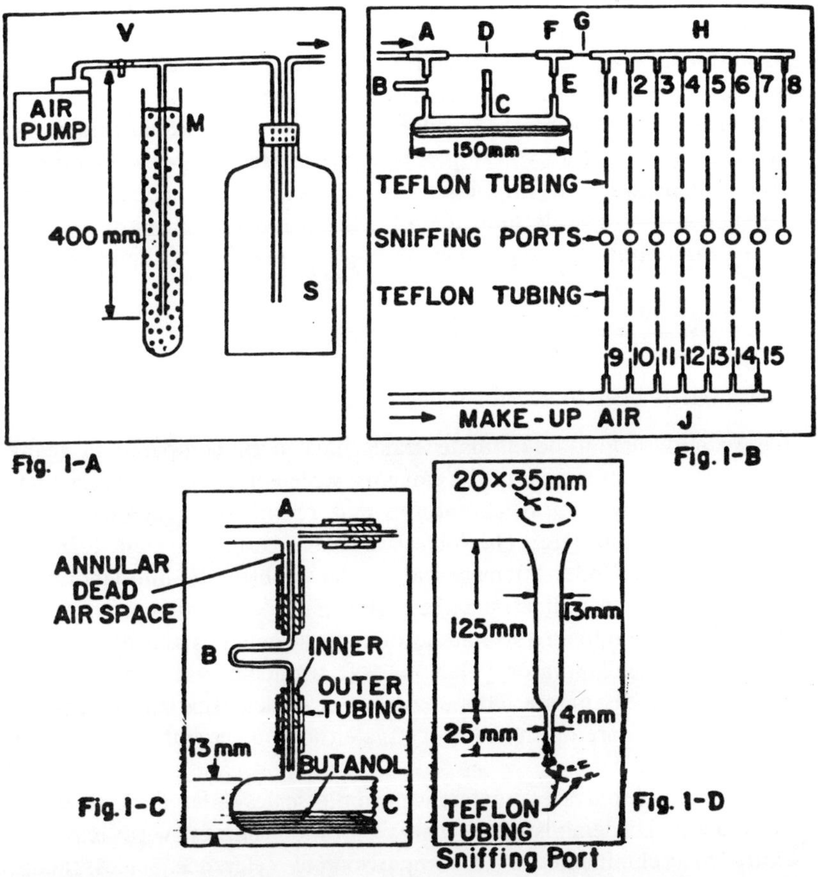

FIGURE 4-3 Butanol-scale olfactometer. The l-butanol is in the vessel, C. Air enters a tee, A, where the air flow is split: one part passes through capillary B, the butanol vessel, picks up its vapor, and exits at E; the other part passes through D and mixes with the saturated vapor of butanol in tee F. The splitter, H, divides the mixture into appropriate fractioned flows delivered to sniffing ports. A separate air flow, the makeup air, is supplied to splitter J and then to the sniffing ports in such amounts that the total flow rate from each port is the same. The odor of a sample, not shown, is compared with the odor from the sniffing ports, to locate the port or a position between the ports that smells as strong as the sample. Reprinted with permission from ASTM.[6]

where S is the perceived odor intensity of the sample that is matched against butanol and C is the concentration, in ppm, of 1-butanol (in air emitted at 0.16 L/min) that matches the sample's odor intensity. The coefficients were derived from magnitude-estimation data by three laboratories and from a definition that the odor of 1-butanol at 250 ppm has an odor intensity, S, of 10. Such odor corresponds to a point between categories 3 and 4 on the 0–5 category scale. The S scale has the following characteristics: at S below 1, an odor is close to its threshold intensities; the ratio of S values for two samples is approximately equal to the ratio of the odor intensities of the samples; the scale has no definite end points; and experience with the scale indicates that the standard deviation for S in repeated evaluation of a given odor by a given panel is of the order of 2 units.

Comparison of Odor-Intensity Scales

Category and magnitude-estimate scales have been compared in many studies.[24,40,120] A comparison of the category scale and the butanol scale has been conducted by several laboratories in a round-robin exercise by the ASTM E-18 Committee.[9] The objective was to find which form of communication of odor-intensity data is less subject to interlaboratory variability. Figure 4-4 illustrates the result. Each segment represents one laboratory. The point in the middle of each segment is the mean value reported by the laboratory; lengths of segments indicate standard deviations for the separate laboratories. It is evident that the means are dispersed much more on the category scale than on the butanol reference scale.

Figure 4-5 compares two reference-sample scales—the Turk *et al.* kit and butanol. It demonstrates that calibration curves with expected linearity are obtainable in a comparison of reference scales. Some compositional changes seem to have occurred in the several-year storage of the mixture imitating diesel-exhaust odor.

The principal advantage of the reference scales is that they may be, and are, used in two different modes. In one, samples of odorous air are directly compared with the series of references. In the other, the scale is used to train panelists, and the panelists then use it by memory, e.g., walking through a neighborhood and recording their judgments of ambient-odor intensities at different locations and times, with reference to the scale.

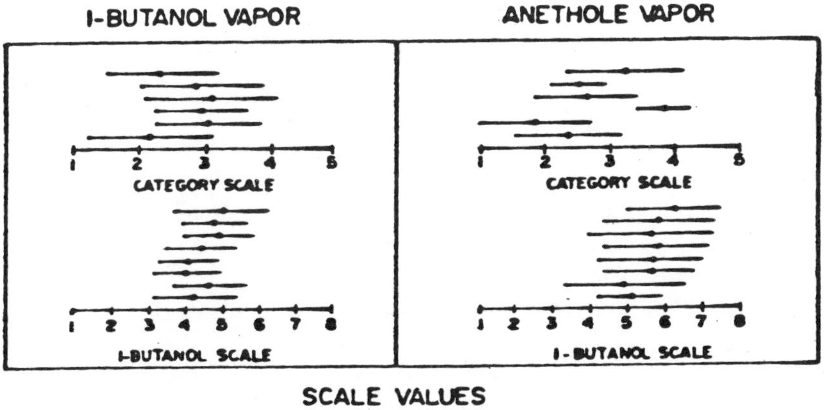

FIGURE 4-4 Odor-intensity evaluation versus category scale and versus 1-butanol scale in multilaboratory exercise by ASTM E-18 Committee. Left, matching butanol to its own scale. Right, matching anethole to butanol scale. Each line segment represents data from a separate laboratory. The center points in the segments are means. Lengths of segments indicate standard-deviation range on each side of mean for panelists of that laboratory. Reprinted with permission from ASTM.[9]

THE DOSE-RESPONSE FUNCTION (CHANGE IN INTENSITY WITH DILUTION)

Small Changes in Concentration

When the concentration of an odorant in an emission is reduced by some form of odor control or by dilution in the atmosphere, the perceived odor intensity decreases. For a small change in the odorant concentration, a particular decrement of concentration is needed to produce a just-noticeable change in the perceived odor intensity. For a given odorant, this fractional change in concentration tends to be roughly constant and is known as the Weber fraction. In the nineteenth century, Gustav Fechner used the Weber fraction to derive a logarithmic law of sensory magnitude. The law, known as the Weber-Fechner law, states that sensory magnitude grows as a logarithmic function of stimulus magnitude.[12] It is now considered an obsolete and inadequate expression of intensity change with dilution. "Odor-difference threshold" is another term used to describe the minimal concentration change needed to produce a just-noticeable difference (jnd) in the perceived odor intensity. Earlier researchers reported that a concentration change of 15–30% was typically needed to produce a jnd in intensity. More exact recent work[22] has indicated that,

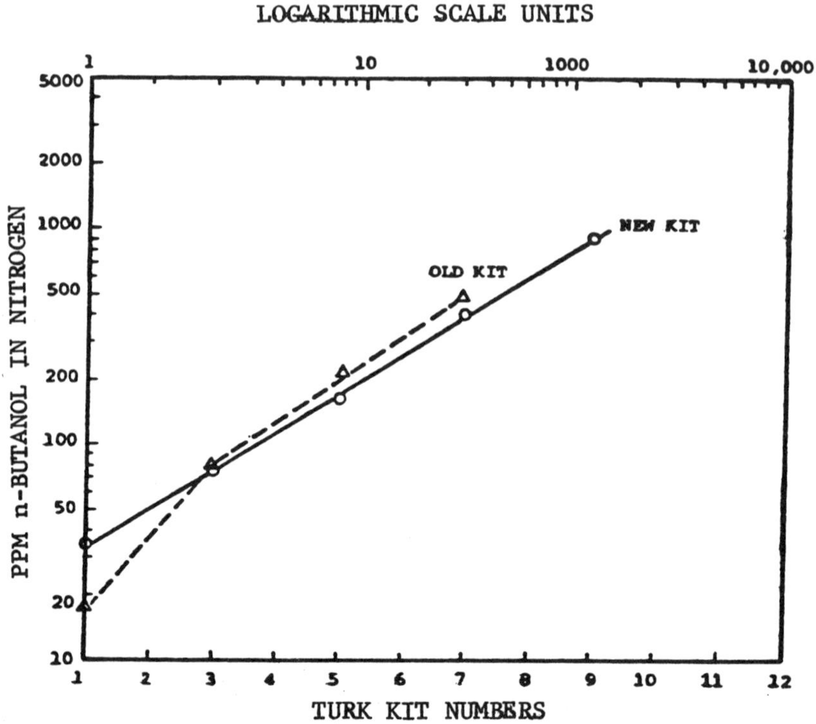

FIGURE 4-5 Comparison of *n*-butanol and Turk *et al.* kit scales ("old kit" is several years old). Storage of the kit resulted in some decrease in odor intensity for the weakest-odor samples of the scale. From IITRI files, Project 6183, Chemical Species in Engine Exhaust and Their Contribution to Exhaust Odors. Project supported by NAPCA and CRC.

with increased precision in odorant concentrations and better control of sample presentation, the jnd is as low as 5% for some odorants.

Large Changes in Concentration

Most modern work on odor intensity uses the relation known as the psychophysical power law, sometimes referred to as Stevens's law,[24,105,120] which is operative for all kinds of sensations. One of its principal formulations is $S = kI^n$ where S is the perceived intensity of sensation, I is the physical intensity of the stimulus (odorant concentration, C, for odor stimulus), and k and n are constants. The values of n are different for different senses. For odors, n is less than unity and is different for different odorants,[23,24,105] typically in the range of 0.2–0.7. Factors that influence judgments of odor intensity have recently been exhaustively reviewed.[74]

In the logarithmic form, the relation for odors becomes

$$\log S = \log k + n \log C,$$

where C is the concentration of the odorant and n is the slope of the psychophysical function. The value of n is measured by applying an appropriate sensory intensity measurement method to odorous-pollution samples of several dilutions. Typically, either a magnitude-estimate scale or a reference scale, such as the butanol scale, is used. A plot of log S against log C produces a straight line if the above equation is obeyed. In some cases, a more complicated intensity-concentration relationship is observed.[37] Figure 4-6 shows examples of normal and abnormal relationship. Such plots for individuals may differ considerably, but plots obtained by averaging group responses can be statistically stable, with n values agreeing to within 0.1 unit.[23,105]

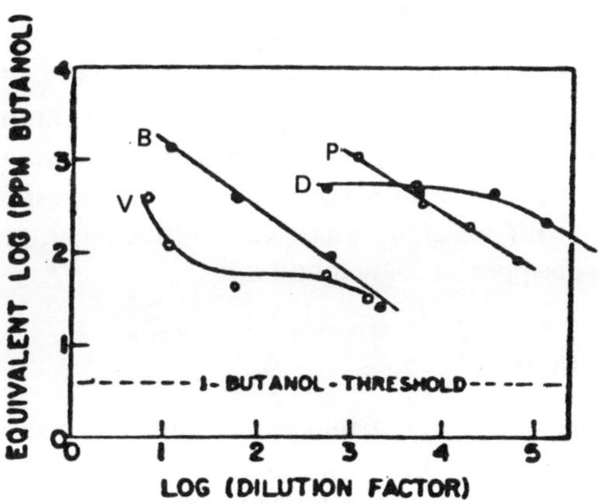

FIGURE 4-6 Normal and abnormal forms of odor-intensity change with dilution. Dilution factor refers to dilution from saturated vapor of the odorant. B, benzaldehyde; D, diacetyl; P, pyridine; V, vanillin. Reprinted with permission from Dravnieks.[37]

Significance of Dose-Response Function in Odor Control

The usefulness of measuring the *S–C* relationship in odorous-pollution control is in estimating the extent of the decrease in odor intensity, with emission treatment, for near-source locations where the odor remains above its threshold. Thus, for an emission with $n = 0.33$, a decrease in the odorous emission by a factor of 8 (equivalent to an 88% reduction in the odorant concentration) would decrease the perceived odor intensity by only a factor of $8^{0.33} = 2$ and would not be very apparent by memory to a casual observer.

In dilution in the atmosphere, an emission odor is converted to a weaker ambient odor. Again, the dose-response function indicates the rate of intensity decrease with dilution. The *S–C* functions have been used in situations where many different odorous emissions occur, and a priority for emission treatments needs to be established. Those with a lower n value respond to dilution by atmospheric dispersion less than others with a larger n. Odorants that smell stronger than others at emission concentrations, but have larger n values, will decrease in odor intensity faster than those with lower n values. At some dilution, a crossover may occur, where the initially stronger odor ceases to dominate and the initially weaker odor becomes dominant. Such possibilities must be taken into account in planning for reductions in odor complaints.

Odors with smaller slopes approach their odor thresholds more slowly as dilution is increased. The shape of *S–C* curves in the vicinity of the odor threshold is disputed.[91] If one assumes that the function essentially continues its course beyond the threshold, an odor with a lower n will decrease in its potentially detectable intensity with dilution more slowly, and the statistical probability of its detection below the threshold will be higher than for an odor with the same dilution threshold but a greater slope.[37] Odors with lower n values are considered more pervasive—less susceptible to disappearance by dilution.[37]

ODOR THRESHOLD

The entire range of dilution of an odorous emission or odorous ambient air can be subdivided into three regions. At sufficiently large dilution (low concentration of odorants), odor detection by most people is, with respect to statistical probability, impossible. At insufficient dilution, odor can be easily detected by most people. Between these two regions is an intermediate one in which odor detection may or may not be possible, depending on the sensitivity of the sense of smell and the mode of sample presentation. It is in this intermediate region that odor detection and recognition threshold are observed, and it is this region that presents considerable difficulty in defining precise threshold values.

Threshold is not an exact property of a substance, as is, for example, vapor pressure. However, if experimental conditions are controlled, if panelists are appropriately selected, and if presentation protocols are adhered to strictly, reasonably reproducible values of dilution thresholds can be obtained. These will be reliable unless various conditions, such as sample flow rates, or the characteristics of sample presentation are changed. Threshold data obtained by different, but in themselves reproducible, methods usually have significant correlations, even if the numerical values are different,[105] so calibration curves relating one method to another can be successfully developed.[37]

Signal-Detection Theory: Odor Detectability Index

Signal-detection theory originally evolved to treat radar signal detection in the presence of random electric noise and dealt with detectability and recognizability of weak signals—those which are close to the noise levels. It was later introduced in psychophysics,[122] and its application to odorous air-pollution measurement was explored by Lindvall.[86] It is applicable in the near-threshold range to weak odors, such as those in the atmosphere at high dilution.

This theory postulates that, when an observer judges the presence or absence of an odor in a sample, the judgment depends both on the sensitivity of the observer's sense of smell and on the criteria used to decide whether a signal (odor) is perceived on the background of various other spurious signals (noise). For some, a "yes" decision is reported only if they are very sure of odor. Others may be more free in reporting "yes." It is recognized that the reporting of an odor when one smells odorless air and the failure to report an odor when a weak odor is present are inherent effects in judgment and are not to be ascribed to irreproducibility in judgment.

Four outcomes are possible:

- Odor is present and is reported (H, a "hit").
- Odor is present, but is not reported (M, a "miss").
- Odor is absent, but is reported (FA, a "false alarm").
- Odor is absent and is not reported (CR, a "correct rejection").

From the numbers, N, of responses in each of these categories, a statistical detectability index, d', is determined. This is a sensitivity measure separated from the decision criteria effects. Numerically, the procedure is as follows: First, two conditional probabilities are calculated, P (hits) for hits and P (false alarms) for false alarms, such that:

$$P\,(\text{hits}) = \frac{(N \text{ of hits})}{(N \text{ of hits}) + (N \text{ of misses})}$$

$$P\,(\text{false alarms}) = \frac{(N \text{ of false alarms})}{(N \text{ of false alarms}) + (N \text{ of correct rejections})}\,.$$

Tables[122] derived from probability distribution equations are then used to find a d' value for the calculated values of P(hits) and P(false alarms). Such tables are constructed by subtracting the z score (units of standard deviation) that corresponds to P(false alarms) from the z score that corresponds to P(hits). Values approach zero when nonodorous air is repeatedly judged and approach 2 in the range of relatively easy detection of an odorant-containing sample repeatedly presented in a random order with nonodorous air presentations.

The signal-detection theory has been used to measure ambient traffic odors;[87] panelists in a mobile laboratory made judgments on samples of air delivered from outside. In one study on the d' values in detection of hydrogen sulfide,[86] d' was found to increase linearly with the logarithm of the concentration of this odorant.

Calculations of d' require large numbers of responses (preferably in the hundreds). Because of that and because of the relative novelty in its application to odors, this method has not yet found extensive use in routine odor measurements. Forced-choice multiple-sample methods are relatively free from the decision criteria effect, are more economical of panelists' time, and are better understood by odor measurement practitioners; thus, they are increasingly preferred in routine work.

It is possible that the detectability-index method can be used in studies of population response in an open environment. This would require monitoring the ambient air simultaneously by analytical methods, inasmuch as odorant concentrations in air fluctuate, thus precluding experimentation at constant concentrations.

Detection versus Recognition Thresholds

The more usual way to characterize the detectability of an emission or ambient-air odor is through determination of odor thresholds. The odor detection-threshold test measures the lowest concentration of an odorant at which an odorized sample can be distinguished from nonodorous air. In the odor recognition-threshold test, the panelists must be familiar with and readily recognize the character of the odorant; the test establishes the lowest odorant concentration at which this odor can be recognized. The odorant concentration at the recognition threshold exceeds that at the detection threshold by a factor of 1.5–10.[70]

Common Forms of Odor-Threshold Tests

In any common form (one different from the signal-detection-theory approach) of either the detection or recognition odor-threshold determinations, panelists are presented with samples of nonodorous air and with various dilutions of an odorous emission or odorous ambient air. Panelists judge which sample evokes an odor sensation and which does not.

The objective is to find the greatest dilution (the lowest concentration of the odorant) beyond which (i.e., greater dilution, lower concentration) fewer than half the panelists detect (or recognize) the odor. More than half the panelists would detect the odor at a dilution that is smaller (a concentration that is greater). This concept appears simple, but all kinds of interferences occur in such determinations. The experimental design is critical; e.g., the order of sample presentation influences the threshold value determined.

Increasing-Dilution Series (Descending Series) In this design, panelists evaluate an odorous sample first and then evaluate more and more dilute samples until no odor is detected. The problem often encountered with this procedure is that stronger odors fatigue the sense of smell and make the odor of weaker samples presented later more difficult to detect.

An additional problem with increasing-dilution series is physicochemical; it arises when the sample presentation devices are constructed to deliver different dilutions from the same dilution and sniffing system. Odorants are adsorbed (and sometimes absorbed) in the construction materials of the equipment. With increasing dilution, adsorbed odorants desorb and temporarily contaminate the weaker sample.

Decreasing-Dilution Series (Ascending Series) Very dilute and therefore nonodorous samples are evaluated first, and then the concentration is increased until the odor is detected by all panelists. The problem with this approach is anticipation: panelists anticipate that an odorous sample will eventually appear and tend to report detection or recognition of an odor prematurely. In one experiment, nonodorous air samples were repeatedly presented to a well-trained panel, and after the fourth presentation "odor" reports began to occur (A. Dravnieks, personal communication).

In both these methods, the decision criterion used by the panelists operates. At each presentation, each panelist must decide whether odor is present or absent. The responses will depend on the panelists' readiness to be wrong, when they are not yet sure.

Randomized-Dilution Series In this design, weaker and stronger dilutions are presented in a random order. This approach is used in an effort to eliminate anticipation. The problem encountered here is much the same as in the descending series: A temporary desensitization by a stronger odor makes the detection of a weaker odor more difficult. The adsorption-desorption problem also may distort the concentrations of odorants in the samples presented.

In a comparison of several presentation designs, it was concluded[86] that random presentation makes it almost impossible to determine odor-detection thresholds.

Modified Ascending Series (Insertion of Blanks or Out-of-Order Samples) In an attempt to reduce the anticipation observed in the ascending series and to provide some verification of the reliability of panelists' judgments, the ascending series is modified as follows: instead of steadily progressing with increases in concentration, a nonodorous "blank" sample is presented once in a while, or an out-of-order sample is presented (e.g., repeating a previous, more dilute sample). For example, the ASTM 1391-67[3] version of this dilution test, which uses dilution syringes, recommends that each third or fourth presentation be an out-of-order sample.

It appears that this approach is more palliative, rather than curative, of the problem in the ascending series. In a typical test with a limited number of presentations, there will be many variants in the order of presentation of blanks or repeat samples; each set of sequentially presented dilutions will in essence consist of short segments of the ascending or descending series.[15] Usually, a method for dealing with "odor" reports in blanks is not prescribed. Reports of odor in blanks are considered inconsistencies and thus warning signs related to the validity of data. It is known, however, from the signal-detection theory, that such reports are a normal effect in judging weak signals.

Multiple-Sample Forced-Choice Methods In this approach, panelists give their judgments only after smelling and comparing several samples. Some are blanks (nonodorous), others contain odorous samples diluted to identical concentrations. Panelists must smell all samples in the set and only then report which are odorous. They then proceed to the next, usually more concentrated set of odorized and blank samples.

In some multiple-sample methods, the response "no odor in any of the samples" is permitted.[2] This results in judgments based both on sensitivity and on the panelists' decision criterion. The proponents of the non-forced-choice response believe that it is improper to request definite judgment if a panelist is not sure.

However, in the more frequently used multiple-sample methods, a forced-choice response is prescribed. A panelist who cannot decide by smell must simply guess. This approach simplifies the statistics and to a very large extent removes the decision-criterion effect from the response.

The number of samples in a set is usually between two[4,42] and five,[14] with one or two being odorous samples and the rest blanks. The three-sample method is usually termed the triangle method. Several sets are presented one after the other. Usually, each set contains more of the odorous emission or ambient air in the nonblank samples than the previous set; thus, the series is ascending. The typical increase per step is by a factor of 2 or 3.

In two-sample sets, the only feasible question is which sample is odorous. When there are three samples (two of them blanks) per set, the question may be which sample is different from the others (detection threshold).

The multiple-choice samples within a set can be made available all at the same time for the panelist to make mutual comparison. This version is usually preferred by panelists. Otherwise, the samples of a set may be presented sequentially; for example, with a set of three samples, it could be asked whether the first, the second, or the third had an odor. Usually, a precaution is taken to avoid a positional bias—preferring the left to the right sample or, in the three-sample sets, the left, or the center, or the right. This effect is controlled by changing the order of the samples in the set or by arranging the samples in a circular mark-free fashion.[50] However, a recent study[59] indicated that the positional bias does not affect the results of odor-threshold determination and appears only in sets with samples below the threshold concentration.

In 1978, ASTM Committee E-18 on Sensory Evaluation completed development of a recommended practice on odor- and taste-threshold determinations by an ascending-series forced-choice triangle method.[7] This procedure was selected because it is easily standardized, may be applied to odorants added to a background that is already odorous, and is now widely used in flavor research and industrial quality control.[8] The ASTM text does not deal with the physical form of sample presentation or the selection of panelists.

Go–No Go Methods In some applications, mostly in quality control, multiple-sample methods are used to judge whether the odor threshold is within specified limits. In such cases, the emphasis is on establishing with sufficient confidence that an odor is not noticeable at, e.g., a probability of 0.05 (confidence level, 95%), meaning that the odor detection could have occurred by chance in one case in 20. It may be specified that an odorous sample diluted by a factor of 100 should not be detectable by smell. A sample so dilute can be compared with nonodorous air samples in a forced-choice multiple-set design with a statistically sufficient number of panelists[59] or a few panelists in a go–no go design a sufficient number of times.[19] Appropriate statistical tests are then applied to the data to estimate the significance of the result for different numbers of panelists. For example, in the forced-choice triangle test, seven panelists in a panel of nine must make the same and correct choice to reach $p < 0.01$ (confidence level, 99%);[59] such coincidence of identical selections in a test in which none of the three samples were odorized could occur in one case in 100.

Threshold by Extrapolation of Intensities An approach not based on odorous-nonodorous sample selection consists of extrapolation of the odor intensity data from odor intensity measurements above the odor threshold. In a typical application,[71] odorous air samples are presented in an ascending series. The odor intensities at each concentration are rated, e.g., on a 0–4 category scale. In a semilogarithmic plot of dilution factor versus intensity category, a best-fit straight line is constructed. The straight line is extrapolated to the zero category,[71] or some other category, and the dilution at that point is considered to be the threshold dilution.

In the U.S.S.R., a somewhat similar procedure has been reported. Odor intensities at several dilutions are rated on a 0–5 scale, and odor threshold is taken to correspond to category 1 on the scale.

A similar approach uses ratings of the odor intensities according to the ASTM butanol scale.[6] Several suprathreshold concentrations of the odorous sample are compared for their odor intensity by matching parts-per-million butanol-scale values. A plot of logarithm (sample dilution) versus logarithm (ppm butanol) is usually straight, and extrapolation to the butanol threshold concentration yields an estimate of the odor threshold of the odorous sample.

Olfactometers for Threshold Determination

Olfactometers are devices for the presentation of odorous samples to panelists, most usually for odor-threshold determination, and not devices that measure odors themselves.[38] Some olfactometers have been called odorimeters, osmometers, osmoscopes, etc.

Dilution Methods Odorous samples can be diluted with nonodorous air in two different ways. In so-called static dilution (actually better termed batch dilution[4,20,93]), a sample is diluted to the desired fixed volume. In dynamic dilution, a flow of the odorous sample is mixed with a flow of the odorless air in a desired ratio.[50,71] In flowing systems, adsorption effects can be dealt with more easily than in batch dilution, and the same flow can be used to deliver the diluted sample for smelling at a controlled rate. The present consensus of Air Pollution Control Association (APCA) Odor Committee TT-4,[108] concerned with air-pollution odor measurements, is in favor of dynamic methods, excluding the Scentometer.

Sample Presentation to Panelists Odor test rooms in which air is rapidly circulated and an odorous sample is released for evaluation by panelists are occasionally used.[121] Samples cannot be changed rapidly in a test room, and the use of large panels is unwieldy. Odor-room work is usually conducted by a few expert odor judges. Such rooms tend to have a slight background odor, and expert judgment is needed to evaluate the effect of a sample on the room air.

In some olfactometers, hoods (chambers) of appropriate size are used, into which a panelist inserts his head to smell a sample passing in a steady measured flow through the hood.[25] Large samples and high volumetric flow rates are needed.

Mask olfactometers are face or head masks with air supplied at 7 L/min or faster—the flow rate needed to satisfy breathing requirements. Odorous samples are mixed with the breathing air in gradually increasing concentrations,[110] or nonodorized air and odorized air are fed to the mask alternately in accordance with some predetermined schedule.[19] Mask olfactometers can be used directly in odorous air (such olfactometers have odor adsorbers that produce nonodorous dilution air from the odorous ambient air[110]), and some use portable cylinders with nonodorous air or oxygen for sample dilution. The rationale for the mask method[89] is that the dilution of the odorous sample in the air breathed can be tightly controlled; in other types of olfactometers, in which samples are smelled from sniffing ports, the surrounding ambient air may dilute the odorous sample emitted from the port before it enters a panelist's nose. However, the significance of this argument does not seem to be experimentally proved; in a comparison of four odor-measurement systems,[131] the panelists were least sensitive with the mask device. The mask olfactometer is cumbersome if a large number of panelists is required. Even if individual masks are used, there is often a residual odor. As with any single-sample method that requires yes-no judgment, decision criteria and anticipation may complicate the results.

The Scentometer,[18] another type of olfactometer, uses glass tubes that are inserted into the panelists' nostrils; the panelists inhale through the nose.

Another variation of sample presentation, also with nostril inserts, is designed to limit the total amount of diluted sample per sniff to 30 or 50 ml;[84] this is a typical volume of a usual sniff, and it is considered that sensitivity is higher with a short sniff than with full nasal breathing.

In all other forms of olfactometry, diluted samples are presented at sniffing ports, which range from glass nose cones[71,79] to open ends of tubes of various sizes and shapes. An unsettled question in the use of sniffing ports involves the influence of the flow rate from such ports on the determined odor threshold. Rates typically range from 0.5 L/min[133] to 150 L/min. Some studies have indicated that the ease of odor detection increases with increased flow rate.[35] An increase from 0.5 L/min to 10 L/min from a 25- × 35-mm elliptic port increased the dilution needed to reach the detection threshold of *n*-butanol by a factor of 3 in a forced-choice triangle test; the dilution threshold, in terms of the dilution ratio, varied linearly with the logarithm of the flow rate. Despite differences in the flow rates, the different dynamic-dilution olfactometers produce threshold values that correlate through linear plots of different slopes.[35,131] Lower flow rates consume less sample per test, can be maintained with olfactometers that are more portable, and use smaller and less expensive components. They also permit work in normally ventilated rooms, which are more generally available. Specific olfactometers for use with collected samples or in ambient air are discussed below.

At high rates of emission from sniffing ports, the air flow begins to be felt and tends to facilitate odor detection. This leads to the claim that stimulation both by feeling the flow and by the odor may be an artificial situation that complicates data interpretation.

In some methods of sample presentation, the time allowed for sniffing is controlled. Thus, a signal is given when the sample for the evaluation begins to be presented, and another signal directs the panelist to report whether an odor was perceived.[102] The next sample—odorized or a blank—is presented, etc. In such a method, the same port is used to deliver odorous samples and blank air, and some error may be introduced if the dilution and delivery system does not properly follow the intended sample-concentration changes as fast as expected, owing to adsorption, desorption, and some volume retention. This can be corrected by the use of several sniffing ports. These can supply the same sample dilution and thus permit simultaneous evaluation by several panelists;[54,71] or they can supply different concentrations from different ports, with other ports supplying blanks for comparison.[42,69,93]

Obviously, there are large differences in the designs and methods of application of different olfactometers, and at least several dozen different devices have been described in the literature and some patented. Five types of olfactometers are known to be available commercially and in use in more than one laboratory. The ASTM D 1391 syringe dilution test is still another system in multiple use.

D 1391 Syringe Test[3,4] Panelists eject the gas content of the syringe into the nose for smelling. Preparation consists of making various dilutions of the odorous sample with nonodorous room air in glass syringes. Because the original procedure prescribed for this test was ambiguous in some details and somewhat impractical for use with large panels, it was subjected to the Mills modification,[93] to eliminate the trial-and-error aspect of the ASTM and Mills method. The test procedure was further modified by Benforado;[20] the modification was related to the manner in which the sample is administred to the panel from a master syringe and improved the reproducibility. The revised test method has been prescribed for stack emission by many air-pollution control authorities. It has been ill-advisedly extended to weak ambient odors, but apparently only rarely.

More recently, both forms of the syringe test have been criticized[10,108] on the grounds of conflicts with correct sensory-evaluation principles, and another revision was formulated.[4] The salient changes are as follows. Panelists' sensitivities are pretested with an odor relevant to the particular pollution odor, rather than with a vanillin-methylsalicylate discrimination method, as in the earlier procedure. At each dilution, two syringes are compared, one with a diluted odorous sample, another with nonodorous air; this requires that a panelist judge which syringe contains the odorous sample, whereas in the earlier version only one syringe at a time was presented and a yes or no answer was required. Samples are presented in ascending odorous-sample concentration, increasing by a factor of 2 per step; in the earlier version, out-of-order concentrations had to be presented once in a while. Correct selection of the syringe at three consecutive concentrations (this can occur by chance in one case out of eight) is taken to indicate that the panelist's threshold was reached at the lowest of these three concentrations. The panel mean threshold is simply the geometric mean of the panelists' thresholds. Earlier, in the Mills modification, the panelists' responses were pooled on a plot of logarithmic dilution against probability, with 50% probability taken as the threshold datum.

The test is now inapplicable to weak odors (detection thresholds at dilutions by less than 25 times) and usually applicable to stronger odors (detection thresholds at dilutions in excess of 50 times); the original test did not have this limitation. The limitation excludes use of this test for most odors in ambient air.

This recent revision has corrected many defects of the syringe test. However, some problems remain. Adsorption of odorants on the syringe walls interferes with maintenance of intended dilution factors. The rate of expulsion of the sample into the nose is difficult to control, and the total time for sniffing is short. Consequently, APCA Odor Committee TT-4 declined to endorse the revised ASTM D 1391 and has stated a preference for olfactometric methods based on dynamic dilution.[108]

The most recent (1978) revision of ASTM D 1391 abolishes the concept of odor concentration units per cubic foot and adopts the metric system with a dimensionless dilution factor, Z.

Scentometer[18,61,72] This device was developed for measurement of odors in ambient air. It consists of a plastic box, two glass nosepieces, activated-carbon filters, and several calibrated holes in the box. An operator inserts the nosepieces into his nostrils and breathes through the nose, first obtaining only air deodorized by the carbon filters. He then opens selected holes to admit the odorous ambient air into the box, where it is diluted, during inhalation, with the carbon-filtered nonodorous air. Different holes provide different dilution ratios (D/T), or dilution to threshold. The panelist judges whether adding the odorous air resulted in odor. Typical dilution-to-threshold ratio settings of 2, 7, 31, and 170 are defined in the manual, but documentation on the actual dilution ratios that result is not available. The manual states that odors at a D/T of 2 usually do not constitute a pollution problem and those at a D/T of 31 always do.

The device has been used in community odor surveys[57] and is prescribed in some odorous-pollution control regulations.[76]

The Scentometer method has been criticized on several grounds. It is not possible to verify whether the operator actually detected an odor. There is some question about the sensitivities of the test, inasmuch as the Scentometer operator is already exposed to the odorous atmosphere, which produces some olfactory fatigue (adaptation); this effect must be overcome at the beginning of the test, by breathing through the Scentometer's carbon filter for at least a few minutes. With the nasal inserts in nostrils, this is difficult. An alternative way to prevent fatigue is to use the device in an open automobile while moving from an odorfree area into the odorous air, traversing the emission-originated plume to the other side of the plume into nonodorous air. This technique is described in a recently published set of procedures (see the appendix to Flesh and Burns[57]).

The degree of mixing in the Scentometer during inhalation is uncertain. The carbon-filtered air sometimes has a weak odor that may complicate accurate detection of the added odorous air. Consequently, APCA Odor Committee TT-4 arrived at a consensus[108] against the use of the Scentometer in air-pollution control regulations and recommended the selection of some better controlled device that uses dynamic dilution. About half the Committee members thought that measurement of odors in ambient air should not be used in regulations, because such odors are transient and make it difficult to obtain consistent measurements and are expensive to sample and monitor; they recommended that measurements required by the regulations be confined to those made at the source of the odorous emission.

Misco (Sanders) Olfactometer[94] This olfactometer is a mask device, used both in ambient air and on odorous samples in the laboratory by a few expert panelists. Nonodorous air is supplied to the mask. When the Misco olfactometer is used in odorous ambient air, diluted air is generated by filtration through active carbon and silica gel. The odorous sample is gradually added to the breathing air until an odor is reported. At this point, flowmeters are read and the odorous-sample dilution is calculated. An appropriate calibration odorant (for instance, methyl sulfide for measurement of paper- and pulp-mill odors) is used to relate the pollution odor being measured to the odor threshold for the calibration odorant.

The Misco device was originally developed by Sanders at the California Public Health Service. The Misco Company undertook to manufacture the device. (An olfactometer similar to Misco's is made by the Eutec Company.)

It is understood that a similar device, equipped with electronic sample-presentation controls, is in use in Bay Area Air Pollution Control District laboratories. It is operated as a go–no go test method, to determine whether an odor is detectable at some specific sample dilution. In this form,[131] samples of diluted odorous air and nonodorous air are presented in a random order at approximately 15 L/min. Each presentation lasts 5 s, with a 10-s interval between presentations. A total of 10 nonodorous and 10 odorous presentations are administered; all odorous presentations are at the same dilution specified by regulations. During the intervals, only nonodorous air is presented. The sample is considered odorous if odor was reported in eight of 10 odorized presentations. However, a test is considered invalid if odor was reported in three or more nonodorous-air presentations.

Technical performance data on this device and its evaluation method have not yet been published. Anticipation effects are eliminated. Instrumentation artifacts resulting from rapid changes of samples may occur, but the extent of these is not known.

The statistical analysis (chi-squared) of the above limits for positive and negative test results indicates that a sample will be considered odorous only if the probability of a positive result by chance is at $p < 0.05$ (confidence level, over 95%). A nonodorous sample can be found to be odorous by chance in approximately 2.5% of such 20-response tests.

T04 (Mannebeck) Olfactometer[89] This was developed in Germany, and several dozen units are in use there. It is a portable mask device that uses pressurized oxygen from a portable cylinder to supply dilution gas and to drive an ejector that sucks in ambient odorous air and mixes it with the oxygen supplied to the mask. A valve controls the rate of introduction of the ambient-air sample. The dilution ratio is calculated from flowmeter readings. In the simplest form of its application, the procedure is susceptible to the anticipation effect.

Dynamic Triangle Olfactometer[50,51,75,127] This device was developed initially for use in the rendering industry both for emission and for ambient-air odor, but now is in use in various industrial and research laboratories. It was designed to permit use of the forced-choice triangle method in detection-threshold determinations. Such measurements are done over a range of sample dilutions.

The device provides fixed dilutions. For each, three sniffing ports arranged in a circle are attached to the olfactometer by a flexible conduit. Two ports supply nonodorous air, and one supplies the diluted odorous sample; all three form a triangle set. The flow rate from the ports is 0.5 L/min. In the olfactometer used for industrial odorous emission, there are six dilutions. Each next dilution is more dilute than the preceding by a factor of 3. In the olfactometer for ambient odors, five dilutions are provided, up to dilution by a factor of 81.

The sample consumption rate is approximately 0.1 L/min for emission and 1 L/min for odorous ambient air. All flows are continuous and constant during the test. The total odorous emission from the device is low enough to permit work in normally ventilated rooms. However, the low flow rates complicate the decontamination of the dilution and port system if strong odors have been inadvertently admitted.

Each panelist begins with the lowest-concentration set and proceeds toward stronger samples. In each set, the panelist must indicate which of the three ports exhibits an odor (is different from the other two for detection or is characteristic of a reference for recognition). The selection is signaled by pressing a signal button at the port that lights a corresponding bulb on a signal board. An ED_{50} value (effective dosage for 50% of panelists) is calculated by combining the responses of panelists; it is a dilution at which an estimated 50% of the panelists would have difficulty in reliably detecting the sample odor.

The method has the advantages described for the ascending-series forced-choice triangle method.[7,8]

The standard deviation of ED_{50} values from repeated measurement of the same sample by the same panel of nine is typically around 0.1 in $\log(ED_{50})$ units.[35,51,88] For a dilution factor of 100, one standard deviation gives the range 80–125.

Hemeon Olfactometer[71] This device is larger than the others described and uses a flow rate of 150 L/min from the sniffing ports. Judgments consist of rating the perceived intensities of the odors of samples diluted to different extents, and the threshold is estimated by extrapolation to zero odor. Dilution can be increased by an increase in the dilution air flow, a decrease in the odorous-sample flow, or both. For strong odors with very low thresholds, a supplementary predilution stage is provided. Diluted odorous samples are emitted from three glass cone-shaped sniffing ports to three panelists who can judge the odor simultaneously. A test begins with dilution sufficient to make the sample nonodorous. Sample concentration is then increased, and at each concentration the panelists rate the odor intensity on a 0–4 scale. A plot of the logarithm of the dilution ratio against the category-scale rating is prepared, and the best-fit approximately straight-line plot through points is extrapolated to zero intensity, indicating the dilution threshold for detection.

The high flow rate is considered an advantage for a rapid change in dilution, but requires a larger device and better ventilation than the other olfactometers. The odor sensation is accompanied by the feel of air flow, and this may or may not be an advantage; its significance has not been critically explored.

Use of Different Olfactometers The ASTM syringe method and the Scentometer reflected the state of the art when odor-control regulations first became technically more developed. Previously, public-nuisance laws constituted the common approach to odor control.

Dynamic olfactometers, except the Scentometer, later appeared, and many organizations designed their own devices. Today, only one pollution-control jurisdiction, the Bay Area District in San Francisco, specifies *dynamic olfactometry* and uses a specific locally developed apparatus, or the equivalent, in its laboratory. Other jurisdictions specify the ASTM D 1391 syringe test or its Mills modification; some add "or equivalent." The procedures for establishing the equivalence of different olfactometric methods have not been spelled out. Presumably, they are not restricted to an exact numerical equivalence of measured detection-threshold values, but could consist of calibration curves relating data derived with one method to data derived with another.

One reason for specifying the ASTM D 1391 or equivalent is that a data base exists from the early work by the Los Angeles County Air Pollution Control District,[29] in which data were collected on the dilution thresholds of various industrial emissions that seemed to reduce population annoyance caused by emission odors to a tolerable point. It is questionable whether such data, obtained in a limited number of cases in specific topography and meteorology, should be indiscriminately used in all other situations.

Some air-pollution control jurisdictions accept the data obtained by a dynamic olfactometer as a substitute for the syringe-test data, on the basis of reasonably valid technical arguments, even if the regulations have not specifically provided for an alternative method. Thus, the Illinois EPA does not specify use of a dynamic olfactometer instead of the syringe method, but grants permission for the use of a dynamic triangle olfactometer for weak odors; permission needs to be requested separately for each case (A. Dravnieks, personal communication).

Odor Committee TT-4 of the Air Pollution Control Association endorsed dynamic olfactometry over ASTM D 1391 syringe and Scentometer and recommended that specifications be developed for the performance of dynamic olfactometers.[108]

Comparison of Odor-Threshold Measurement Systems

A comparison of the overall performance of the ASTM D 1391 methods and the Misco, dynamic triangle, and Hemeon olfactometers has been made by TRC, Inc., for the Illinois EPA.[131] The odorants used in the comparison were diluted vapors of phenol, butyric acid, dimethylamine, dimethyl sulfide, and hydrogen sulfide. These are typical malodorants that occur in the most annoying emission. Both the values of detection threshold obtained and the operational characteristics were compared. Figure 4-7 shows the relationships between the values of dilution thresholds for odor detection obtained with various pairs of methods.

The following gives estimates of the ratio, *R*, for the highest to the

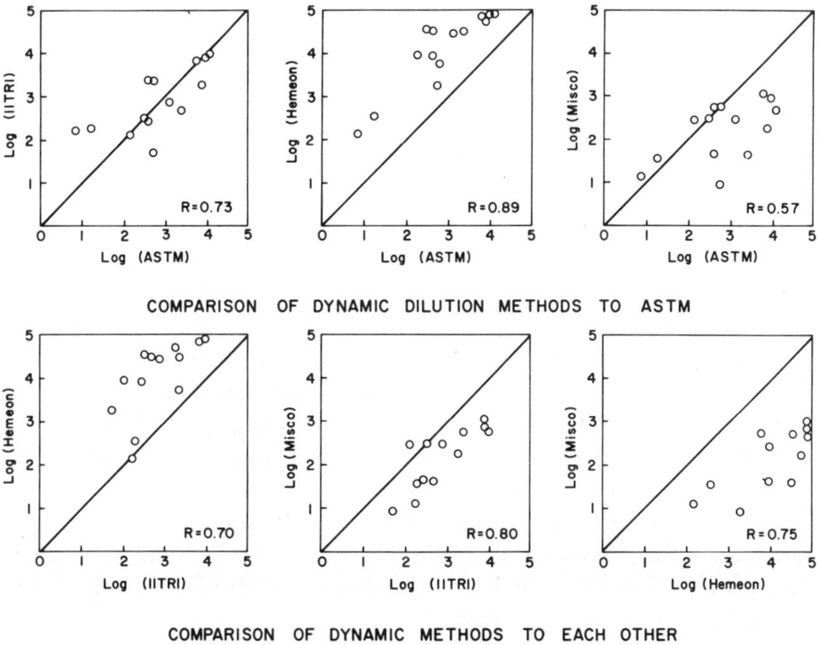

FIGURE 4-7 Comparisons of four odor detection-threshold measurement systems in the laboratory, using various malodorants. The number in the lower right corner of each plot is the correlation coefficient for the least-squares fit line through the actual data points; this line is not shown. Points would be on the drawn diagonal if the two systems yielded exactly the same numbers for detection thresholds. Each data point corresponds to measurements of the threshold for the same odorous sample by the same panel at the same session by two methods (indicated by titles on the coordinates). Reprinted with permission from Dravnieks.[35]

lowest values of odor thresholds for the same odorous samples. The data apply to a typical panel of 8–10 panelists, not specifically selected for the homogeneity of their odor sensitivities.

Same panel, same day, same high-state-of-art system: $R = 2.5$

(Note: based on a range of four standard deviations of panel means.)

Static system, different panels deliberately selected to represent a maximal difference in odor sensitivity in a normal population: $R = 10$

Same panel, same day, same high-state-of-art system, sample flow rate to nose changed from 0.5 to 9 L/min: $R = 4$

Same panel, same day, different commercially available systems: $R =$ up to 200

If the values were identical, points on Figure 4-7 would be on the diagonals. The correlation coefficients indicate the goodness-of-fit of the least-squares straight line that could be drawn through the points in each diagram. Correlations between methods have substantial statistical significance.

The Hemeon method (probably because of the higher sample flow rate and the method of estimating the odor threshold by an extrapolation of the intensity response) gives the highest values; the panelists appear to be more sensitive in this test. Surprisingly, the mask method, which is supposed to conserve odorous vapors by delivering them directly into breathing air, operated as though the panelists were less sensitive than in the other methods. The mask device was considered inconvenient by the panelists, especially when odor rapidly increased in intensity while the mask was on. The triangle method was found to be convenient and desirable. It must be pointed out that the use of the forced-choice triangle principle is not limited to the particular type of olfactometer used.

The TRC work found the dynamic triangle olfactometer to be the most suitable for the odor detection-threshold determination, with a comment that a higher flow rate might facilitate deodorization of the device during changes from one sample to another.

The six-dilution dynamic triangle olfactometer for odorous emission was compared with the ASTM D 1391 method with rendering emission and some single odorants.[50] Data from this work are combined with the TRC data in the upper left part of Figure 4-8. A reasonable agreement between the two studies, but not exact correspondence between the values of odor-detection thresholds obtained by the two olfactometric methods, is evident.

Comparison of Dynamic Triangle Olfactometer and Scentometer The five-dilution dynamic triangle olfactometer for odorous ambient-air samples was compared in the laboratory under closely controlled conditions with the Scentometer, but for rendering odors only.[51] In another experiment,[51] Scentometer readings at different positions around a rendering plant were made simultaneously for each position by three Scentometer operators, and bag samples were taken at the same time for a laboratory evaluation by dynamic triangle olfactometer. The odor-detection thresholds for experienced and inexperienced panelists agreed very well. Scentometer data correlated with the dynamic-olfactometer data for experienced panelists, but not for novices. The lower part of Figure 4-8 shows the relation between the Scentometer D/T settings and the dynamic-triangle data.

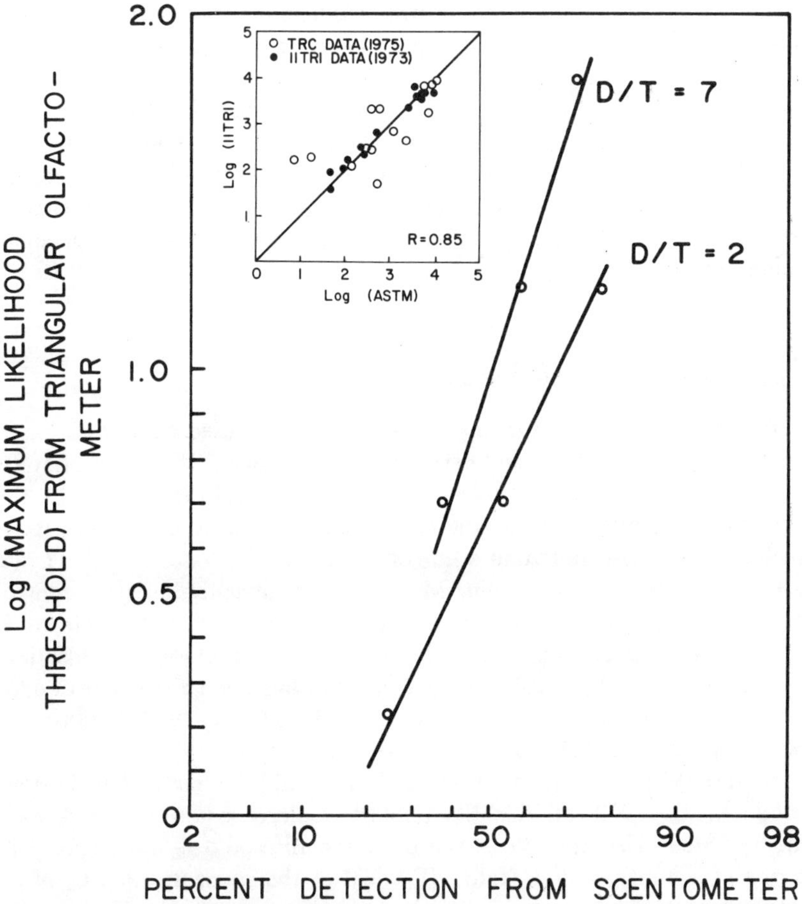

FIGURE 4-8 Comparison of dynamic triangle olfactometer results with ASTM D 1391 test results (upper left corner) and with Scentometer results.[35] The data points in the insert represent rendering odors, some single odorants, and the malodorants of Figure 4-7. The Scentometer D/T setting signifies the nominal dilution factors in accordance with the Scentometer manual. The left coordinate is equivalent to log ED_{50} or log Z, as defined in ASTM D 1391, 1978 revision.[4]

APCA *TT-4 Test* In 1977, several laboratories participated in an interlaboratory detection-threshold measurement exercise organized by APCA Odor Committee TT-4. Explicit directions were given on how to prepare the controlled concentrations of odorous-air samples in plastic bags. Each participating laboratory measured the odor-dilution thresholds of such preparations with whatever olfactometric method it commonly used. The measurement results from different laboratories scattered substantially. It was speculated that a part of the problem was in sample-preparation irreproducibility, and a repetition of the cooperative exercise with a gaseous odorous sample that could be circulated between laboratories and periodically verified analytically is being considered.

A similar interlaboratory test is planned for Europe and was discussed at a Society of German Engineers (VDI) meeting on odor-active substances in September 1977 in Dusseldorf, Germany (A. Dravnieks, personal communication).

Units for Expression of Dilution

Before 1978, ASTM D 1391 and ASTM D 1356-73a used definitions based on "odor units." Odor was likened to a "substance." If 1 ft^3 of odorous sample had to be diluted to N ft^3 to reach its odor-dilution threshold, the "odor concentration" in the original sample was said to be N units/ft^3. If such a sample was emitted at a rate of V ft^3/min, the "odor emission" rate was said to be NV units/min. Most existing air-pollution odor-control regulations, except those based on the Scentometer, refer to such units.

There were several objections to these definitions. Odor is a sensation, not a substance. The so-called odor concentration is not a measure of odor intensity. With wider use of metric units, the situation with "odor units per cubic foot" was becoming even more confusing.

In the 1978 revision of D 1391,[4] syringe-dilution method definitions proposed in 1970 at the Stockholm Conference on Odors[123] were incorporated. The dilution to the detection threshold is now defined in terms of a dimensionless factor, Z, equal to the concentration, C, of an odorant in the sample, divided by its concentration, C_{thr}, at the detection threshold; then $Z = C/C_{thr}$. Numerically, Z is equal to odor units per cubic foot. Thus, if 1 volume of odorous sample has to be brought to a total of 1,000 volumes to reach the dilution threshold, $Z = 1{,}000$. If such a sample is emitted at 50 m^3/min, the odorous emission rate is $1{,}000 \times 50 = 50{,}000$ m^3/min. This is the volume of a nonodorous atmosphere that can accommodate emission for 1 min without becoming odorous, as measured by a threshold test. In essence, it measures the clean-air demand of the odorous emission. Numerically, it is equivalent to odor units per minute after conversion from cubic meters to cubic feet.

ODOR CHARACTER

Odor character is the array of the odor notes of the odor sensation that permit one to distinguish odors of different substances on the basis of experience. It is said then that odors are different if they differ in character notes.

At sufficiently high concentration, most substances with molecular weights up to about 200 or 300 have odors. Some exceptions are substances that have two or more –OH groups or combinations of several other polar groups. However, odorous substances very rarely smell so much alike that they cannot be distinguished by odor in a direct comparison. Thus, there are many tens of thousands of different characteristic odors, even disregarding the odors that result from mixtures of odorants.

Odor character is evaluated by a comparison with other odors, either directly or through use of descriptor words. The ASTM E-18 Sensory Evaluation Committee canvassed 31 contemporary lists of odor descriptors.[33] The result was a composite list of 830 descriptors, which may not have included special descriptors used in some industries. Later, about 100 people in various laboratories cooperatively screened this list and concluded that approximately 160 odor descriptors are considered useful and important.

Although there are many distinguishable odors, some can be grouped by similarity. Serious attempts at odor classification began with Linnaeus in 1752; since then, dozens of classification systems have been proposed, and these were reviewed by Harper *et al.* in 1968.[65] It has been suggested that studies of specific hyposmias—decreased, or below-normal, sensitivity to selected odors—may lead to a physiologic classification[13] based on 30–40 classes.

No odor classification has yet been universally accepted. For practical purposes, aside from an open-ended wordy description of an odor, it is now understood that each odor may have a combination of character notes of different applicability. The odor character is then best described by methods known as multidimensional scaling or profiling.

In multidimensional scaling, the odor is characterized by either the degree of its similarity (or dissimilarity) to a set of reference odors or the degree of applicability of various descriptors (sometimes termed attributes) to it. The result is an odor profile.

Direct-Comparison Profiling

In direct comparison, the odor of the sample is compared with a series of odors of reference odorants or odorant mixtures. The sample odor is scored for the degree of its similarity to each of the reference odors; scales of scores may be bipolar, with "very similar" at one end and "very dissimilar" at the other end. Scales for scoring commonly consist of five to nine categories.[65,66,115,134,135]

The selection of reference odors for comparison depends on the odor character of the expected odorous samples. For example, in characterizing the odors of diesel exhaust,[125] four odorous mixtures were designed, to represent burnt-smoky, oily, pungent-acid, and aldehydic-aromatic odor characters.

The more diverse the odor characters of the expected samples, the larger the number of reference odors needed to characterize their odors. One typical scale has nine reference odorants:[115,135]

Odorant	*Odor Character*
1-heptanol	oily
vanillin	sweet
methylsalicylate	fragrant
benzaldehyde	spicy
guaiacol	burnt
butyric acid	rancid
1-hexanol	metallic
1-propanol	etherish
ethyl disulfide	sulfurous (or sulfidic, to avoid confusion with sulfur dioxide odor)

Even the nine-odorant scale is insufficient for more complete characterization of many odors that may occur. Many odors, such as that of pyridine, may be quite dissimilar to any of the nine, and hence not readily characterized. A list of chemicals to represent 45 different odor-character notes has been published.[64]

Thus, some basis for selecting reference odors is available, but, with the exception of the diesel-exhaust odors,[125] the use of reference odors for characterization of air-pollution odors has been very limited. In a multisource environment, it is often possible to select some specific composite materials from industrial processes for use as reference odors

TABLE 4-1 Odor Reference Standards, Kearney, NJ[a]

Odor Type	Odor Reference Standard
1. Foundry core	Core mixture made up in laboratory to formula used by the foundry, then kept in a warm sand bath for odor presentation to panel
2. Baking exhaust	Exhaust gases from oven sampled in evacuated stainless steel bomb
3. Sour fat	Sour fat from skim tank of rendering plant mixed with a little mineral oil
4. Inedible protein	Dried inedible protein product from rendering plant
5. Varnish cooking	Raw aged linseed oil
6. Fishy-varnish	Mixture of linseed and menhaden oils, 4:1
7. Fatty acid	Linseed oil fatty acids

[a] Reprinted with permission from Turk and Mehlman.[124]

and compare the odors in the ambient air with them. Table 4-1 is an example of such a collection.[124]

Reference odors can be used either for direct one-by-one comparison with an odorous sample or for odor-recognition training. In the latter case, the evaluation of an odorous sample, such as ambient air, may be conducted by memory.

In comparing any two sample odors, the evaluation result may change when the order of presentation of the two samples is changed. The cause is a temporary adaptation of the sense of smell to the first odor, which results in a selective desensitization to some odor notes in the second sample. The usual remedy for this difficulty is to reverse the order of the presentation in a later experiment and then calculate the mean of both ratings.

In direct comparison, an odor rated against itself is not always reported as identical. Temporary distortion of a panelist's sensitivity to various odor notes may change odor perception, and the recovery of the sensitivity may occur at different rates for different odor notes. In a cooperative experiment by the ASTM E-18 Sensory Evaluation Committee,[44] in which 50 panelists participated in rating mutual dissimilarities of 10 odors on a 0–7 scale (0 = no difference), the exercise included comparing two samples of the same odor. The dissimilarity of identical odors was scored at 0.33 for the clearly aniselike anethole and 1.05 for the less distinct odor of 1-butanol.

Odor Descriptor Profiling

In this approach, a sample odor is evaluated by scoring the degree of applicability of various descriptor terms (odors). The descriptors refer to various character notes. The judgments may use a category scale or some graphic scale. In one form of multidimensional scaling, descriptors are paired and shown at opposite ends of a scale, to represent antonyms, and the panelists mark the in-between position that best characterizes the odor of the sample.[113] One of the simplest and most frequently used systems of grading the degree of applicability of a descriptor is a 0–5 scale[44,65,66] (0 = not applicable; 5 = applicable to a very high degree).

The descriptors should include all pertinent odor notes. There are two approaches to the selection. Some researchers would like to have as simple a scale as possible. Others argue that too simple a scale will produce profiles that are similar for some odors even if the odors are quite different and that additional descriptors need to be provided to differentiate such odors.

An example of a simple scale that did not work is that of Crocker and Henderson.[28] They believed that a four-digit system can describe all odors. In a four-digit number, the first digit indicated the fragrant note (on a 0–9 scale); the second, the acidic; the third, the burnt; and the fourth, the caprylic (goaty). This system has only historical significance; it was found to be too simplistic to deal with the complexity of odor character. An attempt to apply the Crocker-Henderson scale to industrial odors in Louisville (cited in U.S. EPA[126]) was not successful.

Harper *et al.*[66] developed a scale of 44 descriptors for the food industry. This scale quite broadly covers many kinds of odor notes. In work on air-pollution odors, it was found that profiles for some clearly different odors were similar. A much more elaborate list of descriptors was needed to differentiate odors in practice.[45] Therefore, Harper's scale was expanded[44] by using lists of odor descriptors collected by the ASTM E-18 Committee,[133] and some British descriptors were Americanized (Figure 4-9). The descriptors for index 001–022 and 031–052 were taken from the list of Harper *et al.*

A slightly abridged version of this scale (with 136 descriptors)[36] was evaluated in an interlaboratory test (four laboratories) implemented by the ASTM E-18 Committee.[44] Ten odorants were used. Replicate descriptor profiles were developed several weeks apart in each laboratory and evaluated; they correlated at $p < 0.05$ (confidence level, higher than 95%) in 93% of tests. Mean profiles for the same odorant by different laboratories correlated at $p < 0.05$ in 97% of cases. The average number of descriptors per odorant per panelist was approximately 10, with a range of 8.4–12.2 for different laboratories. Thus, descriptor profiles appear suitable for documenting and communicating odor-character data.

In a previous interlaboratory test, 12 laboratories invited by the ASTM E-18 Committee participated in evaluation of short-duration odors of nine pure chemicals as they were eluted from a gas-chromatographic column.[45] Mixtures of different concentrations were used. The 126 panelists who participated were permitted to choose their own terms to describe odors, and together they used a total of 136 terms. It was possible to cluster these terms in 22 groups—e.g., "anise" with "licorice," etc. For each odorant, a composite descriptor profile for all 12 laboratories was calculated. The descriptor profiles generated by each laboratory correlated with the composite profile at $p < 0.1$ (confidence level, 90%) in 91% of tests; most of the poor correlations were found for data from three laboratories.

Thus, either with a free choice of odor-descriptor terms or with multidescriptor scales, adequate agreement in odor-character evaluation may be possible, even without training of the panelists and despite differences in the use of terms by different panelists. Obviously, training and the presentation of examples of odors could improve correlations.

Some Classifications of Pollution Odors

Classifications have been proposed for pollution odors, as in Tables 4-2 and 4-3. The latter is for odors of water of industrial origin and is a three-level classification scheme.[2] At the most generalized, left end, four descriptors are used: sweetness, pungency, smokiness, and rottenness. These are used on three levels to characterize eight typical odor classes, and each class is subdivided into two to four types of chemicals.

Comparison of Odor-Character Measuring Methods

A reference-sample scale in which the odor of a sample is directly compared with a set of reference odors avoids semantic problems, because no names have to be attached to the various odor notes. The method does require a large number of reference odorants, especially if a large variety of odors must be characterized. Unique groups of odors, such as diesel exhaust, require specially designed odorant mixtures to exemplify the few distinguishable odor notes. The direct-comparison method is logistically more complex than the descriptor-profile method—odorants must be properly stored, handled, etc. Also, experience has shown that identical odors are not rated identically in all cases, and an odor match in a direct comparison is not as nearly perfect as might be expected.[44]

The descriptor-profile method is logistically simple. However, it does

1	2	3	4	5	6	7	8	9	10	11	12
						1					
SAMPLE DESIGNATION							PANEL CODE			PANELIST	

(Col. 1–6, 8–10: letters or numbers. Col. 11–12: numbers only.)

ODOR QUALITY EVALUATION

* SMELL SAMPLE. YOU CAN RE-SMELL IT AS NEEDED FOR EVALUATION.

** GO THROUGH LIST BELOW. FOR EACH DESCRIPTOR, *ENCIRCLE* THAT SCORE NUMBER WHICH BEST CHARACTERIZES THE *DEGREE OF PRESENCE* OF THAT ODOR NOTE IN THE SAMPLE ODOR. IF ABSENT, *DO NOT ENCIRCLE ZERO.*

*** INITIAL OR SIGN, AND DATE.

Initial or Signature . . . . Date . . . .

MEANING OF THE ODOR QUALITY SCALE:

ABSENT	SLIGHTLY		MODERATELY		EXTREMELY
0	1	2	3	4	5

Index	*DESCRIPTOR*	*SCORE*	*Index*	*DESCRIPTOR*	*SCORE*
001	FRAGRANT	0 1 2 3 4 5	031	OILY, FATTY	0 1 2 3 4 5
002	SWEATY	0 1 2 3 4 5	032	LIKE MOTHBALLS	0 1 2 3 4 5
003	ALMOND-LIKE	0 1 2 3 4 5	033	LIKE GASOLINE, SOLVENT	0 1 2 3 4 5
004	BURNT, SMOKY	0 1 2 3 4 5	034	COOKED VEGETABLES	0 1 2 3 4 5
005	HERBAL, GREEN, CUT GRASS	0 1 2 3 4 5	035	SWEET	0 1 2 3 4 5
006	ETHERISH, ANAESTHETIC	0 1 2 3 4 5	036	FISHY	0 1 2 3 4 5
007	SOUR, ACID, VINEGAR	0 1 2 3 4 5	037	SPICY	0 1 2 3 4 5
008	LIKE BLOOD, RAW MEAT	0 1 2 3 4 5	038	PAINT-LIKE	0 1 2 3 4 5
009	DRY, POWDERY	0 1 2 3 4 5	039	RANCID	0 1 2 3 4 5
010	LIKE AMMONIA	0 1 2 3 4 5	040	MINTY, PEPPERMINT	0 1 2 3 4 5
011	DISINFECTANT, CARBOLIC	0 1 2 3 4 5	041	SULPHIDIC	0 1 2 3 4 5
012	AROMATIC	0 1 2 3 4 5	042	FRUITY (CITRUS)	0 1 2 3 4 5
013	MEATY (COOKED, GOOD)	0 1 2 3 4 5	043	FRUITY (OTHER)	0 1 2 3 4 5
014	SICKENING	0 1 2 3 4 5	044	PUTRID, FOUL, DECAYED	0 1 2 3 4 5
015	MUSTY, EARTHY, MOLDY	0 1 2 3 4 5	045	WOODY, RESINOUS	0 1 2 3 4 5
016	SHARP, PUNGENT, ACID	0 1 2 3 4 5	046	MUSK-LIKE	0 1 2 3 4 5
017	CAMPHOR LIKE	0 1 2 3 4 5	047	SOAPY	0 1 2 3 4 5
018	LIGHT	0 1 2 3 4 5	048	GARLIC, ONION	0 1 2 3 4 5
019	HEAVY	0 1 2 3 4 5	049	ANIMAL	0 1 2 3 4 5
020	COOL, COOLING	0 1 2 3 4 5	050	VANILLA-LIKE	0 1 2 3 4 5
021	WARM	0 1 2 3 4 5	051	FECAL (LIKE MANURE)	0 1 2 3 4 5
022	METALLIC	0 1 2 3 4 5	052	FLORAL	0 1 2 3 4 5
023	PERFUMERY	0 1 2 3 4 5	053	YEASTY	0 1 2 3 4 5
024	MALTY	0 1 2 3 4 5	054	CHEESY	0 1 2 3 4 5
025	CINNAMON	0 1 2 3 4 5	055	HONEY-LIKE	0 1 2 3 4 5
026	POPCORN	0 1 2 3 4 5	056	ANISE (LICORICE)	0 1 2 3 4 5
027	INCENSE	0 1 2 3 4 5	057	TURPENTINE (PINE OIL)	0 1 2 3 4 5
028	Cantaloupe, Honey Dew MELON	0 1 2 3 4 5	058	FRESH GREEN VEGETABLES	0 1 2 3 4 5
029	TAR-LIKE	0 1 2 3 4 5	059	MEDICINAL	0 1 2 3 4 5
030	EUCALYPTUS	0 1 2 3 4 5	060	ORANGE (FRUIT)	0 1 2 3 4 5

FIGURE 4-9 An expanded Harper's scale for odor-quality profiles. Reprinted with permission from Dravnieks *et al.*[44]

Index	DESCRIPTOR	SCORE	Index	DESCRIPTOR	SCORE
061	BUTTERY (FRESH)	0 1 2 3 4 5	104	HOUSEHOLD GAS	0 1 2 3 4 5
062	LIKE BURNT PAPER	0 1 2 3 4 5	105	PEANUT BUTTER	0 1 2 3 4 5
063	COLOGNE	0 1 2 3 4 5	106	VIOLETS	0 1 2 3 4 5
064	CARAWAY	0 1 2 3 4 5	107	TEA-LEAVES-LIKE	0 1 2 3 4 5
065	BARK-LIKE, BIRCH BARK	0 1 2 3 4 5	108	STRAWBERRY-LIKE	0 1 2 3 4 5
066	ROSE-LIKE	0 1 2 3 4 5	109	STALE	0 1 2 3 4 5
067	CELERY	0 1 2 3 4 5	110	CORK-LIKE	0 1 2 3 4 5
068	BURNT CANDLE	0 1 2 3 4 5	111	LAVENDER	0 1 2 3 4 5
069	MUSHROOM-LIKE	0 1 2 3 4 5	112	CAT-URINE-LIKE	0 1 2 3 4 5
070	WET WOOL, WET DOG	0 1 2 3 4 5	113	PINEAPPLE (FRUIT)	0 1 2 3 4 5
071	CHALKY	0 1 2 3 4 5	114	FRESH TOBACCO SMOKE	0 1 2 3 4 5
072	LEATHER-LIKE	0 1 2 3 4 5	115	NUTTY (WALNUT, ETC.)	0 1 2 3 4 5
073	PEAR (FRUIT)	0 1 2 3 4 5	116	FRIED CHICKEN	0 1 2 3 4 5
074	STALE TOBACCO SMOKE	0 1 2 3 4 5	117	WET PAPER-LIKE	0 1 2 3 4 5
075	RAW CUCUMBER-LIKE	0 1 2 3 4 5	118	COFFEE-LIKE	0 1 2 3 4 5
076	RAW POTATO-LIKE	0 1 2 3 4 5	119	PEACH (FRUIT)	0 1 2 3 4 5
077	MOUSE-LIKE	0 1 2 3 4 5	120	LAUREL LEAVES	0 1 2 3 4 5
078	BLACK PEPPER-LIKE	0 1 2 3 4 5	121	BURNT MILK	0 1 2 3 4 5
079	BEAN-LIKE	0 1 2 3 4 5	122	SEWER ODOR	0 1 2 3 4 5
080	BANANA-LIKE	0 1 2 3 4 5	123	SOOTY	0 1 2 3 4 5
081	BURNT RUBBER-LIKE	0 1 2 3 4 5	124	CRUSHED WEEDS	0 1 2 3 4 5
082	GERANIUM LEAVES	0 1 2 3 4 5	125	RUBBERY (NEW RUBBER)	0 1 2 3 4 5
083	URINE-LIKE	0 1 2 3 4 5	126	BAKERY (FRESH BREAD)	0 1 2 3 4 5
084	BEERY (BEER-LIKE)	0 1 2 3 4 5	127	OAK WOOD, COGNAC-LIKE	0 1 2 3 4 5
085	CEDARWOOD-LIKE	0 1 2 3 4 5	128	GRAPEFRUIT	0 1 2 3 4 5
086	COCONUT-LIKE	0 1 2 3 4 5	129	GRAPE-JUICE-LIKE	0 1 2 3 4 5
087	ROPE-LIKE	0 1 2 3 4 5	130	EGGY (FRESH EGGS)	0 1 2 3 4 5
088	SEMINAL, SPERM-LIKE	0 1 2 3 4 5	131	BITTER	0 1 2 3 4 5
089	LIKE CLEANING FLUID (Carbona)	0 1 2 3 4 5	132	CADAVEROUS, Like Dead Animal	0 1 2 3 4 5
090	CARDBOARD-LIKE	0 1 2 3 4 5	133	MAPLE (AS IN SYRUP)	0 1 2 3 4 5
091	LEMON (FRUIT)	0 1 2 3 4 5	134	SEASONING (FOR MEAT)	0 1 2 3 4 5
092	DIRTY LINEN-LIKE	0 1 2 3 4 5	135	APPLE (FRUIT)	0 1 2 3 4 5
093	KIPPERY (SMOKED FISH)	0 1 2 3 4 5	136	SOUPY	0 1 2 3 4 5
094	CARAMEL	0 1 2 3 4 5	137	GRAINY (AS GRAIN)	0 1 2 3 4 5
095	SAUERKRAUT-LIKE	0 1 2 3 4 5	138	CLOVE-LIKE	0 1 2 3 4 5
096	CRUSHED GRASS	0 1 2 3 4 5	139	RAISINS	0 1 2 3 4 5
097	CHOCOLATE	0 1 2 3 4 5	140	HAY	0 1 2 3 4 5
098	MOLASSES	0 1 2 3 4 5	141	KEROSENE	0 1 2 3 4 5
099	ALCOHOL-LIKE	0 1 2 3 4 5	142	NAIL POLISH REMOVER	0 1 2 3 4 5
100	DILL-LIKE	0 1 2 3 4 5	143	FERMENTED (Rotten) FRUIT	0 1 2 3 4 5
101	CHEMICAL	0 1 2 3 4 5	144	CHERRY (BERRY)	0 1 2 3 4 5
102	CREOSOTE	0 1 2 3 4 5	145	VARNISH	0 1 2 3 4 5
103	GREEN PEPPER	0 1 2 3 4 5	146	SOUR MILK	0 1 2 3 4 5

FIGURE 4-9 *Continued*

TABLE 4-2 Community Odor Nuisance Classification[a]

Odor Type	Typical Occurrence
Organic nitrogen compounds	Reduction of animal matter, including fish; typical representative compounds would be trimethylamine and skatole; animal and fishy odors are different from each other and could be two classes, instead of one
Phenolic odors	Curing of phenolic resins, creosoting operations, and the like; typical components are phenol, cresols, xylenols, and carvacrol
Organic sulfur	Petroleum refinery emission, emission from pesticides manufacturing, gas odorant leakage, and the like; typical components are mercaptans, sulfides, and disulfides
Organic acidic odors	Emission of acids like butyric, valeric, and phenylacetic
Burnt odors	Burning operations; tarry and asphaltic odors
Fragrant or floral	Manufacturing or compounding of flavors and perfumes
Solvent odors	Dry-cleaning exhaust; solvent drying emission
Camphoraceous	Naphthalene or *p*-dichlorobenzene odors
Oily	Varnish cooking; foundry core oven emissions; typical odor of linseed oil
Gassy unsaturated	Pungent, gassy odors like those of acrylic materials, diesel exhaust, etc.

[a] From A. Turk (personal communication).

suffer from semantic problems with respect to the meanings of descriptors. One solution may be in the selection of panelists and in their training in the uniform use of descriptors, as is done in milk-tasting. Nevertheless, experience with collaborative testing of the descriptor-profile method[44] has indicated that the same odors produce significantly correlated descriptor profiles in different laboratories, despite the fact that both inexperienced and experienced panelists were used, without training in the uniform use of descriptors.

In the ASTM E-18 Committee study,[44] a special form of a statistical treatment of odor descriptor profiles, based on a chi-squared statistic, was found to yield overall odor-dissimilarity indexes that correlated highly with the direct dissimilarity rating of the same odor pairs. Thus, the semantic problems in using the descriptors are not as critical.

Odor Classification Spaces

In the multidimensional scaling of odors, various reference odors or odor descriptors are interdependent to various degrees. The question arises of whether some parameters can be derived to classify odors in a mathematical way.

For instance, a tridimensional space (or one with more dimensions) may be possible in which each odor is represented by a point, and the distances between the points are proportional to the differences between the odors. The development of a map from given distances between cities is analogous. A multitude of distances results in a two-dimensional map, or a tridimensional one if the distances are sufficiently accurate to include elevation.

Several attempts have been made to develop such odor space models.[32,65,95,98,111-113,115,136] In most attempts, a few dimensions have been sufficient to systematize the data. However, the number of different odors used in each instance has been relatively low—up to 45.[32] Usually, some important odor types are missing. A generally applicable classification based on odor spaces has not yet been developed.

HEDONIC TONE (PLEASANTNESS-UNPLEASANTNESS)

A distinction must be made between the acceptability and the hedonic tone of an odor. Acceptability is usually a judgment made by a specific person in the context of a specific situation and specific expectations. For example, an otherwise pleasant odor may be unacceptable if it is a part of persistent odorous air pollution in a residential area and originates from a fragrance-manufacturing factory, and not from a flower garden.

When an odor is evaluated for its hedonic tone in the neutral context of an olfactometric presentation in the laboratory, the relative pleasantness or unpleasantness is determined by each panelist's experience and emotional associations. The response may be quite individualistic; thus, an odor pleasant to many, such as musk, may be declared highly unpleasant by some.

Hedonic Measurements

Hedonic judgment is both a category judgment (odor is pleasant, unpleasant, or neither) and a magnitude judgment, inasmuch as odors may differ in their relative unpleasantness or pleasantness.

Methods used to evaluate hedonic tone are similar to the odor-intensity

TABLE 4-3 Odors (in Water) Classified by Chemical Type[a]

Odor Characteristics[b]						
Sweetness	Pungency	Smokiness	Rottenness	Odor Class	Chemical Types	Examples
100	50	0 to 50	50	Estery	Esters Ethers Lower ketones	Lacquer, solvents, most fruits, many flowers
100	50 to 100	0 to 100	50	Alcoholic	Phenols and cresols Alcohols Hydrocarbons	Creosote, tars, smokes, alcohol, liquor, rose and spicy flowers, spices and herbs
50	50	0 to 50	50	Carbonyl	Aldehydes Higher ketones	Rancid fats, butter, stone fruits and nuts, violets, grasses and vegetables
50	100	0 to 50	50	Acidic	Acid anhydrides Organic acids Sulfur dioxide	Vinegar, perspiration, rancid oils, resins, body odor, garbage

100	50 to 100	50 to 100	0 to 100	Halide	Quinones Oxides and ozone Halides Nitrogen compounds	Insecticides, weed killers, musty and moldy odors, husks, medicinal odors, earth, peat
50	50	100	100	Sulfury	Selenium compounds Arsenicals Mercaptans Sulfides	Skunks, bears, foxes, rotting fish and meat, cabbage, onion, sewage
100	50	50	100	Unsaturated	Acetylene derivatives Butadiene Isoprene	Paint thinners, varnish, kerosene, turpentine, essential oils, cucumber
100	50	0 to 50	100	Basic	Vinyl monomers Amines Alkaloids Ammonia	Fecal odors, manure, fish and shellfish, stale flowers such as lilac, lily, jasmine, and honeysuckle

[a] Reprinted with permission from ASTM.[2]

[b] The degree of odor characteristic perceived is designated as follows: 100 indicates a high level of perception, 50 indicates a medium level of perception, and 0 indicates a low level of perception.

rating, except that there are two directions: negative for unpleasant odors, and positive for pleasant odors.

Category Scales These may be defined purely semantically (e.g., dislike very much, dislike, dislike slightly, neither like nor dislike, like slightly, like, like very much) or numerically (e.g., -3, -2, -1, 0, $+1$, $+2$, $+3$). Various pictorial hedonic scales have been developed, such as that in Figure 4-10. This scale was used for rating attitudes to diesel-exhaust odors[118] in public-opinion surveys; it does not require numerical or lingual expertise.

Magnitude-Estimate Scales Here, the panelist first judges whether the odor is pleasant, unpleasant, or neither. If the odor is pleasant or unpleasant, the panelist indicates by some magnitude response how pleasant or unpleasant the odor is. This may be done by assigning a number in a free open-ended selection of numbers. Another method consists of marking the length of a line or a bar on paper.[42,97] Still another uses a flexible self-rewinding tape that the panelist pulls out of a container as much as will indicate the degree of unpleasantness or pleasantness.[43,49] The tape is positioned with its graded side toward the panel leader, to prevent use of divisions by panelists as memorizable guides. Ratios of numbers or lengths permit the estimation of how much more unpleasant one odor is than another. Sometimes, it is useful to calibrate the panelists' responses for their numerical or length range immediately. For example, panelists may be asked to mark off their responses to an imaginary "moderately unpleasant" odor. This length is assigned 100 units, and each panelist's rating of the odorous scale is divided, for normalization, by the length indicated for "moderate."

Comparison of Hedonic-Measurement Methods Although hedonic judgments are individualistic, they are not as arbitrary as sometimes believed. Three sets of data were compared. Ten odorants were common to a hedonic evaluation conducted in 1964 at the University of California by 20 panelists who used category ratings[134] and to an evaluation in 1975 in Chicago by nine panelists who marked the length of a bar on paper.[97] The hedonic values were ranked in order of unpleasantness-pleasantness; both sets of data correlated at $p < 0.05$ (confidence level, higher than 95%). Another group of 10 odorants from the 1964 data set was compared with hedonic evaluation of their odors by nine panelists in 1977 in Chicago by the tape-extension method.[49] The order of hedonic rating correlated at $p < 0.001$ (confidence level, higher than 99.9%). Thus, the relative order of hedonic-tone values of odors in a bland laboratory context rated by panels is reasonably stable, and a hedonic-value scale may be possible for odors. There has been an attempt at this.[49]

Change in Hedonic Tone with Odor Intensity

Unpleasantness usually increases with odor intensity in a complex way.[97] Pleasant odors may increase in pleasantness with odor intensity when the intensity is low, but begin to become less pleasant, and may become unpleasant at higher intensities. A comparison of hedonic tone at the same odor intensities should be more useful than a comparison without regard for intensity, and comparisons over a range of intensities may be even more informative.

The hedonic tones of air-pollution odor samples from many different industrial sources have been measured by marking lengths of a bar on paper.[42] To relate the measured values to odor intensity, the intensities derived by the ASTM E 544 method were also measured. Figure 4-11 represents the plot of the data. Overall, unpleasantness decreases with a decrease in intensity, but at each intensity value there is a large dispersion in unpleasantness of various odors. A new sample may be compared for its relative unpleasantness with the industry average at the same intensity, or with a cluster of points for related emission from the same type of industry.

Thus, relative ratings of hedonic tone of odorous air-pollution samples may be possible.

PANELS AND PANELISTS

Panelist Selection

The type of sensory evaluation that will be made on odors and the objective of the evaluation will determine the rationale for panelist selection.

If the objective is to measure, for some odor, the odor-sensitivity distribution and the mean odor-detection threshold of the population at large, no selection of panelists is necessary, and as many panelists as possible should be used. Figure 4-12 is a plot of distribution of sensitivity of 100 panelists to the odor of isobutyric acid.[14] It is typical in the sense that the distribution is normal (Gaussian) with respect to the logarithm of the odorant concentration.

For some odors, people can be found who are significantly less sensitive than the average population.[13,14] For isobutyric acid, 10 of 400 people were one-fiftieth as sensitive as the average of the rest[14] and formed their own

FIGURE 4-10 A hedonic scale used for evaluation of diesel-exhaust odors in population surveys. Reprinted with permission from Springer.[118]

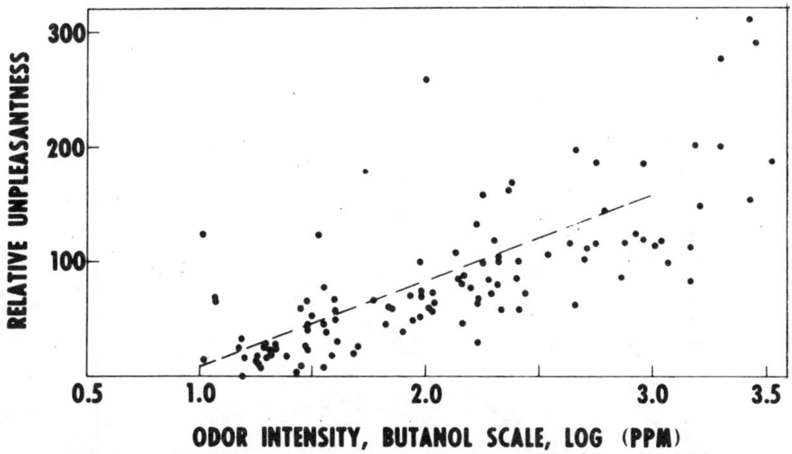

FIGURE 4-11 Scattergram of unpleasantness values of various industrial odorous emissions. Each point represents one emission sample. Its measured intensity can be read at the horizontal coordinate, and the unpleasantness on the vertical coordinate. Data represent many types of emissions. The dashed line is the least-square fit plot through all points. At the intensity coordinate of 2, the highest point is for a paper- and pulp-mill emission. The cluster of points at the same coordinate around the dashed line is for emissions from auto-body finishing operations. Reprinted with permission from Dravnieks.[42]

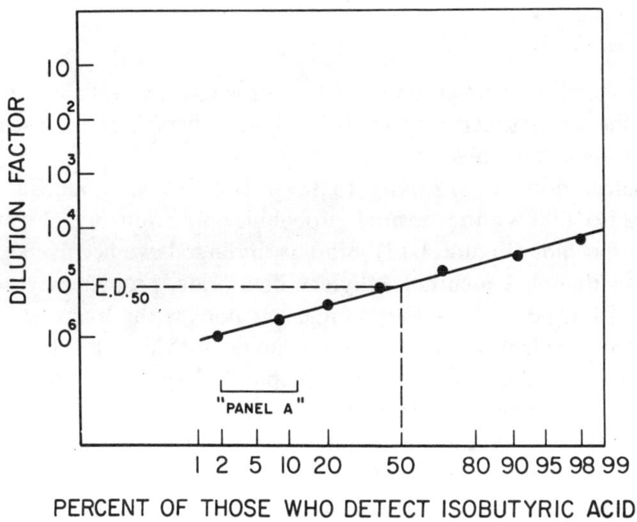

FIGURE 4-12 Odor-detection threshold distributor for 100 people. Based on data from Amoore.[14]

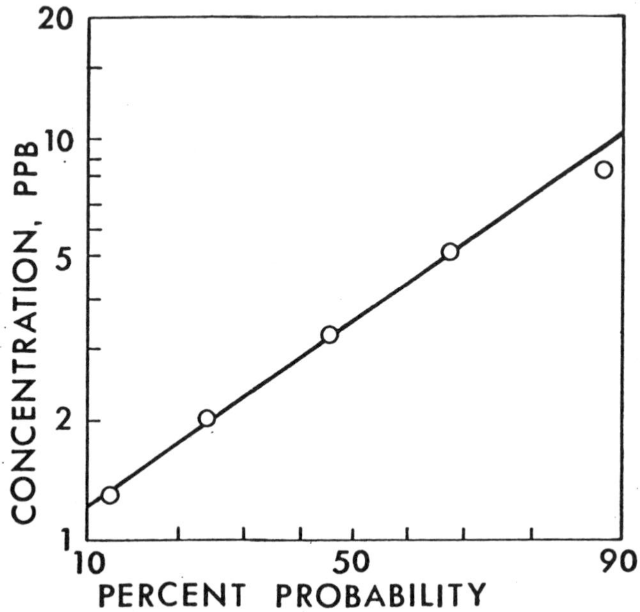

FIGURE 4-13 Distribution of individual odor-recognition thresholds for ethyl sulfide in air. Plot is in log (concentration) versus probability (%) coordinates and is based on responses by 33 people. Based on data from Wilby.[133]

normal probability-distribution cluster. This is termed "specific anosmia," although the term "specific hyposmia" might be better, because sensitivity is merely lessened, not absent.

The distribution of sensitivity to several sulfur compounds[133] among panelists also follows the normal probability function in a logarithmic concentration plot (Figure 4-13). Similar findings have been reported for 2-mercaptoethanol, 1-menthol, ethylene dichloride, propylbenzyl alcohol, *d*-camphor, formic acid, and ammonia, but not for the hydrocyanic acid resulting from hydrolysis of potassium cyanide; for this acid, the threshold range appeared much larger and trimodal.[21] Very little information is available on the range of sensitivity to various odors in the population.

Lower sensitivity in a given person to one odor does not automatically imply lower sensitivity to all odors, so it is important to test sensitivity with an odor appropriate to the odors that will be measured. Thus, sensitivity to the typical malodorants butyric acid and methyl disulfide (components of rendering odors) had no relation to sensitivity to vanillin and methylsalicylate.[132] It is obvious that persons with specific anosmia should not be included.

Figure 4-12 further illustrates the effects of various rationales for panelist selection. Some advocate using the more sensitive fraction of the panelists, such as panel A, to provide a safety factor in the results.[3] Others would select a homogeneous group—those whose sensitivity is average and quite similar, such as those represented in the center of Figure 4-12. Such a group would produce data with a high degree of reproducibility, which in reality is an artifact generated by panelist selection. A less arbitrary approach is the selection of a few panelists to approximate the normal distribution of sensitivity of a larger group (Brown *et al.*[21] and Bethea, cited in U.S. EPA[126]). A large group of prospective panelists is tested in duplicate or triplicate with the type of air-pollution odor sample that is to be evaluated. Those who are inconsistent in their judgments are not considered for panel membership. The rest form the panelist pool and are classified on the basis of their sensitivity. Each smaller panel is assembled from this pool of qualified panelists in proportion to the number of members of their sensitivity groups in the entire panelist pool. Panels so composed would closely reflect the approximate distribution of sensitivity of the panelist pool.

Familiarity with the relevant odors is required in evaluation of subtle differences in odors. This is especially important in surveying an area for occurrence of pollution odors, to identify their sources and decide which sources should be the prime candidates for control. If panelists are selected for this purpose, they must be tested for consistency, ability to discriminate odor qualities, and sensitivity to all pertinent odors, e.g., to specific odor notes in evaluation of diesel-exhaust odors.[125]

For evaluating the hedonic tone and acceptability of odors, the exact opposite applies.[90] Here, a thorough familiarity with the odor produces bias. Although some familiarity cannot be avoided, evaluation of the odor should preferably be conducted away from its source without informing panelists about the origin of the odor.

Panel Size

The principles and methods for panelist selection for various sensory tasks include many considerations,[90,125] including panel size. For most routine tasks, a panel of nine or 10 is about the smallest, inasmuch as data obtained with smaller panels cannot be statistically tested with sufficient resolution of probabilities. Larger panels, 15–100, are needed for hedonic judgments.[90]

Panelist Training

For such tests as odor-threshold and odor-intensity measurements, in which population response is to be estimated, the only training needed is in how to proceed with smelling and responding. For tests that are to establish finer differences between odors, training in procedures is accomplished by conducting evaluation trials with many typical samples that encompass the range of samples of odorous pollution expected. Detailed protocols for panelist training are available.[12,125]

SAMPLING

Samples can be taken into an analytical device or an olfactometer, or in some cases (e.g., ambient air) smelled directly. The advantage of sampling into an evaluation device is that the degradation of the sample is minimized. In other cases, it is advantageous to take a batch sample of ambient odorous air or emission at the source for evaluation in a laboratory by methods that are impractical in the field.

SAMPLING FOR ANALYTICAL EVALUATION

Methods differ, depending on whether a specific odorant or a group of odorants are to be measured.

Field Monitoring

The concentration of some odorants either in the ambient air or in stack emission can be measured and monitored in a semicontinuous manner by devices that can operate in the field. For instance, hydrogen sulfide and some highly volatile organic sulfur compounds can be determined quantitatively in emission-gas samples or, in ambient air, taken periodically by portable gas analyzers. Such analyzers use batch sampling into a gas-chromatographic separation column to separate components of the sample. A photometric hydrogen-flame ionization detector selectively detects elution of sulfur compounds from the GC column[67]* Another type of equipment measures total reducible-sulfur content.

*Portable gas chromatograph Model 513, Analytical Instrument Developments, Inc., Avondale, Pa.; Rotorod gas sampler Model 721, Metronics Associates, Inc., Santa Clara, Calif.

Other simple and less precise direct-sampling analytical devices use color reactions. Hydrogen sulfide in air is detected semiquantitatively by darkening of a lead acetate-impregnated paper.[99] Colorimetric tubes that contain packing with reagents that change color in the presence of specific groups of compounds are available for a large variety of chemical groups,[116] but their performance has not been widely discussed. For emission, a gaseous odorous sample is pulled through the tube by a syringe that can be attached to the tube.

Sampling for Analysis in Laboratory

Techniques for sampling emission and odorous ambient air have advanced substantially in recent years, owing to the advent of thermally resistant porous organic polymeric adsorbent materials with a large surface area.

Freezeout and Active-Carbon Methods Previously, freezeout and active-carbon methods were used.[5] In the freezeout method, all volatile vapors in air are condensed by passage through cold traps cooled to the temperature of liquid nitrogen; or through stages of cooling with ice, dry ice, and liquid nitrogen. In the active-carbon method, vapors are adsorbed on the extremely large surface of fine pores in carbon.[116] The carbon method is in extensive use in monitoring personal exposure of workers to NIOSH-controlled organic vapors, such as chlorinated hydrocarbons. A device with a small battery-operated pump and a carbon adsorber is carried by the worker. Adsorbed vapors are later eluted from the carbon adsorber with a solvent and analyzed to calculate the cumulative exposure to the potentially harmful vapors.

Neither of these methods is well suited to sampling for odorant analysis. Because water vapor is a substantial portion (several percent) of the atmosphere and may be a very substantial portion of emission, freezeout samples contain much water. Water interferes with many forms of analysis and is a complicating factor especially when ambient air has to be analyzed for parts-per-billion concentrations of odorants. Concentrating the odorants from aqueous samples with solvent extraction and evaporative concentration is a laborious procedure, and it is complicated by possible losses in the extraction and evaporation of the solvent and by interference caused by impurities in most solvents.

In the carbon method, sample recovery from the carbon poses problems. Recovery by thermal desorption requires a considerable increase in temperature and readily results in thermal and catalytic changes in the composition of the sample. Elution with a solvent is less harsh, but some materials may be incompletely recovered. Impurities in the solvents are difficult to avoid, and the volume of the solvent required results in a sample that needs to be concentrated by vaporization.

Polymeric Adsorbents Porous organic polymers that are relatively thermally stable and have large surface areas are available.[17,41,92,104,106,109] These can adsorb most organic vapors quite efficiently, but retain little water vapor. The principal materials have trade names of Porapak, Chromsorb 100 Series, and Tenax. A gaseous sample is pulled through a tubular element packed with such materials. Organic vapors are adsorbed, and most of the water vapor passes through. A sample is recovered by heating in an inert-gas flow. The surface of such materials is noncatalytic and the temperatures needed are lower than for recovery from active carbon. Therefore, sample degradation is minimal, although with some unstable compounds it may be a potential problem and has to be critically explored.

For some analytical methods, a sample can be recovered from such polymers with a solvent. For most purposes dealing with very low concentrations of odorants, problems with impurities in the solvent and the undesirable dilution of the sample by the solvent make the extraction method less attractive; it is rarely useful below parts-per-million (vol/vol) concentrations of organic vapors in air, if the analysis is by gas chromatograph (A. Dravnieks, personal communication).

Of the three materials named, Tenax GC is increasingly used. It is based on 2,6-diphenyl-*p*-phenylene oxide, as probably the most thermally stable of the available materials. Although the material is the same, there are almost as many mechanical arrangements for sample collection as there are laboratories that use it.[41,92,104,106,109] They differ in the amount of adsorbent used, the geometry of its placement, sampling rates, and mechanisms of the sample recovery. Performance is characterized by breakthrough volumes for various vapors, because the collector operates very much like a GC phase in frontal adsorption analysis. Each substance has its characteristic retention volume. Those which are more volatile and are adsorbed less continue to accumulate in the collector until this volume is exceeded; then, as much vapor enters as leaves, and the amount retained remains constant, although proportional to the concentration of the odorant in the air. Heavier odorants may be totally retained until quite a large volume of the gaseous sample has passed. In accordance with the effects inherent in gas chromatography, for a sample of constant air volume, the content of each component in the collector is proportional to its content in air, regardless of whether the specific retention volume has been exceeded; the proportionality factors are different for compounds that exceeded their retention volumes and for those which did not.

Some samples at high temperatures may contain water vapor that would condense when the sample is cooled to the ambient collector temperature. Such samples are usually prediluted to a controlled extent in a flowing system with inert gas until such condensation is prevented. The collector temperature can also be increased. Both methods decrease collection efficiency.

The volume of air to be sampled depends on the expected concentration of odorants; it may be from a fraction of a liter to many tens of liters. Sampling times are from a few minutes to an hour, or even more. Sampling rates are from 30 ml to several liters per minute.

The polymer method is rapidly replacing freezeout methods. The use of grab samples in inflatable bags or preevacuated cylinders for sample collection has little or no advantage. Such devices are bulkier than small polymeric collectors, are difficult to keep free of impurities, and still require additional concentration when samples are brought to the laboratory for analysis.

Samples collected in polymeric adsorbents can be closed, transported, and stored for hours and sometimes days, although early analysis is desirable. Before reuse, collectors usually need reconditioning by heating in an inert-gas flow. This is done shortly (hours or a day) before sampling, because impurities tend to build up even in closed collectors, in ways that are not entirely clear.

SAMPLING FOR SENSORY EVALUATION

The integrity of a sample of odorous air or emission is best preserved if the sample is evaluated without storage and transportation. However, such evaluation requires that panelists be brought to the site of odor occurrence. This is difficult if the odors are transient or change with location or if a large number of panelists is needed.

Evaluation in the Field

In-the-field sensory evaluation is undertaken in two types of situations.

One type is in a form of *odor survey.* When odorous emission disperses in the environment, it reaches different locations at different concentrations, which fluctuate with atmospheric conditions and winds. Taking samples in such a situation for laboratory evaluation is prohibitively expensive. The most suitable form of survey involves the use of trained odor experts or highly trained panelists.[58] They travel extensively in the community exposed to odorous air pollution and make notes on the intensities and characters of odors at different locations and different times. It is highly desirable that additional odor information be collected to support and verify these observations. Documentation of meteorologic conditions is essential in estimating the frequency of occurrence of the ambient odors, provided that the frequencies of occurrence of different meteorologic conditions for the specific source location are known from climatic data. If some samples are available for sensory evaluation in the laboratory, verification or calibration of the field measurements against more exact measurements made by larger panels under more favorable odor-testing conditions is possible. If analytical samples of ambient air and emission are taken, the analytical data can be used to reconcile walking-survey data with emission-dispersion equations. All such supplementary information increases the credibility of the walking survey by experts and permits better estimation of the benefits to be derived from odor-control measures.

Another type of in-the-field evaluation uses *mobile laboratories.* These are trailers or motor homes equipped with air deodorization equipment to maintain odorless air inside and with instrumentation to deliver a flow of odorous air or emission samples for measurements inside.[87,132] Panelists conduct odor evaluations in the comfortable environment of the mobile laboratory, and the odorous samples delivered undergo a minimum of composition change. The operation of such laboratories is expensive, but they may be useful for critical evaluation of odor-complaint disputes and for evaluation of sampling techniques (for comparison with results obtained on samples taken to a stationary laboratory).[132]

The reverse of the mobile-laboratory method is used in testing of odors from vehicular sources. Here, the laboratory is stationary, but the odor sources are brought to some testing stand for monitoring of operational variables, and emission is piped to the laboratory for evaluation.

Sampling for Sensory Evaluation in Laboratory

Sampling of odorous ambient air or emission for later evaluation by panels in a properly equipped odorless space requires the proper choice of sampling equipment, containers, and sample-recovery methods.

Sampling requires a device to withdraw a batch sample of an emission or ambient air and a container for sample storage (Figure 4-14).

In some cases, a simple ejector[110] is used in which a nonodorous-gas flow sucks in the gaseous sample, dilutes it to a known ratio (which depends on ejector characteristics), and pushes it into a storage container. This method is particularly suitable for hot, moist gas streams.

A frequently used sampling device for emission is a peristaltic pump that uses an essentially nonodorous plastic tubing—e.g., food-grade Tygon tubing[50]—and kneads it by rollers or other mechanical elements to deliver

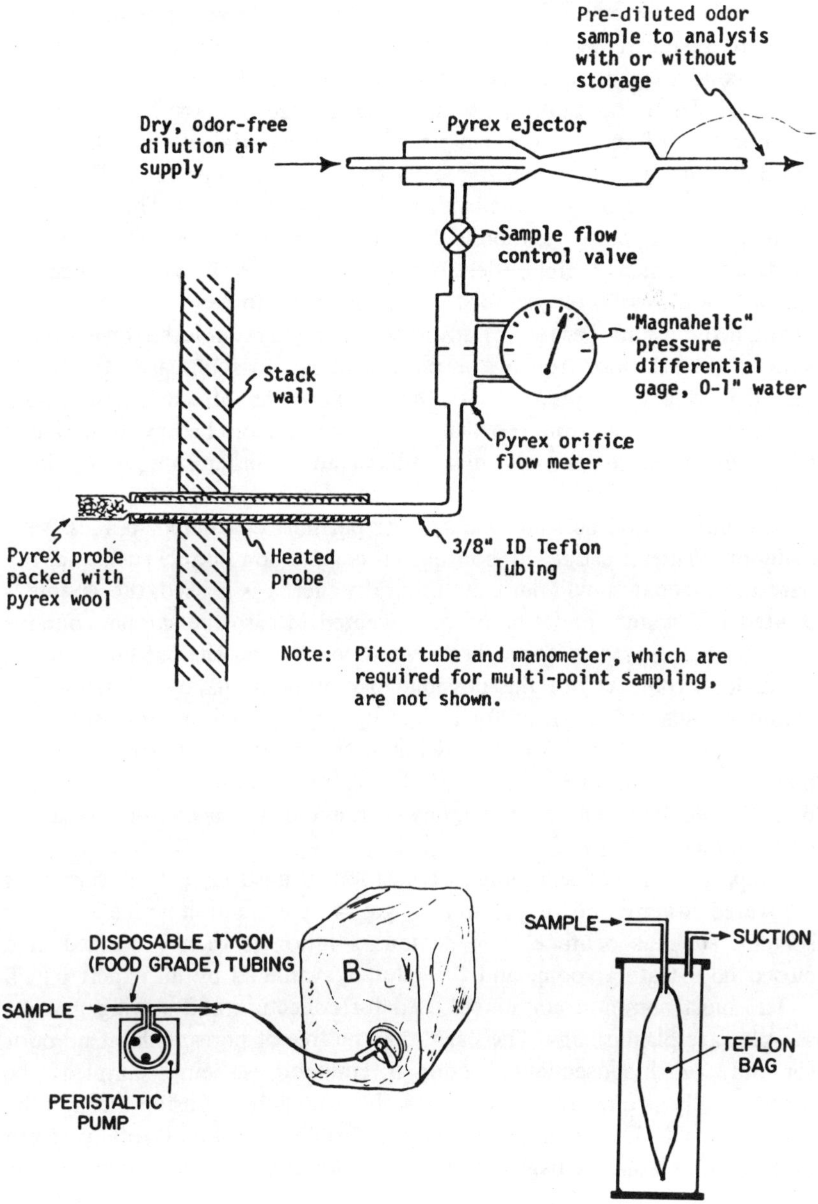

FIGURE 4-14 Methods for sampling odorous emission and ambient air. Top, reprinted from Wahl *et al.*[132] Bottom left, reprinted with permission from Dravnieks and Prokop.[50] Bottom right, reprinted with permission from Dravnieks *et al.*[51]

the sample into a container. A gaseous sample can be sucked either from a negative- or positive-pressure duct and is pushed into the container. Some slight condensation of excess water vapor does not interfere with the pumping. The initial emission sample is discarded to allow for adsorption equilibration of the flowing sample with the material of the tubing and sampling line. The container, typically a plastic bag, is preflushed to allow for the initial loss of odorants by sorption in the bag walls. The advantage of the peristaltic-pump method is that a new piece of tubing element can be used for each new sample; therefore, no cleaning of the pump is needed.

For a weakly odorous sample, it is undesirable to have the sample pass through the pump, because of adsorption and sorption losses or contamination by weak odor from warm plastic tubing. In such cases, the bag is placed in a larger drum,[51,110] with the bag-filling spout outside the drum. When air is pumped from the drum—e.g., with a hand bellows pump—the bag is inflated with the odorous ambient air or other sample. Again, a preflush is used.

Another method uses an evacuated or partially evacuated stainless-steel cylinder.[5] When it is opened, odorous air or emission sample rushes in. If a partially evacuated cylinder containing dry inert gas is used, the sample is diluted and water condensation is prevented in case the sample contains condensable moisture. The cylinder method obviates the use of pumps in the field. Problems are the possible loss of odorants by adsorption at cylinder walls and the need for a thorough cylinder-cleaning between the samples. For recovery, the cylinder is pressurized by introducing excess inert gas, so dilution of the sample before its olfactometric evaluation is unavoidable; its extent is known, however, and can be accounted for in the calculations.

A special form of sampling is collection in adsorbents.[121] A sample is recovered either by heating in an inert gas or by extraction with a low-odor solvent, such as pentane. The pentane solution is then vaporized in a closed odor-testing room, and the odor is evaluated by an expert panel.

The most common containers used for collecting and storing gaseous samples are plastic bags. The bags must be free of pores, exhibit no odor (or only an inconsequential odor if emission is being sampled), be mechanically strong, and prevent loss of odorants by diffusion through the wall material. Diffusion is almost impossible to prevent totally. Even multiple-film laminate bags used in the food industry to prevent flavor loss from packaged foods do not totally prevent it. The shelf-life of samples in bags is limited and may be different for different types of samples.

In the United States, bagged samples can usually be transported and evaluated within 24 h after sampling. Several studies of the storage of odorous or other gaseous samples have been conducted,[50,51,73,114,132] some with inconsistent results that indicate artifacts in sample preparation or analysis. The most advisable course is to pretest the storage of the particular type of odorous sample in the proposed bags, on the basis of the odor-detection threshold tests with nine or more panelists. A statistical analysis of the panel data would show whether a statistically significant odor degradation had occurred. Degradation can occur not only by diffusion through the walls, but also by reaction between the components of a sample. Thus, weak mixtures of hydrogen sulfide in air of normal humidity rapidly degrade in glass containers or syringes, but remain stable for 1–2 days in thick-walled (0.5-mm) polyethylene containers (A. Dravnieks, personal communication).

Popular containers for emission samples are thick-wall cubic collapsible polyethylene 18-L (5-gal) "Cubutainers."[50] They have a very slight background odor, but are inexpensive and are discarded after use. They have survived air-freight shipping in cardboard boxes and even by themselves.

For weaker odors, much more expensive Tedlar[68] or heat-sealable FEP Teflon[51] bags are often used. Such bags can be deodorized by repeated flushing with odorless air.

Recovery of Samples

Recovery is the reverse of sampling. The sample is removed from the bag either by a peristaltic or other type of odorless pump at a controlled rate into an olfactometer or to sniffing ports[50] or by being pushed into the olfactometer by compression of the bag in a cylinder.[51] In either case, a few minutes are allowed for overcoming adsorption losses in the tubing that connects the bag to the olfactometer.

Sample Size

The flow-rate requirements of the olfactometer determine the sample size. The size typically ranges from 15 to 200 L.

OTHER FORMS OF SAMPLING

The Mills version of ASTM D 1391[93] proposes sampling into a glass container, initially filled with mercury, by lowering the level of mercury. This method has now been almost universally rejected because of the toxicity of mercury.

Another modification[93] for sampling hot and humid emission uses a 250-ml glass sampling bulb filled with dry air. One end of the bulb has a long glass capillary tube that is inserted into the sample stream and adjusts to its temperature. The other end of the bulb is connected to a glass syringe, and 5 or 10 ml of sample is pulled into the glass bulb; this results in a 1:50 or 1:25 dilution of the moist gas sample with dry air. Experience with adsorption on glass surfaces makes it questionable whether at the high surface-to-volume ratios in such collectors adsorption losses can be ignored, especially for large odorant molecules, which strongly adsorb at glass surfaces (particularly if the molecules contain polar groups and double bonds).

These methods of small-scale sampling are more and more being replaced by bag sampling and dynamic dilution techniques.

COMPILATIONS OF DATA ON ODORS

The ASTM E-18 Committee compiled and published tabulations of odor and taste detection and recognition threshold values from the literature and from unpublished laboratory data.[119] A new enlarged edition of this compilation was published in 1978,[55] and the ASTM has established a procedure for a periodic updating of the compilation. Another compilation was published in Holland in 1977.[128]

Data in these compilations show large differences in the thresholds of the same odorants reported by different authors. These differences are probably caused by differences in methods, panelists, and purity of compounds.

Perhaps the mean values of the thresholds reported in these compilations can be used as tentative benchmarks of odor thresholds for specific chemicals.

Odor-intensity data on 110 substances have been tabulated.[105]

Data on odor character, which in the scientific literature is usually termed "odor quality," are scattered in the literature and have not been collected and systematized.

CURRENT APPROACHES TO THE MEASUREMENT OF INDUSTRIAL POLLUTION ODORS

EMISSION

Most of the data base on emission odors is in terms of odor-detection limits, usually in the form of odor-detection thresholds—the dilution that is needed to make the odor of the emission nondetectable is measured. Such measurements are made to satisfy the applicable regulations of particular air-pollution control jurisdictions.

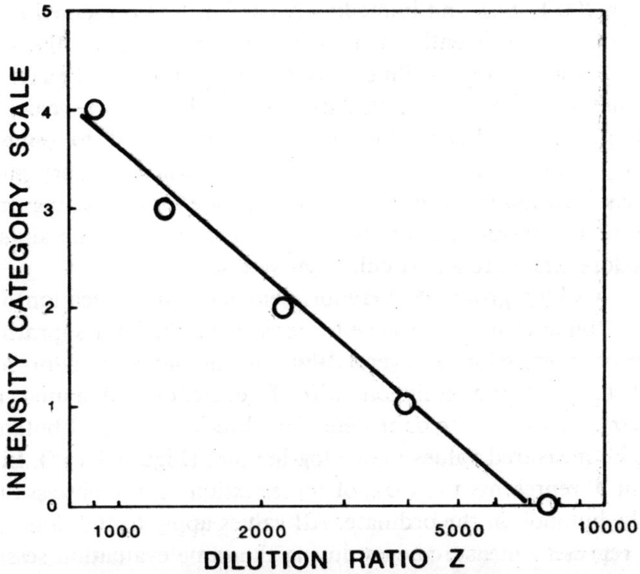

FIGURE 4-15 Dose-response plot for odorous-emission dilution test using category scale for measurement of odor intensity. Based on data from Hemeon.[71]

However, limited use is also made of measurements of odor intensities at one or several dilutions of the emission. Because the odor intensity of an odorous sample decreases with dilution at different rates for different odorants, such measurements provide more useful information on the degree of dilution (by reducing the output of odorants at the source or by atmospheric dispersion) that is needed to reduce odor to zero intensity or to some intensity specified as low enough.

In one case, an emission sample is continuously withdrawn with immediate dilution to a controlled ratio, and the odor intensity of the diluted emission is measured with the ASTM E 544 butanol scale.[103] The values obtained are used to estimate to what degree (e.g., after the atmospheric dilution) the odor may be a cause of complaints.

The use of dose-response functions for an extrapolation to some odor intensity has already been mentioned. This is the procedure proposed for the Hemeon olfactometer,[71] with extrapolation to zero odor. Figure 4-15 shows the principle.

One working group (J. J. Franz, personal communication) selects the "very slight odor" as the target value for odor control. Emission samples are taken in 200-L bags and immediately (within 1 h) presented in diluted form at several consistently increased concentrations of the emission sample. The flow rate for smelling is over 30 L/min. A small panel (four to five persons) rates the odor intensity on a 0–4 category scale, with 1 representing a very slight odor. The dilution needed to obtain such intensity is then estimated from the dilution-intensity category plot. This dilution has been found to serve as a useful guide for estimating, from the atmospheric dispersion equations, the possibility of complaints about ambient odors related to a particular emission source.

Another working group (A. Gaynor, personal communication) uses the ASTM E 544 butanol-intensity scale for measuring the odor suprathreshold intensities of an emission at several dilutions, measures the odor-detection threshold ED_{50} of the emission with forced-choice dynamic triangle olfactometry, measures the odor-detection threshold ED_{50} of butanol, and combines all measured values in one log-log plot (Figure 4-16[42]). In Figure 4-16, point T represents the ED_{50} of the emission on the abscissa and the ED_{50} of the butanol on the ordinate. All values apply to the same panel of nine and represent measurements during the same evaluation session; the threshold point is usually on the continuation of the straight line connecting the suprathreshold intensity points. The vertical line marks the dilution characterizing the odor-detection threshold of the most sensitive panelist, the approximate ED_{10} value for a panel of nine. The right-hand scale is the S scale[96] (see the discussion of odor-intensity measurements) and consists of numbers proportional to the perceived odor intensity of the various locations on the dose-response plot. Such plots permit calculating odor intensities of emission at different degrees of atmospheric dispersion, so that the perceived odor intensities in various locations under different meteorologic conditions can be estimated. No exhaustive data bases exist to prove or disprove definitely that the predicted ambient-odor intensities agree with the observed values. However, in application to the emission odors from auto-body finishing operations, where several types of odors occur, those which had a smaller slope in Figure 4-13 were found to dominate the character of the ambient odor at greater distances from the plant, even if their intensities at the source were lower.

AMBIENT AIR

Recent approaches to measurement of industrial odors in ambient air have extended beyond simple measurements of odor-detection thresholds at some selected locations and times.

In a Dutch study, the statistical distribution of the detection thresholds within a population was considered to relate the odor impact on the

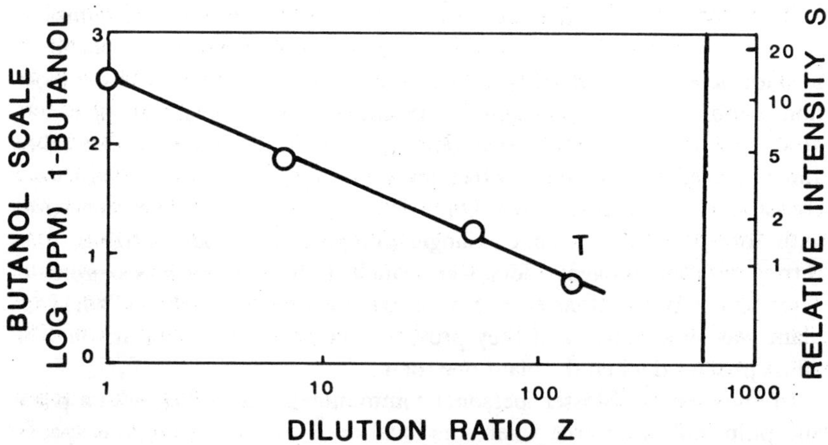

FIGURE 4-16 Dose-response plot for odorous-emission dilution test using ASTM E 544 1-butanol scale for measurement of odor intensity. Data from A. Gaynor (personal communication).

community to the atmospheric dispersion equations.[27] The odor-impact indexes were telephone-complaint frequency and response to distributed questionnaires. The statistical distribution of odor-detection thresholds for specific air-pollution odors was not measured, but was assumed to be a typical logarithmic normal probability function. An example[133] of the distribution function is shown in Figure 4-14: with ethyl sulfide at 4 ppb in air, 50% of the people would be able to detect its odor; at 1.5 ppb, only 15% would be able to do so. Atmospheric concentrations of emission odors were estimated for different locations in the community with dispersion equations. It was found that functions like that in Figure 4-13 quite satisfactorily predict the distribution of the frequencies of complaints in the various areas in the community.

In another work, extensive ambient-odor detection-threshold measurements and chemical analytical data were obtained to calibrate the local atmospheric dispersion model (A. Gaynor, personal communication). Detection thresholds were found to correlate with the concentrations of emission measured analytically. The industrial odorous-emission source was an auto-body finishing plant with several hundred stacks that was in a city and surrounded by residential, school, and commercial buildings. The dispersion model was developed by placing a scale model of the plant and community in a wind tunnel with equipment to imitate various meteorologic conditions. However, it was questioned whether dilutions of the ambient odorous air along a typical street would follow the values of dilutions predicted from the model. Emission was characterized in terms of odor-detection threshold, total emission rate, and analytical composition. Ambient odors were sampled simultaneously at 10 sites along streets downwind from the plant; analytical samples at all sites were also taken and analyzed to provide a direct measure of the dilution of the source emission at the various sites. Odor-detection thresholds were measured with forced-choice dynamic triangle olfactometry;[51] local citizens, concerned with the ambient odors, were panelists. In total, air was sampled in three sets of 10 locations each. Similar measurements were made when the plant was shut down, and they provided the data for a comparison with odors produced when the plant operated.

In one case (R. Blosser, personal communication), dealing with a paper and pulp mill in an area where the local odor-control regulations specify maximal permissible emission of several sulfur compounds,[56] analytical monitoring instruments are used to measure the ambient concentrations of these compounds at the property line. The maximal permissible concentrations are set on the basis of expected dilution by the time the odorants in the emission reach the property line, so that at this location the odors would be at an estimated detection threshold. Direct analytical monitoring at the property line is used by the plant to ascertain that the detection threshold there is not exceeded.

Direct perceived-intensity measurements in ambient air are increasingly used to survey the distribution of ambient odors in the community.

In a Danish study,[1] panelists were trained to estimate the ambient-odor intensities by reference to the Turk *et al.* odor-intensity scale developed originally for diesel-exhaust odor-intensity measurements.[125] The ambient-odor survey was conducted by several panelists who wore plexiglass hoods supplied with air deodorized by active carbon. At various locations in the community, they opened the hoods and estimated the ambient-odor intensity, by reference (by memory) to the intensity scale.

In connection with an auto-body finishing plant in the United States, panelists selected from the company's personnel served as a periodic "odor patrol" (A. Gaynor, personal communication). They walked in the community and characterized, by reference to a memorized ASTM E 544 butanol scale, the ambient-odor intensity at different locations. This permitted a recording of the odor-intensity distribution under different meteorologic conditions.

In still another study,[78] a mask device was adapted to measure the perceived odor intensities in ambient air. The intensity of an ambient odor was compared with the intensity of butanol odor (as in the ASTM E 544 test), which could be supplied to the mask at various concentrations.

DISCUSSION

GENERAL NEEDS

When odorous emission creates a community odor problem, there is assumed to be a need for technologic control of odor. The extent of the needed control and the success of control cannot be determined unless odors are measured.

Measurements on the emission combined with some empirical knowledge provide support for calculations on the degree of odor control needed. Similar measurements performed after the implementation of odor control yield data on the efficiency of the control. After dilution in the atmosphere, the odorous emission reaches various locations in the community; the odors are then ambient odors. Ambient odors vary with location and time, because atmospheric dispersion depends on meteorologic conditions, distance from the emission source, and topography. Measurements of ambient odors at some selected sites and times are methodologically possible and may be useful; some air pollutants are already similarly monitored at fixed sites. However, such limited information does not reflect the broad scope of the impact of odor on the community. Time, location, meteorologic, topographic, and demographic variables require that the adverse response of the community to the ambient odor be measured by other means. Community response is shown in a variety of ways, such as by complaints to industry, press, public officials, and other public figures and by answers to odor-survey questions. There is a need to develop satisfactory indexes that can measure, at least on a relative basis, the extent of public annoyance with an ambient odor and the response to reduction of the annoyance as technologic odor control is implemented.

Because the potential for odor annoyance depends on odor intensity and character, and these in turn depend on the type of emission and its atmospheric dispersion within specific topography, it is unlikely that some generally applicable target values could be defined for all emission that would eliminate the impact of odor on the community. Differences in the sensitivity of individuals to ambient odors also complicate the selection of such a target. Only elimination of an odor would eliminate the impact. In most cases, this would entail excessive cost, usually with a high penalty in energy. Odorous emission is diluted by the atmosphere before it has its impact on the community, so elimination of emission odors themselves is not necessary to eliminate ambient odors, except under very unusual atmospheric conditions (e.g., a heavy inversion).

A compromise that in many cases may be technologically possible at tolerable economic and energy penalties is a reduction of emission odors to the point where some appropriate community response index reaches a judiciously selected low value or indicates a virtual cessation of complaints. In such an approach, the following chain of actions is visualized:

1. Odorous properties of the emission and the emission rate are measured.
2. Community response to the existing ambient odors is characterized through an odor-response index under several types of atmospheric conditions.
3. An engineering judgment is made on the target value for reduction of the emission odor and the means of achieving this target.
4. Technologic odor control is implemented.
5. The emission odor is measured to ascertain that odor reduction has occurred and to estimate the efficiency of odor control.
6. Community response is monitored under various atmospheric conditions.
7. If the adverse community response continues to be substantial, further improvement in odor control is considered by returning to step 3 and continuing until the odor impact is no longer of measurable dimensions.

Although community response to an ambient odor under a variety of atmospheric conditions should be the most important dimension for measuring the impact of the odor on the community, it may often be more expeditious to survey ambient odors in a community by using estimates based on responses of a small group of adequately trained persons.[58]

This discussion has used many vague terms that pinpoint the need for research on methods to measure the impact of ambient odors on a community and to arrive at agreed-on target values for permissible impact. However, the problem of setting specific targets for pollutants is not peculiar to odors. It occurs with many pollutants, when the degree of the physiologically measured health effects is a continuous function of the concentration and duration of exposure and the criteria for permissible exposure must be set as a specific point on a continuous dose-response curve.

SAMPLING

Change in an odor during sampling and sample storage is a possibility that needs consideration. The change is least when the odorous sample is delivered for evaluation directly in a continuous flow. This requires special logistics, such as the use of mobile odor-evaluation laboratories. A common alternative is batch-sampling with evaluation at a reasonably odor-free location as soon as possible after sampling. With this method, comparisons should be conducted on the odor threshold of the batch sample and the threshold of a sample obtained by direct flow of the odorous emission (or ambient odor) into the same threshold-measuring device. Batch-sampling would be considered valid if the two thresholds were statistically indistinguishable. Acceptability of storage of batch samples must be validated by similar means, including comparison of odor thresholds of identical samples after storage for specified durations. Different materials may need to be used in the storage equipment (bag or adsorbent) for different types of odorous samples. Batch-sampling must be conducted with a preflush of the sampling equipment with the sample, to avoid sample losses by adsorption and absorption.

Odorous samples are diluted to measure odor thresholds or to evaluate the dose-response function (change in odor intensity with dilution). One form of dilution, that in the ASTM D 1391 syringe-dilution test, uses odorless air to bring a smaller volume of the sample to a selected fixed volume. Adsorption of the odorant on the container walls is a source of difficulty in such a method, especially if the final volume is small. This form of dilution should be replaced by the so-called dynamic-dilution methods, in which measured sample and dilution air flows are continuously mixed to produce needed dilution ratios. In the dynamic methods, the initial depletion of the odorants by adsorption on the dilution-system walls ceases rapidly when equilibrium is reached between vaporous and adsorbed odorant.

ESTIMATING THE ODOR CONTROL NEEDED

When a need for odor control is evident from an adverse reaction of the community, the extent of odor control that will significantly reduce or eliminate the ambient odor must be estimated. This can be done on the basis of measurements of the emission-odor threshold, from experimentally determined dose-response functions, or from measurements of the thresholds of the diluted, ambient odors. The last of these approaches requires decisions as to the time and location of sampling and is therefore less straightforward.

The emission-odor detection threshold can be used for engineering estimates in two ways. First, there is considerable information, including that reflected in some odor-control regulations, on the emission-odor thresholds below which the corresponding ambient odors seem to cease to be a problem. Second, threshold values can be used in atmospheric-dispersion models to estimate the odor thresholds expected in ambient air under various dispersion conditions at different locations. If the estimated ambient odors are too high, a reverse calculation will indicate what emission-odor threshold target will reduce or eliminate ambient odors. Such dispersion calculations for multisource situations may be questioned, because additivity of thresholds cannot necessarily be expected.

In the use of dose-response functions, the perceived odor intensities of the emission are measured at several dilutions. The measurement may use any odor-intensity reference scale consisting of odor samples of graded intensities or categories. Category scales are logistically simpler, but the data obtained by different groups on the same samples are more difficult to compare, because of the differences in the meaning of the categories for different people. From the plot of the dose-response function, the dilution needed to reach the odor-intensity target can be estimated. Atmospheric-dispersion models are then used to determine whether the dilution targeted from the response plots will occur naturally in the atmosphere. If not, the degree of reduction in odor intensity can be estimated, and a point on the intensity-dilution plot can be selected so that further atmospheric dilution will bring the odor intensity to the desired target value.

Ambient-odor threshold measurements can also be used directly. If the ambient odor is 5 times above the threshold ($Z = 5$) in the particular location at the particular time, a reduction in the odorous emission by a factor of 5 would presumably bring the ambient odor to an approximate threshold value. This would apply, however, only to the same location under identical atmospheric conditions. The problem in this approach is to select the appropriate sampling site and atmospheric conditions. Many ambient-odor threshold measurements at a number of locations under different atmospheric conditions may be required to obtain a balanced estimate of the needed emission-odor control, and the procedure would then be long and cumbersome.

None of the three approaches produces an exact solution to the odor-control problem. The data obtained by such methods provide only guidance for the selection of an odor-control method, in addition to a baseline for evaluation of the efficiency of later odor control. Comparison based on the resulting ambient-odor thresholds is very cumbersome and uncertain; this is an added disadvantage of the ambient-odor method in dealing with emission-odor control.

ODOR-THRESHOLD MEASUREMENTS

Methods in which one sample at a time is presented and a panelist reports whether it has an odor have one important disadvantage: validation that the panelist indeed perceived an odor is not possible unless many presentations are made at appropriate dilutions. This is a particular disadvantage when there are conflicting opinions on the presence or absence of the odor problem, e.g., in arguments between the organization responsible for an odorous emission and the clean-air enforcement authority or concerned citizens. A practical and efficient procedure for solving this difficulty uses a forced-choice multiple-sample technique. One or more samples contain diluted odorous emission or ambient air, the others are odorless air. The panelist smells all the samples and indicates which is (are) odorous. An incorrect choice signals that the odorous sample at the dilution used cannot be reliably detected by its odor by a particular panelist. Judgments of several panelists at several odorous-sample dilutions are combined to obtain the mean panel threshold.

Methods in which one sample at a time is presented should be considered to yield advisory data only. Critical values should be obtained by the forced-choice multiple-sample methods. The one-sample method has use in scientific research and yields significant data when many judgments are obtained and the values are analyzed by statistical techniques based on signal-detection theory.

Odorous samples in threshold determinations can in principle be presented in ascending, descending, or randomized order, with respect to the concentration of the odorous materials in the diluted samples. Ascending order is preferred in the forced-choice multiple-sample method. An anticipation error—reporting an odor when there is none but one may be expected—cannot occur in the multiple-sample presentation. Descending order and random order suffer from difficulties caused by adaptation (when a weak odor cannot be detected after a stronger one has been smelled) and desorption hysteresis effects (odorants adsorb rapidly when a higher concentation is handled in the dilution system, but desorb slowly and contaminate the weaker concentration when it is handled).

Because most of the odor-threshold data are applied on a relative basis—comparing one emission with another, or treated and untreated emission, or ambient and emission odors—the argument as to which technique produces the most useful results and would reflect performance of an odor in open air is not critical. Criteria for performance of the measurement method should be based principally on the reproducibility of the mean panel values, ease of application, relative freedom from errors inherent in various sensory methods (adaptation, anticipation, and judgment criterion used by the panelist), cost, and economy in time, both per evaluation and per panelist. Developing such performance criteria for dynamic-dilution odor-threshold measurement methods is an obvious need in dealing with pollution odors.[25]

If ambient-odor threshold is defined as nondetection of odor in 50% of attempts to detect, then the value of such a threshold is not a useful criterion for estimating the extent to which the odor should be reduced to eliminate complaints. Nondetection of odor in even a much higher proportion of presentations—inhalations—may still imply detection in a substantial number of presentations; this would usually be interpreted as evidence of the presence of an odor in the ambient air.

Another use of the ambient-odor threshold measurements is in verification of atmospheric-dispersion models. In such an application, odor thresholds of the emission and of the ambient air are measured. It is important that the same method, and preferably the same panelists, be used to measure both; different methods produce different thresholds, even with the same panelists.

ODOR-INTENSITY MEASUREMENTS

Measurement of perceived intensities of ambient odors is a more satisfactory method for the evaluation of the impact of odors on a community. This and the odor character are the only properties of an ambient odor of which a person in the ambient air is aware. Perceived intensities of ambient odors cannot be estimated from dilution thresholds, because two different odors at the same multiple of their threshold concentrations may have different odor intensities.

The following short table, based on data of Katz and Talbert,[20] indicates, for several highly unpleasant odors, how much an odorous-air sample with a "faint" odor must be diluted to reduce the perceived odor intensity to "very faint" and to "none."

	Dilution factor to reduce to:	
Odorant	*"very faint"*	*"none"*
Allylmercaptan	3	9
Allylamine	5	25
Methyl sulfide	23	530
Thiophenol	52	2,500
Allyl sulfide	70	5,000

In reverse, allylmercaptan, for example, smells faint at a concentration ninefold above its threshold, and allylsulfide smells faint only at a concentration 5,000-fold above its threshold.

Because multiples of the odor-detection threshold are not an appropriate measure of the suprathreshold intensities, there is a need to measure the intensity directly.

Three principal methods for measuring odor-suprathreshold intensity have been discussed: category scales, magnitude-estimate scales, and reference-sample scales. The last of these is preferred, because it does not require special training and it avoids problems as to what various categories or magnitudes should mean. The ASTM E 544 butanol-odor reference scale is a good potential candidate for determining suprathreshold intensities.

The current trend in method development is to use intensity measurements in odorous-pollution studies;[1,6,37,78,82] this indicates an increasing concern with the perceived intensity, rather than detection threshold.

ODOR CHARACTER

One of the principal uses of odor-character evaluation is in tracing an ambient odor to its source. In such an application, the use of emission samples as odor-character reference scales is preferred.

For a broader range of odor character, descriptor scales and the resulting profiles are useful. These produce data on the influence of process changes and odor-control devices on the character of odor. The character of an odor is closely related to its hedonic tone, but hedonic tone can be evaluated directly and independently of character.

HEDONIC TONE

The hedonic tone of an odor is probably its most important property in the determination of its impact on a community. Its measurement, by either a category scale or a magnitude-estimate scale, is simple. Relative judgments on hedonically negative (unpleasant) odors can probably indicate which odor is potentially more annoying. However, it is not yet possible to predict from hedonic data the impact of an ambient odor on a community.

CONCLUSIONS

GENERAL

1. Odorous emission and odorous ambient air have various analytical and sensory properties. Knowledge of the numerical values of some of these properties is essential in any approach to the control of odorous pollution.
2. The analytical properties of odorous emission and odorous ambient

air are characterized by the chemical identities and concentrations of the odorants present.

3. The sensory properties of odorous emission and odorous ambient air are perceived odor intensity, change in intensity with dilution (dose-response function), odor detectability (including detection and recognition thresholds), odor character, and hedonic tone (which refers to the pleasantness and unpleasantness of the odor).

4. Determination of the sensory properties of odors from analytical data on odorous samples in most cases is not yet possible.

5. Several methods are available for measurement of every analytical and sensory property of odorous emission and odorous ambient air. (Methods for estimating the impact of pollution odors on a community from data on the analytical or sensory properties of these odors are discussed in Appendix B.)

6. Limited information is available on the performance of the various measurement methods and is sufficient for a preliminary selection of those which are most suitable and appropriate for further development; in particular, there is a need for more comprehensive investigations of their reproducibility, of means of improving reproducibility, and of applicability to various types of emission and ambient-air odors.

7. Existing odor-control regulations, wherever they attach numbers to odor properties, almost exclusively prescribe some form of measurement of odor-detection threshold as a basis for determining the severity of odorous pollution; by far the dominant measurement methods are the ASTM D 1391 syringe-dilution and the Scentometer methods. Odor-measurement methods are available that are free of several shortcomings of these two and that can yield more useful information on pollution odors, especially on the dose-response function for specific odors.

SELECTION OF PROPERTIES AND METHODS

8. The most suitable properties for the evaluation of odors in ambient air are measurements of perceived intensities, of change in intensity with dilution, and of detection. Measurement of recognition thresholds is useful for specific purposes, such as verification of atmospheric-dispersion models.

9. The most suitable properties for measuring and monitoring emission odors are detection thresholds and perceived intensities at one or more dilutions, including dose-response function plots. Such data permit calculation of the effect of odor control and estimates of the perceived reduction in odor complaints. Experience accumulated by odor-control technologists on the relation between emission-odor threshold and

incidence of population complaints permits crude estimates of the degree of odor control that will be needed.

10. Direct evaluation of odors of emission or ambient air is preferable wherever practical and economically feasible; such evaluation precludes changes in the samples that may be caused by sampling and storage. However, in many cases this approach may be logistically and economically prohibitive; an appropriate form of batch-sampling for later off-site evaluation is a valid option.

11. It is likely that different types of odorous emission and ambient odors will require specific adaptations of the odor-measurement methods, especially with respect to sampling and sample storage, because loss of odorants by adsorption on and diffusion through the container walls depends on the properties of the odorants, as well as on the container-wall material.

12. In the selection and training of panelists, two rationales coexist. Where the relative changes effected by odor control are to be monitored, panelists with similar sensitivity to the particular odor in question may be useful. However, such a panel provides little information on the distribution in the population of different sensitivities to the odor being measured. A randomly selected panel of statistically sufficient size provides at least limited information on such distribution and permits crude estimates of the percentage of the population that will be annoyed when a given degree of odor control is practiced.

13. An important factor in the measurement of emission and ambient-air odors is the selection of a practical method for measuring the appropriate property in a way that is easily standardized. In the case of sensory measurements, the selected procedures should be essentially free of various specific effects inherent in sensory evaluation, so that reproducible results may be obtained when identical odorous samples are evaluated by different working groups.

MEASUREMENT OF ODOR THRESHOLD

14. Odor threshold is a function of the measurement-method variables and may be defined only by reference to specific measurement systems. The most useful functional value ought to be based on the detection or recognition threshold of an odor in free ambient air, when one enters such air from an essentially nonodorous environment; if this value for the particular odor were known, the measurement system that produced the threshold value closest to such open-air value, either directly or through some calibration plot, would be a preferred system for odor-threshold measurement. Such open-air work is unknown.

15. At least four systems for odor-threshold measurement are commercially available in the United States. They differ widely in design and use. In a study of limited scope, they have produced different numerical values for the same odorous samples with the same panelists on the same day; however, the values obtained by different systems could be related by calibration plots.

A number of important factors, apart from sampling and sample storage, influence the results of odor-threshold measurement:

- *Individual sensitivities* for odors may differ by a factor of 20, or even more—still within the limits of normal probability of distribution in the population.
- Many *statistical designs of sample presentation* in current use suffer from differences in personal judgment criteria, in addition to differences in individual sensitivities; from anticipation effects; from the effect of olfactory fatigue; and from a lack of verification that an odor was perceived when the panelist stated that it was. The pre-1978 version of the ASTM syringe-dilution test and the Scentometer method both suffer from these problems. The multiple-sample forced-choice ascending-concentration design is essentially free of most of these problems and economical of the panelists' time.
- Measurement methods that use *dynamic mixing* and controlled rate of sample delivery are more suitable than the *static methods* (such as the ASTM syringe-dilution test) for sample-handling, in that they permit elimination of the effect of adsorption losses on the sample odorant concentration and better control of the rate of sample delivery to the nose.
- On the basis of scanty experimental evidence, the mask *design for the sample-nose interface*, compared with a shifting-nozzle design, does not appear to increase panelist sensitivity or reproducibility of odor-threshold measurements. There are no reliable data on the effect of the various sample-nose interface designs on the values of odor threshold or intensity.
- The *rate of sample delivery to the nose* has a great effect on the odor threshold of an odorous sample. This rate is not standardized, beyond the values specified in the instructions of manufacturers of specific odor-measurement systems.

Unless all the factors that affect the values of odor thresholds are standardized, widely different thresholds are likely to be reported for the same samples by different groups. Until then, data obtained with the same system, preferably by the same panelists, can be used in monitoring the efficiency of odor control on a relative basis.

MEASUREMENT OF PERCEIVED ODOR INTENSITY

16. Measurement of perceived odor intensity on the basis of a series of concentrations of a reference odorant requires a minimum of panelist training and gives, from a limited set of experiments, more reproducible results than the method of semantic or numerical category scales, according to interlaboratory comparisons. The ASTM E 544 odor supra-threshold-intensity scale based on *n*-butanol is an adequate candidate method for most odor-intensity measurements.

17. As in the case of odor-threshold measurements, the design of the sample-nose interface and the rate of flow of the sample in this interface influence the perceived odor intensity of the sample. However, because dilution of the sample is not required, direct measurement of perceived intensity of ambient odorous air is feasible.

MEASUREMENT OF OTHER PROPERTIES OF ODOROUS SAMPLES

18. There are methods for evaluation of odor character, and some have been shown to produce satisfactory inter-laboratory agreement in studies of limited scope. Such methods may assist in investigation of ambient odors in a multisource environment. Relating odor-character data to the impact of an odor on a community is complex and not well understood.

19. The hedonic tone (pleasantness and unpleasantness) is widely recognized as a very important factor in determining the relative annoyance of odorous pollution. Limited experience had demonstrated that measurement of hedonic tone in different laboratories with similar sets of odorants produces similar results with respect to the relative pleasantness and unpleasantness of their odors. A broadly accepted hedonic reference-sample scale is not yet available, but appears feasible. The relation between hedonic tone and the annoyance that results when an odor is encountered in the context of odorous pollution is poorly understood, especially in the case of pleasant odors.

20. Analytical measurements are applicable to the monitoring of the content of specific odorants in emission and in ambient air. In a few cases, where relations have been found between the odor threshold (or odor intensity) of odorous samples and the content of specific odorants, analytical measurements can be a valid tool for monitoring odor control. Analytical data may assist in relating an ambient odor to its sources and in verifying atmospheric-dispersion models. In most odorous-pollution cases, many odorants are present, and analytical data cannot substitute for sensory data.

RECOMMENDATIONS

1. Pollution-odor control technology should be supported by better odor-measurement methods than those specified in existing odor-control regulations.

2. Measurement of odor-detection or -recognition thresholds should use dynamic-dilution techniques and statistically based sample-presentation design that reduces the influence of various effects inherent in sensory evaluation, including the role of decision criteria, olfactory fatigue, and the anticipation effect. The use of multiple-sample forced-choice ascending-concentration series is recommended as suitable and efficient. All factors that may influence the results—such as the rationale of panelist selection, sample flow rates, design of the sample-nose interface, and method of calculating results—should be standardized to optimize the reproducibility of results between laboratories.

3. Odor thresholds measured by the optimized method of recommendation 2 should be compared with the existing regulation-prescribed methods and with threshold values that apply to odor detection in open, naturally breathed ambient air.

4. Perceived odor intensity should be considered, with hedonic tone, as the most important property of ambient odors. Methods for measuring intensity should be adapted; the ASTM E 544 scale may serve as a base.

5. The hedonic tone (pleasantness and unpleasantness) of a pollution odor should receive major attention as an important property related to the impact of the odor on the population. A hedonic reference-sample scale should be developed for such measurements.

6. In the adaptation and development of these recommended methods, a broad data base should be generated on their performance before they are incorporated into odor-control regulations. The data base should include an analysis of the influence of the various operational factors on the measured values, analysis of sources of error, findings on the applicability of the various methods, and findings on the degree of agreement that can be expected between measurements conducted in different laboratories.

7. Work on the use of analytical measurements for various types of odorous emission and ambient odors should be pursued so that instrumental monitoring of pollution odors can eventually replace the more cumbersome sensory measurements.

REFERENCES

1. A. L. Smidth & Co., A/S. Skagen Miljo, 1971. Copenhagen: A. L. Smidth & Co., A/S, 1972. 116 pp. (in Danish)

2. American Society for Testing and Materials. ASTM D 1292-65, Standard Method of Test for Odor in Water. Philadelphia: ASTM, 1973.

3. American Society for Testing and Materials. ASTM D 1391-67, Standard Method for Measurement of Odor in Atmospheres (Dilution Method). Philadelphia: ASTM, 1976.

4. American Society for Testing and Materials. ASTM D 1391 (1978), Standard Test Method for Measurement of Odor in Atmospheres (Dilution Method). Philadelphia: ASTM, 1978.

5. American Society for Testing and Materials. ASTM D 1605-60 (1973), Standard Recommended Practices for Sampling Atmospheres for Analysis of Gases and Vapors. Philadelphia: ASTM, 1973.

6. American Society for Testing and Materials. ASTM E 544-75, Standard Recommended Practices for Referencing Suprathreshold Odor Intensity. Philadelphia: ASTM, 1975.

7. American Society for Testing and Materials. Standard Practice for the Determination of Odor and Taste Threshold by the Forced-Choice Ascending Concentration Series Method of Limits. (in press)

8. American Society for Testing and Materials, Committee E-18. Support document to Standard Practice for the Determination of Odor and Taste Threshold by the Forced-Choice Ascending Concentration Series Method of Limits. Philadelphia: ASTM Committee E-18, 1977.

9. American Society for Testing and Materials Committee E-18. Development of odor suprathreshold intensity referencing method. Memorandum. Philadelphia: ASTM Committee E-18, 1973.

10. American Society for Testing and Materials E-18.04.06. Comments on the sensory evaluation aspects of ASTM D 1391-57 (1967), Standard Method for Measurement of Odor in Atmospheres (Dilution Method). Philadelphia: ASTM Committee E-18, 1973.

11. American Society of Heating, Refrigerating and Air-Conditioning Engineers. Odors. The sense of smell; sensory measurement; analytic measurements; odor sources; odor removal, chapt. 12, pp. 12.1–12.8. In ASHRAE Handbook and Product Directory. 1977 Fundamentals. New York: ASHRAE, 1977.

12. Amerine, M. A., R. M. Pangborn, and E. B. Roessler. The Weber fraction. (The Fechner equation, pp. 256–266.) In Principles of Sensory Evaluation of Food. New York: Academic Press, 1965.

13. Amoore, J. E. A plan to identify most of the primary odors, pp. 158–171. In C. Pfaffmann, Ed. Olfaction and Taste. Proc. of the Third Internatl. Symposium on Olfaction and Taste, Rockefeller Univ., 1968. New York: The Rockefeller Univ. Press, 1969.

14. Amoore, J. E. Specific anosmia: A clue to the olfactory code. Nature 214:1095–1098, 1967.

15. Arthur D. Little Inc. Chemical Analysis of Odor Components in Diesel Exhaust. Final report to Coordinating Research Council and U.S. Environmental Protection Agency. No. ADL 74744-5. Cambridge: A. D. Little Inc., 1973.

16. Baker, R. A. Response parameters including synergism-antagonism in aqueous odor measurement. Ann. N.Y. Acad. Sci. 116:495–503, 1964.

17. Ballou, E. V., Ed. Second NIOSH Solid Sorbents Roundtable. Publ. No. NIOSH 76-

193. Cincinnati: U.S. Department of Health, Education, and Welfare, Public Health Service, National Institute for Occupational Safety and Health, 1976. 237 pp.

18. Barnebey-Cheney Co. Scentometer: An Instrument for Field Odor Measurements. A manual. Columbus, Ohio: Barnebey–Cheney Co., 1970. 2 pp.

19. Bay Area Air Pollution Control District. Procedures for Sampling and Evaluating any Odorous Substance (Dynamic Olfactometry). Chapt. 3, regulation 2, paragr. 15301–15303. San Francisco: Bay Area Air Poll. Control District, 1976.

20. Benforado, D. M., W. J. Rotella, and D. L. Horton. Development of an odor panel for evaluation of odor control equipment. J. Air Poll. Cont. Assoc. 19:101–105, 1969.

21. Brown, K. S., C. M. MacLean, and R. R. Robinette. The distribution of the sensitivity to chemical odors in man. Hum. Biol. 40:456–472, 1968.

22. Cain, W. S. Differential sensitivity for smell: "Noise" at the nose. Science 195:796–798, 1977.

23. Cain, W. S. Physicochemical characteristics and supraliminal odor intensity: Reply to Mitchell. Percep. Psychophys. 9:478–479, 1971.

24. Cain, W. S., and H. R. Moskowitz. Psychophysical scaling of odors, pp. 1–32. In A. Turk, J. W. Johnston, Jr., and D. G. Moulton, Eds. Human Responses to Environmental Odors. New York: Academic Press, 1974.

25. Cederlöf, R., M.-L. Edfors, L. Friberg, and T. Lindvall. Determination of odor thresholds for flue gases from a Swedish sulfate cellulose plant. TAPPI 48:405–411, 1965.

26. Cederlöf, R., L. Friberg, and T. Lindvall. The organoleptic evaluation of odors with special reference to the Kraft pulp industry, pp. 111–122. In International Conference on Atmospheric Emissions from Sulfate Pulping, Sanibel Island, Fla., April 28, 1966. Gainesville, Fla.: University of Florida, 1966.

27. Clarenburg, L. A. Penalization of the environment due to stench. A study of the perception of odorous air pollution by the population. Atmos. Environ. 7(3):333–351, 1973.

28. Crocker, E. C., and L. F. Henderson. Analysis and classification of odors. An effort to develop a workable method. Amer. Perf. Essent. Oil Rev. 22:325–327, 356, 1927.

29. Danielson, J. A., Ed. Air Pollution Engineering Manual, Los Angeles County Air Pollution Control District. Public Health Service Publication No. 999-AP-40. Cincinnati: U.S. Department of Health, Education, and Welfare, 1967. 919 pp.

30. Degobert, P. H. Diesel exhaust odor measurements, pp. 566–581. In Proceedings of the Fourth International Symposium on Automotive Propulsion Systems, April 18–22, 1977. Washington, D.C.: U.S. Department of Energy, 1978.

31. Doty, D. M., R. H. Snow, and H. G. Reilich. Investigation of Odor Control in Rendering Industry. EPA Report EPA-R2-72-088, by Fats and Proteins Research Foundation, Inc., to Office of Research and Monitoring, U.S. Environmental Protection Agency, Washington, D.C. 1972. 142 pp.

32. Dravnieks, A. A building-block model for the characterization of odorant molecules and their odors. Ann. N.Y. Acad. Sci. 237:144–163, 1974.

33. Dravnieks, A. A composite list of odor descriptors. Report prepared for American Society for Testing and Materials, Committee E-18, 1974. 33 pp.

34. Dravnieks, A. Approaches to subjective/objective correlations in flavor, pp. 5–25. In J. J. Powers and H. R. Moskowitz, Eds. Correlating Sensory Objective Measurements—New Methods for Answering Old Problems. ASTM Spec. Tech. Publ. 594. Philadelphia: American Society for Testing and Materials, 1976.

35. Dravnieks, A. Development of a Dynamic Forced-Choice Method for the Measurement of Emission and Ambient Odors. IIT Research Institute Project No. C8217 for Fats and

Proteins Research Foundation, Inc. Des Plaines, Ill.: Fats and Proteins Research Foundation, Inc., 1976. 96 pp.

36. Dravnieks, A. Evaluation of human body odors: Methods and interpretations. J. Soc. Cosm. Chem. 26:551–571, 1975.

37. Dravnieks, A. Fundamental considerations and methods for measuring air pollution odors, pp. 429–436. In J. Le Magnen and P. MacLeod, Eds. Proceedings of the Sixth International Symposium on Olfaction and Taste. London: Information Retrieval, Ltd., 1978.

38. Dravnieks, A. Instrumental aspects of olfactometry, pp. 1–61. In D. G. Moulton, A. Turk, and J. W. Johnston, Jr., Eds. Methods in Olfactory Research. New York: Academic Press, 1975.

39. Dravnieks, A. Measuring industrial odors. Chemical Engineering/Deskbook issue, pp. 91–95, Oct. 21, 1974.

40. Dravnieks, A. Odor perception and odorous air pollution. TAPPI 55:737–742, 1972.

41. Dravnieks, A. Organic Contaminants in Indoor Air and Their Relationship to Outdoor Contaminants. Phase One, ASHRAE Project RP 183. IIT Research Institute. New York: American Society of Heating, Refrigerating and Air-Conditioning Engineers, February 1977. (unpublished)

42. Dravnieks, A. Threshold of smell and measurement, pp. 27–44. In P. N. Cheremisinoff and R. A. Young, Eds. Industrial Odor Technology Assessment. Ann Arbor: Ann Arbor Science Publishers, Inc., 1975.

43. Dravnieks, A., and F. C. Bock. Odor Characterization of Odorant Mixtures. IITRI Report C8237. Chicago: IIT Research Institute, 1977. 112 pp.

44. Dravnieks, A., F. C. Bock, J. J. Powers, M. Tibbetts, and M. Ford. Comparison of odors directly and through profiling. Chem. Sens. Flav. 3:191–225, 1978.

45. Dravnieks, A., H. C. McDaniel, and J. J. Powers. Comparison by 12 Laboratories of the Odor Qualities of Nine Chemicals Sniffed from the Bottle and as GLC Effluents. Report to American Society for Testing and Materials, Committee E-18, 1978. 33 pp.

46. Dravnieks, A., and A. O'Donnell. Principles and some techniques of high-resolution headspace analysis. J. Agr. Food Chem. 19:1049–1056, 1971.

47. Dravnieks, A., A. O'Donnell, and H. G. Reilich. Determination of odor components in tobacco smoke: Design of mixture to simulate odor. ASHRAE Trans. 81(Pt.2):200–212, 1975.

48. Dravnieks, A., A. O'Donnell, R. Scholz, and J. D. Stockham. Gas-Chromatographic Study of Diesel Exhaust Using a Two-Column System. Presented at meeting of the American Chemical Society, Los Angeles, Calif., April 1971. (unpublished)

49. Dravnieks, A., and H. J. O'Neill. Annoyance Potentials of Air Pollution Odors. Presented at American Industrial Hygiene Association conference, Los Angeles, Calif., May 11, 1978.

50. Dravnieks, A., and W. H. Prokop. Source emission odor measurement by a dynamic forced-choice triangle olfactometer. J. Air Poll. Cont. Assoc. 25:28–35, 1975.

51. Dravnieks, A., W. H. Prokop, and W. R. Boehme. Measurement of ambient odors using dynamic forced-choice triangle olfactometer. J. Air Poll. Cont. Assoc. 28:1124–1130, 1978.

52. Dravnieks, A., H. G. Reilich, J. Whitfield, and C. A. Watson. Classification of corn odor by statistical analysis of gas chromatographic patterns of headspace volatiles. J. Food Sci. 38:34–39, 1973.

53. Dravnieks, A., J. Whitfield, and J. Shah. Determination of odor components in tobacco smoke. ASHRAE Trans. 80(1):76–91, 1974.

54. Duffee, R. A., J. P. Wahl, W. Marrone, and J. S. Nader. Defining and measuring

objectionable odors, pp. 191–201. In Pollution Engineering Techniques—An Overview Introduction to the Problem. Ann Arbor: Ann Arbor Science Publishers, 1974.

55. Fazzalari, F. A., Ed. Compilation of odor and taste threshold values data. Sponsored by ASTM Committee E-18 on Sensory Evaluation of Materials and Products. ASTM Data Series DS 48A. Philadelphia: American Society for Testing and Materials, 1978. 497 pp.

56. Feldstein, M., D. A. Levaggi, and R. Thuillier. Odor Regulation by Emission Limitation at the Stack. Preprint 73-273. Presented at the 66th Annual Meeting, Air Pollution Control Association, Chicago, June 1973.

57. Flesh, R. D., and J. C. Burns. Procedures for the Identification and Assessment of Community Odor Problems. Report to the U.S. Environmental Protection Agency. La Jolla, Calif.: Copley International Corporation, 1978. 71 pp.

58. Flesh, R. D., J. C. Burns, and A. Turk. An evaluation of community problems caused by industrial odors, pp. 33–44. In A. Turk, J. W. Johnston, Jr., and D. G. Moulton, Eds. Human Responses to Environmental Odors. New York: Academic Press, 1974.

59. Frijters, J. E. R. The effect of duration of intervals between olfactory stimuli in the triangular method. Chem. Sens. Flav. 2:301–311, 1977.

60. Fuller, G. H., R. Steltenkamp, and G. A. Tisserand. The gas chromatograph with human sensor: Perfumer model. Ann. N.Y. Acad. Sci. 116:711–724, 1964.

61. Gruber, C. W., G. A. Jutze, and N. A. Huey. Odor determination techniques for air pollution control. J. Air Poll. Cont. Assoc. 10:327–330, 1960.

62. Guadagni, D. G. Requirements for coordination of instrumental and sensory techniques, pp. 36–48. In ASTM Tech. Publ. 440, Correlation of Subjective-Objective Methods in the Study of Odors and Taste. Philadelphia: American Society for Testing and Materials, 1968.

63. Hall, H. J., and R. H. Salvesen. Evaluation of Measurement Methods and Instrumentation for Odorous Compounds in Stationary Sources. Vol. I—State of the Art. Publ. No. APTD-1180. Washington, D.C.: U.S. Environmental Protection Agency, 1972. 92 pp.

64. Harper, R. Some chemicals representing particular odour qualities. Chem. Sens. Flav. 1:353–357, 1975.

65. Harper, R., E. C. Bate Smith, and D. G. Land. Odour Description and Odour Classification; A Multidisciplinary Examination. London: J. & A. Churchill, Ltd., 1968. 191 pp.

66. Harper, R., E. C. Bate Smith, D. G. Land, and N. M. Griffiths. A glossary of odour stimuli and their qualities. Amer. Perf. Essent. Oil Rev. 59:22–37, 1968.

67. Harrison, G. A. F., and C. M. Coyne. Flame photometric determination at the parts per thousand million level of some volatile sulphur compounds. J. Chromatog. 41:453–455, 1969.

68. Harrison, W., A. Dravnieks, R. Zussman, and R. Goltz. Abatement of Malodors at Confined Dredged Material Disposal Sites. Report by Argonne Natl. Lab. to Office, Chief of Engrs., U.S. Army Contract D-76-9. Washington, D.C.: U.S. Army Corps of Engrs., 1976. 146 pp.

69. Hellman, T. M., and F. H. Small. Characterization of petrochemical odors. Chem. Engr. Prog. 69:75–77, 1973.

70. Hellman, T. M., and F. H. Small. Characterization of the odor properties of 101 petrochemicals using sensory methods. J. Air Poll. Cont. Assoc. 24:979–982, 1974.

71. Hemeon, W. C. L. Technique and apparatus for quantitative measurement of odor emissions. J. Air Poll. Cont. Assoc. 18:166–170, 1968.

72. Huey, N. A., L. C. Broering, G. A. Jutze, and C. W. Gruber. Objective odor pollution control investigations. J. Air Poll. Cont. Assoc. 10:441–446, 1960.

73. Humecki, H. Comparison of Containers for Collection of Odorous Air Samples. Report to Illinois Environmental Protection Agency. Walter C. McCrone Associates, Inc., June 1975. 65 pp.

74. Hyman, A. M. Factors influencing the psychophysical function for odor intensity. Sens. Proc. 1:273–291, 1977.

75. Illinois Institute of Technology Research Institute, Chemistry Department. Instructions for Dynamic Triangle Olfactometer. A manual. Chicago: ITT Research Institute, 1977. 18 pp.

76. Illinois State Environmental Protection Agency. Air Pollution Control Regulations. Rule 802 (d), 1972.

77. Johns, T., and J. C. Sternberg. Gas chromatographic detectors, pp. 179–207. In J. Krugers, Ed. Instrumentation in Gas Chromatography. Eindhoven, Netherlands: Centrex Publishing Co., 1968.

78. Johnson, R. A. A System for Field Measurements of Environmental Odor Intensities and Odorant Concentrations Based on the ASTM Odor-Referencing Method. Preprint #78-68.1. Paper presented at the 71st annual meeting, Air Poll. Cont. Assoc., Houston, Tex., June 1978.

79. Johnston, J. W., Jr. Quantification of olfactory stimuli, pp. 192–220. In C. C. Brown, Ed. Methods in Psychophysiology. Baltimore: Williams & Wilkins Co., 1967.

80. Karolinska Institute. Methods for Measuring and Evaluating Odorous Air Pollutants at the Source and in the Ambient Air. The Third Karolinska Institute Symposium on Environmental Health, Stockholm, 1970. Nord. Hyg. Tidskr. 51:1–77, 1970.

81. Katz, S. H., and E. J. Talbert. Intensities of Odors and Irritating Effects of Warning Agents for Inflammable and Poisonous Gases. Bureau of Mines Technical Paper 480. Washington, D.C.: U.S. Department of Commerce, 1930. 37 pp.

82. Kendall, D. A., and P. L. Levins. Odor Intensity and Characterization of Jet Exhaust and Chemical Analytical Measurements. A. D. Little Inc. NASA CR-121159 ADL 74443. NASA Lewis Research Center. Cambridge, Mass.: A. D. Little Inc., 1973.

83. Kendall, D. A., and A. J. Neilson. Correlation of subjective and objective odor responses. Ann. N.Y. Acad. Sci. 116:567–575, 1964.

84. Köster, E. P. Modified and automated olfactometer. Parf. Cosm. Savons 10:509–513, 1967. (in French)

85. Kruger, L., A. N. Feldzamen, and W. R. Miles. A scale for measuring suprathreshold olfactory intensity. Amer. J. Psychol. 68:117–123, 1955.

86. Lindvall, T. On sensory evaluation of odorous air pollutant intensities. Nord. Hyg. Tidskr. Suppl. 2. Stockholm: Karolinska Institute and National Institute of Public Health, 1970.

87. Lindvall, T. Sensory measurement of ambient traffic odors. J. Air Poll. Cont. Assoc. 23:697–700, 1973.

88. Lundberg, C. A. Dynamic Triangle Olfactometer Evaluation Project. Final Report. 3M Company. St. Paul, Minn., 1975. 31 pp.

89. Mannebeck, von H. Ein tragbares Geruchsmebgerat zur Direktmessung. Landtechnik 28:287–288, 1973.

90. Manual on Sensory Testing Methods. ASTM Spec. Tech. Publ. 434. Philadelphia: American Society for Testing and Materials, 1968.

91. Marks, L. E., and J. C. Stevens. The form of the psychophysical function near threshold. Percept. Psychophys. 4:315–318, 1968.

92. Mieure, J. P., and M. W. Dietrich. Determination of trace organics in air and water. J. Chromatog. Sci. 11:559–570, 1973.

93. Mills, J. L., R. T. Walsh, K. D. Luedtke, and L. K. Smith. Quantitative odor measurement. J. Air Poll. Cont. Assoc. 13:467–475, 1963.

94. Misco Company. Dynamic Olfactometer, Model 7250. Data sheet. Berkeley, Calif.: Misco Microchemical Specialties Company, 1974.

95. Moskowitz, H. R., and C. D. Barbe. Psychometric analysis of food aromas by profiling and multidimensional scaling. J. Food Sci. 41:567–571, 1976.

96. Moskowitz, H. R., A. Dravnieks, W. S. Cain, and A. Turk. Standardized procedure for expressing odor intensity. Chem. Sens. Flav. 1:235–237, 1974.

97. Moskowitz, H. R., A. Dravnieks, and L. A. Klarman. Odor intensity and pleasantness for a diverse set of odorants. Percept. Psychophys. 19:122–128, 1976.

98. Moskowitz, H. R., and C. L. Gerbers. Dimensional salience of odors. Ann. N.Y. Acad. Sci. 237:1–16, 1974.

99. National Draeger, Inc. System of detecting and measuring hazardous gases and vapors, p. 2. Bulletin ND-101. In Gas and Vapor Detection Products. Pittsburgh: National Draeger, Inc., 1977.

100. Nwankwo, J. N., and A. Turk. Electrochemical Analysis of Sulfidic and Amine Odorants. Publ. No. EPA-600/2-76-021, June 1976. 37 pp.

101. O'Donnell, A., and A. Dravnieks. Chemical Species in Engine Exhaust and Their Contribution to Exhaust Odors. Final Report to National Air Pollution Control Administration and Coordinating Research Council, Inc. Report No. IITRI C6183. Chicago: IIT Research Institute, 1970.

102. Ough, C. S., and H. Stone. An olfactometer for rapid and critical odor measurement. J. Food Sci. 26:452–456, 1961.

103. Owens-Corning Fiberglass Corp. Report on Incinerator Odor Evaluation of Minneapolis Roofing Facility. March 13–14, 1978. 10 pp.

104. Parsons, J. S., and S. Mitzner. Gas chromatographic method for concentration and analysis of traces of industrial organic pollutants in environmental air and stacks. Environ. Sci. Tech. 9:1053–1058, 1975.

105. Patte, F., M. Etcheto, and P. Laffort. Selected and standardized values of suprathreshold odor intensities for 110 substances. Chem. Sens. Flav. 1:283–305, 1975.

106. Pellizzari, E. D., B. H. Carpenter, J. E. Bunch, and E. Sawicki. Collection and analysis of trace organic vapor pollutants in ambient atmospheres. Thermal desorption of organic vapors from sorbent media. Environ. Sci. Tech. 9:556–560, 1975.

107. Powers, J. J. Experiences with subjective/objective correlation, pp. 111–122. In J. J. Powers and H. R. Moskowitz, Eds. Correlating Sensory Objective Measurements: New Methods for Answering Old Problems. A symposium sponsored by ASTM Committee E-18 on Sensory Evaluation of Materials and Products. Philadelphia, 11–12 November 1974. ASTM Spec. Tech. Publ. 594. Philadelphia: American Society for Testing and Materials, 1976.

108. Prokop, W. H. Developing odor control regulations: Guidelines and considerations. J. Air Poll. Cont. Assoc. 28:9–22, 1978.

109. Russell, J. W. Analysis of air pollutants using sampling tubes and gas chromatography. Environ. Sci. Tech. 9:1175–1178, 1975.

110. Sanders, G. R., R. A. Umbraco, S. Twiss, and P. K. Mueller. The Measurement of Malodor in a Community by Dynamic Olfactometry. Presented at Third Karolinska Institute Symposium on Environmental Health, Stockholm, June 1–5, 1970.

111. Schiffman, S. S. Contributions to the physicochemical dimensions of odor: A psychophysical approach. Ann. N.Y. Acad. Sci. 237:164–183, 1974.

112. Schiffman, S. S. Physicochemical correlates of olfactory quality. Science 185:112–117, 1974.

113. Schiffman, S., D. E. Robinson, and R. P. Erickson. Multidimensional scaling of odorants: Examination of psychological and physicochemical dimensions. Chem. Sens. Flav. 2:375–390, 1977.

114. Schuetzle, D., T. J. Prater, and S. R. Ruddell. Sampling and analysis of emissions from stationary sources. I. Odor and total hydrocarbons. J. Air Poll. Cont. Assoc. 25:925–932, 1975.

115. Schutz, H. G. A matching-standards method for characterizing odor qualities. Ann. N.Y. Acad. Sci. 116:517–526, 1964.

116. SKC, Inc. Sample Tubes for Chemical Hazards in Air. Eighty Four, Pa.: SKC, Inc., 1976.

117. Snow, R. H. Investigation of odor control in the rendering industry, pp. 147–174. In P. N. Cheremisinoff and R. A. Young, Eds. Industrial Odor Technology Assessment. Ann Arbor: Ann Arbor Science Publishers, Inc., 1975.

118. Springer, K. Combustion odors—A case study, pp. 227–262. In A. Turk, J. W. Johnston, Jr., and D. G. Moulton, Eds. Human Responses to Environmental Odors. New York: Academic Press, 1974.

119. Stahl, W. H. Ed. Compilation of Odor and Taste Threshold Values Data. ASTM Data Series DS 48. Philadelphia: American Society for Testing and Materials, Committee E-18, 1973. 250 pp.

120. Stevens, S. S. The psychophysics of sensory function, pp. 1–33. In W. A. Rosenblith, Ed. Sensory Communication. Cambridge: MIT Press, 1961.

121. Sullivan, F., and G. Leonardos. Determination of odor sources for control. Ann. N.Y. Acad. Sci. 237:339–349, 1974.

122. Swets, J. A., Ed. Signal Detection and Recognition by Human Observers; Contemporary Readings. New York: J. Wiley & Sons, 1964. 702 pp.

123. Turk, A. Expressions of gaseous concentration and dilution ratios. Atmos. Environ. 7:967–972, 1973.

124. Turk, A., and S. Mehlman. Correlations between instrumental and sensory characterizations of atmospheric odors, pp. 27–35. In ASTM Spec. Tech. Publ. 440, Correlation of Subjective-Objective Methods in the Study of Odors and Taste. Philadelphia: American Society for Testing and Materials, 1968.

125. Turk, A., J. T. Wittes, L. R. Reckner, and R. E. Squires. Sensory Evaluation of Diesel Exhaust Odors. Report AP-60. Raleigh: U.S. Department of Health, Education, and Welfare, Public Health Service, National Air Pollution Control Administration, 1970. 67 pp.

126. U.S. Environmental Protection Agency. Air Pollution Aspects of Odors. 2 vols. February 1971. (unpublished)

127. U.S. Patent 3,902,851. Methods for detecting odors and apparatus therefor. Issued to Fats and Proteins Research Foundation, Inc., 1975.

128. van Gemert, L. J., and A. H. Nettenbreyer, Eds. Compilation of Odour Threshold Values in Air and Water. TNO, Zeist, Netherlands: Central Institute for Nutrition and Food Research, 1977. 79 pp.

129. VDI Guidelines. VDI #2457. In VDI-Handbuch Reinhaltung der Luft. Düsseldorf: Verein Deutscher Ingenieure, 1970.

130. Venstrom, D., and J. E. Amoore. Olfactory threshold in relation to age, sex or smoking. J. Food Sci. 33:264–265, 1968.

131. Wade, W. A., III, K. C. Tower, and S. Cha. Technical Report to the Illinois EPA on the Evaluation of Four Odor Measurement Systems. TRC Project No. 32390. Wethersfield, Conn.: The Research Corporation of New England, 1974. 61 pp. + app.

132. Wahl, J. P., R. A. Duffee, and W. A. Marrone. Evaluation of Odor Measurement Techniques: Vol. I. Animal Rendering Industry. EPA No. 650/2-74-008-a, Contract No. 62-02-0662. Washington: U.S. Environmental Protection Agency, 1974. 92 pp.

133. Wilby, F. V. Variation in recognition odor threshold of a panel. J. Air Poll. Cont. Assoc. 19:96–100, 1969.

134. Woskow, M. H. Multidimensional Scaling of Odors. Ph.D. Thesis, University of California, Los Angeles. No. 64-6198. University Microfilms, Inc., Ann Arbor, Mich., 1964. 75 pp.

135. Wright, R. H., and K. M. Michels. Evaluation of far infrared relations to odor by a standards similarity method. Ann. N.Y. Acad. Sci. 116:535–551, 1964.

136. Yoshida, M. Studies in psychometric classification of odors (6). Jap. Psychol. Res. 14:70–86, 1972.

5 Characterization of Odorant Transport in the Atmosphere

Odorants are brought into contact with people by atmospheric transport. In short, the atmosphere is the "connecting link" between the sources of odorous emission and human (or other) receptors. To understand the nature of odor problems or complaints in the United States, it is necessary to consider the changes in odorant composition and concentration at the receptor that result from atmospheric transport. This, in turn, requires an understanding of the mechanisms involved in atmospheric transport of odorants.

GENERAL TRANSPORT AND DIFFUSION

Odorous materials released into the atmosphere are redistributed and diluted and react in ways that may be described by means of mathematical and physical models. The atmosphere disperses these materials rapidly, because the atmosphere is turbulent. Consequently, turbulent diffusion is reviewed here, before the discussion of odorants, themselves.

Turbulence, a process of chaotic motion possessed by almost all natural fluid flows, is easy to recognize, but difficult to define. The dispersion of material in the atmosphere is more rapid than molecular diffusion by many orders of magnitude. Gifford[5] has outlined the areas of diffusion in the lower layers of the atmosphere and discussed the average conditions of wind in the lower atmosphere, describing shearing stress, mechanical turbulence, the Reynolds number of the atmosphere, viscosity, the surface-wind profile, effects of buoyancy, the Richardson number (which is associated with the stability of the atmosphere), turning of the wind with height, and the appropriate diffusion theories that follow from considerations of these atmospheric characteristics.

There are two fundamental theories of atmospheric diffusion: the Fickian or K theory and the statistical theory. The gradient-transport approach was developed by Fick[4] in a paper published in 1855. Fick related atmospheric diffusion to the diffusion of heat in a conducting body. The general case of diffusion in three dimensions—in which the diffusion coefficients, which are not necessarily equal, can vary with the three spatial coordinates—is shown in Equation 1. If the dispersion coefficients K_x, K_y, and K_z are constant, the diffusion is called Fickian. The more general case is called K-theory diffusion. The solution of Equation 1 requires specification of appropriate boundary conditions and, in the atmosphere, the addition of appropriate transport terms.

$$\frac{d\bar{q}}{dt} = \frac{\partial}{\partial_x}\left(K_x \frac{\partial \bar{q}}{\partial_x}\right) + \frac{\partial}{\partial_y}\left(K_y \frac{\partial \bar{q}}{\partial_y}\right) + \frac{\partial}{\partial_z}\left(K_z \frac{\partial \bar{q}}{\partial_z}\right) \tag{1}$$

The second theory, which is in more general use, is the statistical theory of turbulent diffusion. The model most often used to describe atmospheric dispersion statistically is that of Gaussian or normal distribution. It has been demonstrated that repeated averaging of experiments in the atmosphere yields a distribution that is nearly normal in the vertical and cross-wind directions for average concentration. In 1953, Sutton[14] described his model of averaged plume diffusion. He reasoned that the Lagrangian single-particle autocorrelation function must depend only on the intensity of atmospheric turbulence and viscosity. This was the begining of the use of the Gaussian plume-diffusion model.

Strictly speaking, the Gaussian diffusion model applies only in the limit of large diffusion time for homogeneous stationary conditions. Batchelor,[1] in 1949, reasoned, by analogy with a central limit theorem of probability, that the Gaussian function may provide a general description of average plume diffusion, because of the essential random nature of this phenomenon. Pasquill,[12] in 1962, pointed out that for very small diffusion times the distribution particles should take the same form as the wind fluctuation distribution, because the particle trajectories coincide with instantaneous wind. In the atmosphere, this approximates a Gaussian distribution fairly closely.

Hilst,[7] in 1957, and Gifford,[6] in 1959, independently introduced the concept of a fluctuating plume mode. Figure 5-1 demonstrates the fundamental concept as explained by Gifford. Plume-dispersion models of the Gaussian form use elementary components to represent instantaneous Gaussian puffs of material. The steady-state plume is merely an assemblage of superpositions of those puffs, each emanating from a fixed origin translated by the mean wind direction. It is assumed that the distribution of material within the puff is described fully by its second moment, whose change with dispersion times is obtained as a consequence either by Fickian theory or by a statistical theory of turbulence. For mathematical convenience, dispersion in the direction of the mean wind (the X direction in Figure 5-1) is neglected in practice; this leads to the spreading-disk dispersion model for plumes, which is portrayed in Figure 5-1(b).

Plumes from actual sources operate under more complicated processes. This is illustrated in Figure 5-1(c), which includes a superposition of elementary puffs. These puffs are wandering or meandering about the center of the time-averaged plume. Following Gifford, we may consider a one-dimensional problem and write the expression for the relative concentration of material at any point in a particular cloud undergoing dispersion as shown in Equation 2, where f is the material distribution function, centered at a distance Dy from the origin; t is time, counted from some initial instance t_0; and Q is the total amount of material in the cloud. If the entire system is moved with a constant windspeed in the X direction, then Dy will become a distance from the X axis. Gifford[6] wrote Equation 3 to represent the mean value of the relative concentration distribution, which is really an average over many trials; g is the frequency function associated with the variability of Dy over all the trials. From this basic formulation, Gifford then demonstrated that the mean relative concentration from dispersion proceeding at different rates in the horizontal and vertical directions may be written as shown in Equation 4, where Y and Z are the average variances of the spreading puff, D_y and D_z are the variances of the center of the puff about the axis of travel, and y and z represent the fixed point in space at which the mean concentration is being calculated.

$$\frac{X}{Q} = f(y - Dy, t, y_0, t_0) \tag{2}$$

$$M\left\{\frac{X}{Q}\right\} = \int_{-\infty}^{\infty} f(y - Dy, t, y_0, t_0)\, g(Dy)\, dDy \tag{3}$$

$$M\left\{\frac{X}{Q}\right\} = (2\pi u)^{-1}\, (\overline{y^2} + \overline{Dy^2})^{-1/2}\, (\overline{Z^2} + \overline{Dz^2})^{-1/2} \tag{4}$$

$$x \exp\left[-\frac{y^2}{2(Y^2 + Dy^2)} - \frac{z^2}{2(Z^2 + Dz^2)}\right]$$

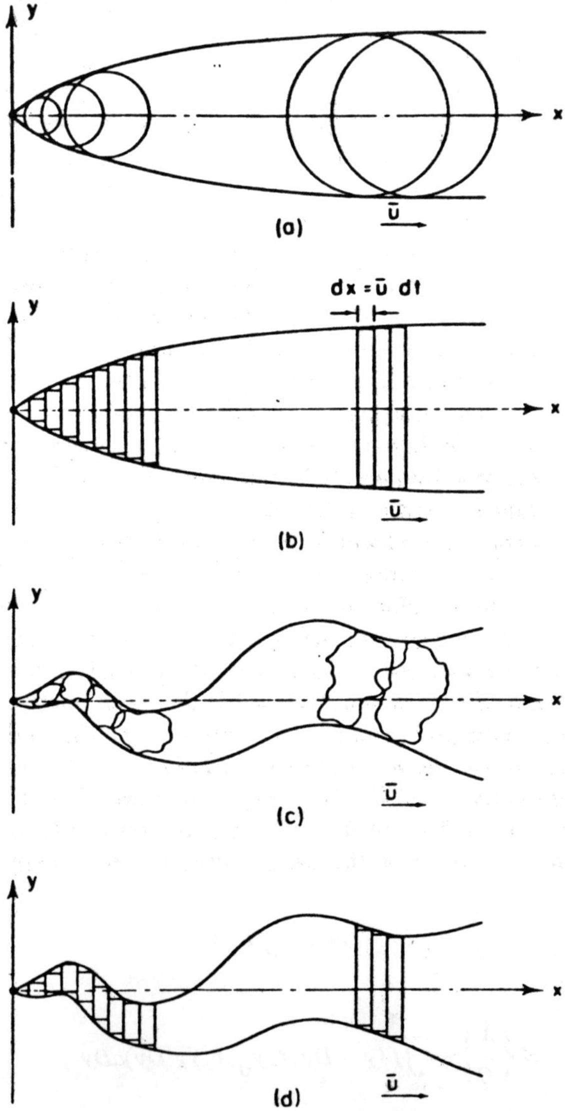

FIGURE 5-1 (a) Schematic formation of plume from superposition of individual averaged elements. (b) Schematic spreading-disk plume model obtained by neglecting x diffusion. (c) Appearance of naturally occurring plumes, with "real" puff elements indicated. (d) Fluctuating-plume model. Reprinted with permission from Gifford.[6]

Finally, Gifford noted that the variance of the point-concentration frequency distribution could be defined as shown in Equation 5.

$$V\left\{\frac{X}{Q}\right\} = M\left\{\left(\frac{X}{Q}\right)^2\right\} - M\left\{\frac{X}{Q}\right\} \tag{5}$$

Högström,[9] in 1964, developed data that could be applied to the fluctuating-plume model described by Gifford by photographing smoke puffs and measuring the diffusion of material about the center of the puff, as well as the meandering of the puff with respect to the mean wind direction for a period of 30–60 min. He further divided it by atmospheric-stability class, on the basis of the lapse rate of potential temperature in the lower layers of the atmosphere. In 1968, Högström[10] applied the data to develop a statistical approach to the air-pollution problem of chimney emission and compared the calculated data with measurements. Finally, in 1972, Högström[8] applied the fluctuating-plume model as a method for predicting odor frequencies from a point source. The Högström model is discussed later.

TRANSPORT AND DIFFUSION OF ODORANTS

For the purposes of the discussion, the odor dilution ratio defined by Equation 6 is the dilution of a sample required to reach a given odor level, detectability point, or other sensory target; Z_t represents the odor dilution ratio, C_0 the odorant concentration of a sample, and C_t the odorant concentration when some sensory target has been reached. Because Z_t is a ratio, it is dimensionless; it is independent of volumes and does not describe the intensity of the perceived odor. Part of the problem of describing concentration of odorous material with classical transport and diffusion models is illustrated in Figure 5-2, taken from Högström.[8] The standard Gaussian plume model predicts an average concentration, as one can see, from the assumed time history. In the example shown, the sensory odor threshold was exceeded three times during the l-h period, but the hourly mean concentration was only about 60% of the odor threshold.

$$Z_t = \frac{C_o}{C_t} \tag{6}$$

We would therefore anticipate that the standard Gaussian plume model would underpredict actual odor levels downwind of a source. However,

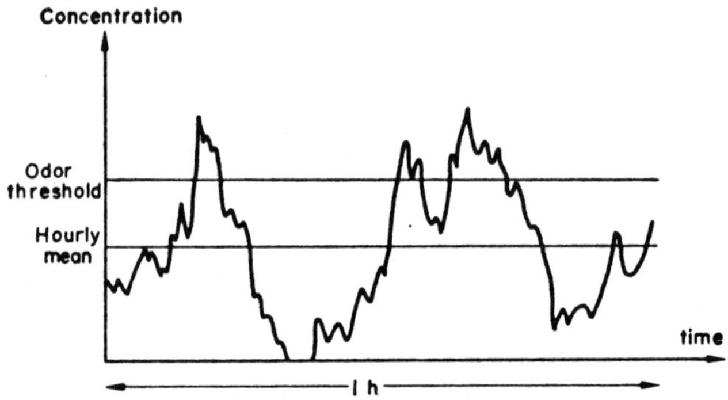

FIGURE 5-2 Schematic diagram showing how the concentration of odorous matter can exceed the odor threshold several times during a period when the hourly mean is much below the odor threshold. Reprinted with permission from Högström.[8]

when such models are applied to multiple point sources—such as various vents on an industrial-plant roof—they frequently overpredict. For example, during a survey of odors associated with an automotive assembly plant, the modeled 200-dilution-factor isopleth corresponded to the observed-nondetectability isopleth for the same atmospheric conditions.[3]

That is why it is necessary to use a model like the fluctuating-plume model to determine the frequency of occurrence of values exceeding the sensory target value, or else to predict the impact of individual puffs to obtain a frequency distribution in that manner.

Up to this point, we have not discussed the potential impact of reaction of odorants during transport in the atmosphere—i.e., the change that could occur through reaction with other compounds in the atmosphere or through disassociation due to sunlight or moisture in the air. If the reaction is one that will decrease the odorant at some specified level, a correction can be made in the source term in the model, similar in manner to radioactive decay, to account for the reaction that occurs. However, if the reaction is not a simple depletion rate, the existing models do not account for that. Such reactions certainly do occur. For example, in a study of odors associated with mixtures of sulfur compounds emitted from viscose processes, variations in odor qualities with distance from the plant were observed. These are summarized in Table 5-1, taken from Polgar *et al.*[13] Although some of this change was due to the dilution of the odorant mixtures and different atmospheric transport because of release-height variations, the change in odor quality beyond 5 km was ascribed by the authors, at least in part, to the photooxidation of carbon disulfide (CS_2) to carbonyl sulfide (COS) during transport in the atmosphere.

A second problem in applying general transport and diffusion models to odors is how to handle different compounds that are released from nearby vents that might have similar odor characteristics. One can simply add the contributions, sum them, and calculate odor impact on the basis of the addition. There are several potential errors in doing this. First, this is not always a true situation for odors, because the compounds that are producing the odor may react to cancel each other or may react to create a new compound that has a lower odor threshold than either of the individual compounds and whose odor threshold cannot be estimated or predicted.

Second, for mixtures that do *not* react, it has been reported that the odor threshold of the mixture is approximated by the geometric mean of the constituent thresholds, provided that the compound with the lowest threshold amounted to less than 10% of the mixture (by volume). Because the resulting mixture of odorants from different stacks is unknown, this can be a major source of error in odor modeling.

Another difficulty of odor transport is the limitation of models to

TABLE 5-1 Odor Quality and Threshold of Synthetic Mixtures[a]

Sample	Odorants at Threshold, ppm	Quality
1. H_2S	0.02	Rotten eggs
2. CS_2	0.67	Medicine, iodine, burnt
3. CS_2	0.45	Sweet, mild, rotten eggs
4. CS_2, aged 3 days	0.78	
5. CS_2, aged 3 days	0.77	
6. COS	0.05	Burnt rubber, carbamate
7. COS	0.12	
8. COS, aged 3 days	0.12	Rotten eggs, burnt rubber
9. 71% H_2S, 29% CS_2	0.01	Rotten eggs
10. 28% H_2S, 72% CS_2	0.05	Rotten eggs
11. 9% H_2S, 91% CS_2	0.13	Rotten eggs
12. 20% H_2S, 80% CS_2	—	6 out of 6: burnt rubber or carbamate, 2 could detect rotten eggs
13. COS/CS_2 (1/1,000)	0.23	Medicine, iodine
14. H_2S/COS (1/70)	—	6 out of 6: rotten eggs
15. $H_2S/COS/CS_2$ (1/3.5/1,250)	0.15	Burnt rubber, shoe wax, sulfur

[a] Reprinted with permission from Polgar *et al.*[13]

distances of somewhat less than 20 km. Odors from individual industrial sources, however, have been detected reportedly at about 80 km from the source.[16] The nature of these mechanisms is unexplained, although Turk *et al.*[15] have suggested that they might be attributed to adsorption (and thus localized concentration) on airborne particles. Obviously, odorants are transported through the atmosphere by mechanisms other than turbulent diffusion. At 20 km or more from the source, these mechanisms dominate.

STATUS OF ODOR MODELING

Högström[8] has developed a method for predicting odor frequencies from a point source on the basis of the fluctuating-plume dispersion model. It is used to give estimates of odor frequencies around a point when the odor threshold of the material emitted is determined by sensory methods. He verified his model by using trained observers who made a large number of instantaneous (yes–no) observations in a variety of locations around a sulfate pulp factory in Sweden. At 2 km, positive odor observations were obtained 10.8% of the time, and the odor frequency predicted by the model was 8.9%. At 5 km, odors were detected 9.8% of the time, and the model calculations indicated 5.6%. At 10 km, odors were predicted 8.5% of the time, and the model predicted 3.1% occurrence. Finally, at 20 km, odors were detected 5.1% of the time, and the model predicted 1.6%. In this case, it is obvious that the prediction of the occurrence of detectable odors was quite good at 2 km, and the validation of the model demonstrated that it was underpredicting as distance from the source increased. The total number of samples varied from 5,000 to 7,500 in this particular experiment.

Högström also conducted other experiments to verify the model with different sensory measurements. In this case, he used automobiles to transport people through the plume. There were five observers in two cars with open windows. They traversed a section of road alternately in north and south directions normal to the plume. The subjects made instantaneous observations of odor every 6 s. The model consistently overpredicted the number of occurrences of odor detection by 30%, possibly because of adaptation or fatigue of the observers. Therefore, there is some disagreement between the two techniques used to observe the odors. However, this does give an idea of the reliability that can be obtained from current odor models.

Clarenburg,[2] in 1973, performed a study of the perception of odorous air pollution with a population in the Netherlands. He developed a mathematical model for the perception of odorous pollution by the population living in the vicinity of a chemical industry in order to describe it quantitatively.

He started with the basic Gaussian plume model and developed a penalization function based on the percentage of the population that would perceive an odor, assuming a log-normal distribution function for such perception. The basic problem was to predict the number of complaints—that is, how many people would perceive odor as a function of the population distribution around the complex in question. The correlation of Clarenburg's model with observed data yielded correlation coefficients above 0.90, except for one case where the value was 0.69.

TRC (The Research Corporation of New England) has developed a puff model that predicts the number of occurrences of specified odor dilution ratios (to detection threshold) during a specified period, such as 1 h. This model was reported by Murray *et al.*[11] at the annual meeting of the Air Pollution Control Association in Houston, Texas, in June 1978. The limited verification data indicate that the highest values of odor level are predicted reasonably well for the test case. That is, the maximal predicted odor dilution ratio was 10 at one point, and that observed was 8. At another point where tested, the maximal predicted odor dilution ratio was 35, and the observed value was 30. No measurements were made of the frequency of occurrence of odors exceeding the threshold at these points in order to test that part of the model.

Odorant transport in the atmosphere is only partially controlled by turbulent diffusion. Reactions of odorants in the atmosphere are largely unknown, and conventional dispersion models may not be capable of handling the complex reactions that do occur. Second- and third-odor closure models may be required in the future. Dispersion models are useful tools in predicting the impact of odorous emission on community odor levels and in developing solutions to odor complaints. However, they must be used with great care and consistency, or serious error will result. Their use is limited to short distances and to nonreactive odorants.

There is a fundamental need for a verification program of the puff model. Högström has verified it for a few cases. In an area downwind of an odor-emitting source, more work needs to be done on the actual frequency of occurrence and to determine why Högström's model apparently underpredicted at greater distances in some cases. Because the TRC model is essentially the same as Högström's model, it follows that the same kind of information is needed there. Clarenburg's model was based on odor complaints to begin with and offers a slightly different set of data, because he was interested basically in developing a penalization function, rather than a control function, to eliminate odor complaints. A subject that needs exploration is the introduction of reactivity into these models, and much more information is needed on potential reaction rates and processes that can occur in the atmosphere for various odorant compounds. Finally, there is a great need for determining the potential additive effects of various compounds, especially at the low concentrations that will be present in the atmosphere.

REFERENCES

1. Batchelor, G. K. Diffusion in a field of homogeneous turbulence. Aust. J. Sci. Res. 2:437–450, 1949.

2. Clarenburg, L. A. Penalization of the environment due to stench. A study of the perception of odorous air pollution by the population. Atmos. Environ. 7:333–351, 1973.

3. Duffee, R. A., and N. E. Bowne. Odor Survey in the Vicinity of an Automotive Assembly Plant. TRC Project Report 61233, January 1973.

4. Fick, A. Über diffusion. Ann. Physik. Chem. 94:59–86, 1855.

5. Gifford, F. A., Jr. An outline of theories of diffusion in the lower layers of the atmosphere, pp. 65–116. In David H. Slade, Ed. Meteorology and Atomic Energy. Washington, D.C.: U.S. Atomic Energy Commission, Division of Technical Information, 1968.

6. Gifford, F. A., Jr. Statistical properties of a fluctuating plume dispersion model, pp. 117–137. In F. N. Frenkiel and T. A. Sheppard, Eds. Advances in Geophysics, Vol. 6. New York: Academic Press, 1959.

7. Hilst, G. R. The dispersion of stack gases in stable atmospheres. J. Air Poll. Cont. Assoc. 7:205–210, 1957.

8. Högström, U. A method for predicting odour frequencies from a point source. Atmos. Environ. 6:103–121, 1972.

9. Högström, U. An experimental study on atmospheric diffusion. Tellus 16:205–251, 1964.

10. Högström, U. A statistical approach to the air pollution problem of chimney emission. Atmos. Environ. 2:251–271, 1968.

11. Murray, D. R., S. S. Cha, and N. E. Bowne. Use of a fluctuating plume puff model for prediction of the impact of odorous emissions. Preprint 78-68.6. Presented at the 71st Annual Meeting, Air Pollution Control Association, Houston, Texas, June 1978.

12. Pasquill, F. Atmospheric Diffusion; The Dispersion of Windborne Material from Industrial and Other Sources. London: D. Van Nostrand Company Ltd., 1962. 297 pp.

13. Polgar, L. G., R. A. Duffee, and L. J. Updyke. Odor characteristics of mixtures of sulfur compounds emitted from the viscose process. Preprint 75-55.2. Presented at the 68th Annual Meeting, Air Pollution Control Association, Boston, June 1975.

14. Sutton, O. G. Micrometeorology. New York: McGraw-Hill Book Co., Inc., 1953. 333 pp.

15. Turk, A., L. R. Reckner, and G. D. Robinson. The Mechanism of Odor Transport. Paper 69-123. Presented at the 62nd Annual Meeting, Air Pollution Control Association, New York, June 1969.

16. Yocom, J. E., and R. A. Duffee. Controlling industrial odors. Chem. Eng. 77:160–168, 1970.

6 Methods of Controlling Odors

Controlling odor is a complex problem. Chief among the complexities, perhaps, is the extremely small amount of odorant that triggers odor perception in humans. It has been estimated that 10^8 or 10^9 molecules of odorant vapor in the nose is sufficient for detection. As an illustration of the minuteness of this volume, a microgram of ethyl mercaptan in the vapor state contains approximately 10^{16} molecules—10^7 or 10^8 times the number of molecules required for detection.[93] Because of our ability to detect odorants at such low concentrations, sources that are of minor concern for criteria pollutants—e.g., leaks from valves and flanges, minor spills, and evaporation from condenser hot wells, sewage systems, or retention ponds—may be the primary sources of odorous emission responsible for community complaints of malodors.

Our olfactory acuity dictates a very high degree of control. In the absence of any standard of acceptable community odor, the only sure way to alleviate complaints of malodors downwind of a source is to preclude detection at all, i.e., to reduce the odorant concentration in the community to less than the detection or recognition threshold. This usually requires control at the source with an efficiency of 95–100%, which is far greater than the efficiency needed for most gaseous emission.

An odorant must be in the form of a gas or vapor if it is to be sensed by the human olfactory system. Although solid particles and liquid droplets in an odorous gas stream may have a pronounced effect on the persistence and other characteristics of an odor, the wide range of methods used for odor control apply to pollutants in the gaseous or vapor state.

Choosing the precise form of control for a specific problem involves consideration of the nature of odors, and this can involve a substantial departure from most engineering principles. An odor is in itself an effect, i.e., a subject's conscious reaction to a stimulation of his olfactory system. Odor control can therefore be directed either at reducing the concentration of an odorant at the receptor or at interfering with the receptor's olfaction process. The approaches to odor control aimed at reducing the concentration at the receptor can be classified according to the engineering operations that are effective: high-temperature oxidation, atmospheric dispersion, absorption and gas-phase reactions, and adsorption. The techniques aimed at interfering with the olfaction process are commonly grouped under the general term "odor modification." The various engineering operations and odor modification are reviewed in separate sections of this chapter. Applications are reviewed later.

Before discussing specific control methods, however, it is appropriate to consider the most effective of all odor-control methods—process change. There is nothing special about the concept of process change as a method of pollution control. This is not definable technology; it is more a state of mind. Engineers, through training and application of this training, have been extremely successful in implementing this state of mind, by developing improved processes when given the incentive to do so. Odor emission presents a sizable challenge to the process engineers involved with odor-producing processes. Approaches to odor control by process modification will vary, but a number of truisms are perhaps worth mentioning with respect to temperature, pressure, volume, maintenance, and housekeeping.

- *Temperature*: The temperature of a process can have a pronounced effect on odorous emission. Simply chilling the water of a vapor condenser during warm weather when water temperature is highest and windows are open may solve an odor problem. Excessive temperatures during drying of a heat-sensitive material may produce odorous decomposition products; this might be avoidable with adequate temperature control. Insufficient temperature in a furnace in which waste gases are burned may produce intermediates more odorous than the original material.
- *Pressure*: Converting a process environment from slightly positive to slightly negative pressure by changing damper positions or fan locations will reduce the number of leak points for odorous materials and make the odor-control job easier.
- *Volume*: A large ventilation volume for an odorous process will tend to dilute odorous emission, but may actually increase the quantity of odorants emitted, if the odorant is a vaporized liquid. Odor intensities downwind from a source are proportional principally to emission rates,

rather than to the concentration of odorants in the effluent. Reducing the ventilation volume may decrease odor downwind and will simplify and reduce costs of additional odor-control equipment required. However, such volume reduction can be achieved only if explosion hazards and other constraints have been adequately considered.

- *Maintenance*: In many odorous processes, the greatest amount of odorant released is from leaks (flanges, pump seals, uncovered vessels, etc.). Controlling such sources of odor is a continuing job for maintenance people.
- *Housekeeping*: Odor-producing materials allowed to accumulate in the open (volatile wastes or putrescible foods and food byproducts) are sometimes the only significant sources of odorous emission from some operations. Good housekeeping can eliminate such problems.

It would be misleading to attempt to present a comprehensive list of odorous air pollutants. Nonetheless, a list of selected typical odorous substances grouped by chemical type can be informative, and such a list appears as Table 6-1.

HIGH-TEMPERATURE OXIDATION FOR THE CONTROL OF ODOROUS WASTE GASES FROM STATIONARY SOURCES

DESCRIPTION

High-temperature oxidation is an air-pollution control process in which odorous waste organic gases or organic particles are converted to odorless gaseous products, such as carbon dioxide and water vapor. The odors are destroyed by exposure of the waste gases to the proper conditions of temperature, time, and turbulence in the presence of air in a combustion chamber. The temperature required depends on the specific contaminants involved and the design configuration of the equipment used. This method will completely destroy the odors in the waste gases at some temperature if the control equipment has been properly designed.

The design of high-temperature oxidation equipment has not been completely standardized, and each manufacturer has units of different design. For this reason, in purchasing this type of equipment, it is well to specify a field test involving sensory odor measurements[3,8,34-36,63] on the stack emission to show that the odor problem is eliminated at the temperature specified by the manufacturer. Odor dilution ratios of less than 25 in the stack outlet (i.e., guideline-acceptable odor level) based on the detection threshold are readily achievable with this type of control equipment fired with natural gas. Oxidation temperatures range mainly from about 600 to 1,500°F (about 315 to 815°C). Residence times can range

TABLE 6-1 Major Odorous Air Pollutants, Olfactory Thresholds, and Related Data

Category and Class	Systemic Chemical Name	Formula	Odor	Mol. Wt.	Odor Threshold, ppm (by vol.)[a]
Sulfur compounds					
Sulfur oxides	Sulfur dioxide	SO_2	Pungent	64	0.47
Sulfides	Hydrogen sulfide	H_2S	Rotten eggs	34	0.0047-0.18
	Carbon disulfide	CS_2	Rotten	76	0.21-0.84
Mercaptans	Methyl mercaptan	CH_3SH	Decayed cabbage	48	2×10^{-5}-0.041
	Ethyl mercaptan	C_2H_5SH	Decayed cabbage	62	3×10^{-5}-0.001
	Propyl mercaptan	C_3H_7SH	Unpleasant	76	0.0016-0.024
	Allyl mercaptan	$CH_2{=}CHCH_2SH$	Garlic	74	0.003-0.017
	Benzyl mercaptan	$C_6H_5CH_2SH$	Unpleasant	124	0.0026-0.04
Thioethers	Dimethyl sulfide	$(CH_3)_2S$	Decayed cabbage	62	0.003
	Diethyl sulfide	$(C_2H_5)_2S$	Foul, garlic	90	0.0048
	Diallyl disulfide	$(CH_2{=}CHCH_2S)_2$	Garlic	146	1.1×10^{-4}-0.012
Nitrogen compounds					
Inorganic	Ammonia	NH_3	Pungent	17	0.47-54
Aliphatic amines	Dimethylamine	$(CH_3)_2NH$	Fishy	45	0.047
	Trimethylamine	$(CH_3)_3N$	Fishy-ammoniacal	59	0.00021
Aromatic nitro compounds	2,4,6-Trinitro-*t*-butylxylene (musk)	$C(C_4H_9)(CH_3)_2(NO_2)_3$	Musk	297	6×10^{-6}-0.005
Heterocyclic amines	Pyridine	C_5H_5N	Empyreumatic	79	0.003-0.23
	Benzo[*b*]pyrrole (indole)	C_8H_7N	Fecal	117	
	3-Methylindole (skatole)	C_9H_9N	Fecal	131	0.05
Cyanides	Hydrogen cyanide	HCN	Bitter almonds	27	0.9
	Allylisocyanide	$CH_2{=}CHCH_2NC$	Sweet repulsive (nauseating)	67	0.18-1.6
	Allylisothiocyanate	$CH_2{=}CHCH_2SNC$	Mustard oil (nose and eye irritant)	99	0.008-0.42

Selenium compounds					
Selenides	Hydrogen selenide	H_2Se	Putrid	81	4×10^{-4}-0.0012
	Ethylselenomercaptan	C_5H_5SeH	Foul, fetid	109	4×10^{-4}-0.0012
	Diethyl selenide	$(C_2H_5)_2Se$	Putrid (nauseating)	137	0.011
Hydrocarbons, Alcohols, and Oxygenates					
Aliphatic hydrocarbons	2-Butene (butylene)	$CH_3CH=CHCH_3$	Gas-house	56	24
	2-Methylpropene (isobutylene)	$CH_2=C(CH_3)_2$	Gas-house	56	20
Phenol	Phenol	C_6H_5OH	Empyreumatic	94	0.047
Aldehydes	Methanal (formaldehyde)	H_2CO	Pungent	30	1.0
	Ethanal (acetaldehyde)	CH_3CHO	Pungent	44	0.066-2.2
	Propenal (acrolein)	$CH_2=CHCHO$	Burning fat	56	0.021-1.8
	4-Hydroxy-3-methoxy-benzaldehyde (vanillin)	$C_8H_8O_3$	Sweet-aromatic	152	1.1×10^{-4}-2×10^{-7}
Ketones	*d*-2-Keto-1,7,7-trimethylnorcamphene (camphor)	$C_{10}H_{16}O$	Aromatic-earthy	152	1.3
Organic acids	Butanoic acid (butyric acid)	$CH_3CH_2CH_2COOH$	Rancid, perspiration	88	0.001-2.2
	2-Methylbutanoic acid (isovaleric acid)	$(CH_3)_2CHCH_2COOH$	Body odor	102	0.015
	Butanediene (diacetyl)	$(CH_3CO)_2$	Sweet butter	86	0.025
Halogen compounds					
Inorganic	Chlorine	Cl_2	Pungent	71	0.31
Aliphatic halogens	Trichloroethylene	$CHCl=CCl_2$	Aromatic	131	0.21
	Triiodomethane (iodoform)	CHI_3	Antiseptic	394	5×10^{-3}
Aromatic halogens	Benzyl chloride	$C_6H_5CH_2Cl$	Aromatic lacrimator	126.5	0.04-0.31
	Chlorohydroxybenzene (chlorophenol)	$C_6Cl_4(OH)Cl$	Medicinal	128.5	0.0036-0.03
Miscellaneous					
Oxygen	Trioxygen (ozone)	O_3	Irritating	48	0.51

[a] Data from Fazzalari.[38]

from less than a second to about 2 s. Turbulence (on the basis of a calculated average velocity at the outlet from the combustion chamber) can be as high as about 30 ft/s (9.1 m/s) at oxidation temperature.

Some odorous inorganic combustibles—such as hydrogen sulfide, ammonia, and cyanides—can be destroyed by high-temperature oxidation, but there is a limit on the concentration of inorganic combustibles in the waste gas stream that can be satisfactorily controlled, because these substances are converted by oxidation to their oxides, which can be objectionable themselves at high concentrations.

Odor problems that ordinarily cannot be satisfactorily controlled by high-temperature oxidation alone are those in which the waste gases contain halogen compounds or compounds (such as phosphates) that form objectionable oxides or acids. When odorous waste gases containing halogens are oxidized, the reaction products include free halogens (fluorine, chlorine, bromine, or iodine), halogen acids, phosgene, etc., all of which are toxic or corrosive and must be removed by chemical scrubbing before discharge to the atmosphere. In the case of phosphates, the treated gases contain phosphorus oxides or acids, which are toxic and also have to be removed.

APPLICABILITY

This method is applicable to a wide variety of industrial processes, such as:

- adhesive-tape curing
- asphalt blow stills and saturators
- brake-lining ovens
- chemical processing
  - soap-making
  - xanthate processes
    - Cellophane
    - rayon
    - sausage casings
- coffee-roasters
- coil-coating lines
- cupola furnace stacks
- fiberglass curing
- foundries
- food-processing
- lithographic ovens
- meat smokehouse
- metal-coating ovens
- paint-baking ovens
- paint-removal facilities
- petroleum-refining
- phthalic anhydride processes
- plastic-curing ovens
- printing presses
- pulp mills
- rendering plants
- resin reactors
- roofing-paper machine hoods
- textile dryers
- varnish burnoff
- varnish kettles
- vinyl-sponge curing
- wire-enameling

HAZARDS

Two potential hazards exist in the high-temperature oxidation of waste gases of some odor-producing industrial processes. These hazards, which could cause malfunction of the control equipment, should be fully evaluated and minimized in the design of any system:

- The odorous organic contaminants may deposit and build up a layer of solid combustible matter or condensed heavy oils in the inlet duct to the control equipment. Either kind of layer could ignite spontaneously when the ductwork heats up and approaches the ignition temperature of the deposited organic matter. Provision should be made by suitable design either to avoid the formation of such a deposit by preventing condensation or to make it easy to clean the inlet duct periodically.
- If the concentration of the contaminant in the waste gas is not constant and can fluctuate into the flammable range, there is a possibility of ignition by the burner in the control equipment and fire-flashback to the source of the emission. This can be prevented by the proper application and selection of control equipment initially and the incorporation of suitable flashback protection, such as a flame arrestor, as required.[48]

CATEGORIZATION OF ODOROUS WASTE GASES

The type of system chosen for controlling odorous waste gases will depend to a great extent on the concentration of the contaminants. A convenient and practical way of categorizing the waste gases by concentration of the contaminants is as follows:

- Nonflammable at concentrations below the lower explosive (flammable) limit (LEL). The LEL is the lowest concentration of a contaminant in air at which the mixture will ignite at room temperature and atmospheric pressure when exposed to a spark or flame.
- Flammable, or potentially flammable, at concentrations greater than the LEL.

NONFLAMMABLE GASES

Several methods of high-temperature oxidation are applicable to controlling odorous waste gases in which contaminants are present in concentrations below the LEL: direct-flame oxidation, catalytic oxidation, and use as the combustion air supply for any plant combustion equipment (such as boilers and air heaters). The maximal concentration of odorous waste gases processed in high-temperature oxidation equipment is usually limited, by insurance underwriters, to 25% of the LEL, to eliminate fire hazards. This limitation ensures that the vapor-air mixture being oxidized will not ignite when exposed to a spark or flame. In some cases with continuous monitoring equipment, concentrations as high as 40 or 50% of the LEL are permitted. In most instances of odor nuisance, the concentration of the contaminating vapors is well below 25% of the LEL—and in many cases, as low as a few parts per million by volume. All the methods listed will completely destroy the odors in the waste gases at some temperature, if the control equipment is properly designed, applied, and operated.

Direct-Flame (Thermal) Oxidation

Direct-flame oxidation has been demonstrated to be one of the most effective and reliable methods for controlling odorous emission. There are many examples of applications of direct-flame oxidation—also referred to as "after-burning,"[2,29,59,86,87] "direct-flame fume incineration,"[5-7,9,10,78] "thermal incineration,"[17,88] "thermal oxidation,"[15,42] "direct-flame combustion,"[2] "direct-flame afterburner,"[81] "direct-gas-flame oxidation,"[30,88,89] and "direct incineration."[60] "Direct-flame oxidation" appears to be the proper terminology for what actually takes place in destruction of the odorants: the odorants are oxidized in the presence of a flame.

In direct-flame oxidation, the odorous emission in concentrations well below the LEL is completely oxidized to nonodorous gases, such as carbon dioxide and water vapor, by exposure to temperatures of 900–1,500°F (480–815°C) in the presence of a flame. The temperature required to do an effective job depends on the specific pollutants involved and the design of the combustion chamber. It has been shown that temperatures of 900–1,500°F, velocities of 15–30 ft/s (4.6–9.1 m/s), and residence times (including flame contact time) of 0.25–0.60 s give satisfactory cleanup,[28,57,59,88] depending on the application and the efficiency required.

Direct-flame oxidation has been used in a large variety of industrial processes that are potentially odor-producing,[8,10,28-30,32,48,61,75,86,87,90] as shown in the following list:

Type of industry	*Process*
Food and drugs	Rendering, smokehouses, fish-processing, coffee-roasting, citrus pulp, dryers, animal-blood dryers, antibiotic fermentation
Chemical	Phthalic anhydride plants, sulfur plants, hydrogen sulfide plants, latex- and silicon-rubber curing, rubber-processing
Petroleum	Asphalt-blowing, petroleum-processing
Municipal	Sewage treatment
Paint and varnish	Varnish and resin cookers
Printing and paper	Rotogravure presses, kraft-mill operations
General manufacturing	Wire-enameling ovens, paint, baking ovens, caster curing ovens, aluminum-chip dryers, glass-fiber curing ovens, can-finishing ovens
Metallurgic	Foundry core, aluminum sweating

A recent development in thermal oxidation involves the use of a regenerative thermal-energy recovery system and achieves oxidation by heating waste gases to the desired temperature by means of a stoneware heat-exchange medium in a packed bed of ceramic castings.[73] This new method appears to offer the potential of destroying organic solvents with very high thermal-energy recovery, but there are no data available on the effectiveness of odor destruction.

Catalytic Oxidation

In catalytic oxidation, sometimes referred to as "flameless combustion," "catalytic after-burning,"[81] "catalytic incineration,"[6] or "catalytic combustion,"[60] the presence of a catalyst allows high-temperature oxidation to take place at a lower temperature and in the absence of a flame. However, a burner is usually required to heat the odorous waste gases to the required temperatures, and the gases are partially oxidized before they reach the catalyst. In catalyst systems, the type of catalyst, the oxidation temperature, the velocity through the bed, and the amount of catalyst are important variables that affect efficiency. Platinum and mixed noble-metal catalysts have been used predominantly in catalytic oxidation equipment for the control of industrial air pollutants. The catalyst does not participate in the reaction. Although the precise mechanism of heterogeneous catalytic oxidation is not well understood, there is general agreement that it proceeds through three necessary steps (and in this order): adsorption on the active surface, chemical reaction (oxidation on surface), and desorption of the reaction products.

The oxidation temperature for a catalyst system is the average temperature of the gases leaving the catalyst bed and ranges from about 600°F to 1,200°F (about 315°C to 650°C). There is a temperature rise across the bed that depends on the amount of the contaminant that is oxidized.

Anything interfering with any of the steps will interfere with the efficiency of the oxidation process. For example, if the waste gases contain inorganic particulate matter that is deposited on the active surface of the catalyst, this adversely affects the adsorption and desorption steps, thereby reducing overall efficiency.

Extensive laboratory tests (Miller and Wilhoyte[62] and Miller and Soward, cited in U.S. EPA[86]) indicate that catalyst systems are capable of achieving efficiencies of over 90% at oxidation temperatures of about 700°F (about 370°C). In actual practice, however, such efficiencies are difficult to achieve, and temperatures of 1,000–1,200°F (540–650°C)[39,57,86] may be required.

The major problems with catalyst systems are the difficulty in obtaining uniform flow and uniform temperature distribution and the susceptibility of catalysts to deterioration due to poisoning, suppression, and fouling (Table 6-2), or due to attrition. Another concern is that catalytic condition is difficult to monitor, compared with the ease of monitoring temperature only in a direct-flame oxidation system. Any high-temperature oxidation system that operates inefficiently for any reason will yield intermediate

TABLE 6-2 Typical Poisons, Suppressants, and Fouling Agents That Affect Catalysts of the Platinum-Group Metals

Type of Agent	Examples
Poison	Heavy metals
	Phosphates
	Arsenic
Suppressant	Halogens (both as elements and in compounds)
	Sulfur compounds
Fouling agent	Inorganic particles
	Alumina and silica dust
	Iron oxides
	Silicones

oxidation products, such as pungent aldehydes and acrylates and burnt odors. With direct-flame oxidation, once the temperature for proper oxidation has been determined in a field test, maintenance of that temperature should ensure proper operation of the system. That is not true for a catalyst system, because higher operating temperatures are required as the catalyst deteriorates.

Catalytic oxidation has been applied with various degrees of success in controlling objectionable odors from phthalic anhydride manufacture, lithographic ovens, wire-enameling ovens, coffee-roasters, hydrogen sulfide tail gases from sulfur production, vulcanization kettles, acrylate monomers in wax-processing, and fabric-coating ovens.[14,16,23,27,39,42,47,52,62,86]

Comparison of Catalytic with Direct-Flame Oxidation

A conclusive evaluation, including a field study with sensory measurements to determine the efficiency of odor destruction by a catalyst system compared with a direct-flame oxidation system at various temperatures, has not been found in the published literature. The main reason is that reliable analytic and sensory methods for measuring the efficiency of field installations have only recently been developed.

It was not until 1965 that a satisfactory analytic method of field-testing the source emission of organic solvents in industrial effluent streams was reported. The first report involved a modified total-combustion analyzer.[16] The detailed method of field-testing for organic solvents, developed in Los Angeles County, was first published in 1966,[55] with the adoption of Rule 66. This method was modified in 1968 on the basis of experience with testing for compliance with Rule 66.[58] The Los Angeles method was adopted by several companies. The results of many tests validating the efficiency of direct-flame oxidation of organic solvents (90–99+%) with this method were reported in the years that followed.[88,89] Similar tests on the catalyst system have not been conducted on a large scale.

Comparative tests[42] showed solvent-hydrocarbon removal effiencies for a direct-flame oxidizer of greater than 90% at temperatures of 1,270–1,370°F (about 690–745°C); for a catalyst system, the maximal efficiency appeared to be 89% at a temperature of 1,100°F (about 595°C). For a platinum-metal catalyst, the actual performance of catalyst systems in reducing many fumes and odors has been found to be inadequate for meeting the Los Angeles standard of 90% efficiency (R. G. Lunche, cited in U.S. EPA[86]). Only one catalyst system had a measured efficiency greater than 90%. This unit is used in an ethylene oxide process and operates at an unusually high temperature for a catalyst system.

Although a sensory method for measuring odors of industrial emission was published[13] by ASTM in 1957, the trial-and-error aspect of this method (a syringe dilution procedure) was a deterrent to its acceptance and general use by industry. The use of this procedure reportedly[19] helped to solve a major odor problem with dimethylamine, a raw material used in the manufacture of synthetic detergent.

In 1963, another group[63] reported on an improved modified ASTM syringe static-dilution technique, which eliminated the trial-and-error aspect. This improved method (Mills/ASTM) has been successfully used by Los Angeles County in evaluating the effectiveness of control equipment in odor-producing industries. With additional improvements, the revised method (Benforado/Mills) has been used successfully by industry in evaluating the odorous emission from plant processes and in determining the effectiveness of control equipment in a variety of applications.[8] Data in Table 6-3 permit an evaluation of the effect of temperature on odor destruction for several applications, on the basis of the Benforado/Mills measurement method.[8] This kind of information is needed to demonstrate the effectiveness of catalyst systems.

Furthermore, there is a definite need for additional comparative testing of catalytic and direct-flame oxidation to demonstrate conclusively the range of applicability of each of these systems for both odor elimination and oxidation of organic contaminants at various temperatures. Any economic comparison of the two methods must be based on the correct operating temperature for equivalent efficiency. Newer methods of odor measurement that use a dynamic olfactometer[34-36] are available for comparative testing.

Use as Combustion Air Supply

It is possible to solve some odor problems by using odorous waste gases as the air supply for a plant boiler, process, or heating furnace. The concentrations of the contaminants must be well below the LEL's, and the volume of the air required for the boiler or furnace should be larger than the volume of waste gases to be oxidized.

The advantages of this method are that large additional capital expenditures are not required and that additional operating expenses for auxiliary fuel may not be required to control the odor problem.

However, the equipment must be fired at all times when the odorous waste gases are being vented, even when firing is not needed to produce steam or heat energy for a process. The resulting overall increase in fuel cost for operation of the boiler or other equipment when it would not otherwise be required may offset any anticipated savings, compared with a properly designed, smaller direct-flame oxidizer used only for odorous waste gases.

This method has been used successfully to eliminate odors from fish-processing plants, to dispose of digester blow gases and relief gas from kraft pulping by oxidizing them in the process lime kiln,[15,44,91,92] and to control emission from smokehouses, rendering cookers, and a variety of odor-producing processes in refineries.[57,59]

FLAMMABLE GASES

When odorous gases are present at concentrations greater than the LEL—which means that they will sustain combustion and burn when ignited in the presence of air—direct combustion, as in a flare, is applicable.[77] Flare systems offer a satisfactory way of disposing of odorous hydrocarbon vapors from refinery petrochemical processes. Operating costs are usually minimal, because the odorous waste gases themselves constitute the fuel.

The greatest hazard is the unanticipated presence of air in the system,[50] which can cause explosions.

A major odor problem in a chemical process that used dimethylamine as a raw material in the manufacture of synthetic detergent was solved with a typical flare stack. In this system, the waste gases—consisting of hydrogen, dimethylamine, ammonia, and nitrogen—varied widely in concentration, so it was necessary to add natural gas to ensure combustion at all times and thus prevent flameout.

ENERGY CONSERVATION

Because high-temperature oxidation involves the heating of odorous waste gases to very high temperatures, the use of heat-recovery equipment to cut down fuel costs should be evaluated.[7,46] If the industrial process operates 24 h/day, 50 weeks/yr, and requires control equipment around the clock, the application of heat recovery is usually easily justified for catalytic or direct-flame oxidation.

Primary heat recovery involves a heat exchanger in which the hot oxidized gases leaving the combustion chamber are used to preheat the cooler odorous waste gases that are entering the combustion chamber. Secondary heat recovery involves a heat exchanger in which the hot oxidized gases leaving the combustion chamber or leaving the primary heat exchanger are used to heat air for plant heating or for a process.

Heat-recovery equipment is normally used to reduce the fuel input to the combustion chamber. A recuperative heat exchanger of the tube type can be used to preheat the odorous waste gases before they enter the combustion chamber where the odors will be destroyed, thereby reducing

TABLE 6-3 Effect of Direct-Flame Oxidation Temperature on Odor Destruction[a]

Summary of Odor Panel Tests
Wire Enameling Oven
Field Test—Portable Direct-Flame Fume Incinerator

Incineration Temperature	Average Odor Strength Odor Units/SCF Inlet	Average Odor Strength Odor Units/SCF Outlet	Effect of Incineration on Odor Strength
1,400°F	2,600	70	97% reduction—acceptable
1,200°F	2,500	350	86% reduction—marginal
1,000°F	1,300	2,100	Increased—unsatisfactory

Summary of Odor Panel Tests
Glass Fiber Curing Oven
Field Test—14,000 scfm Direct-Flame Fume Incinerator

Unit	Incineration Temperature	Average Odor Strength Odor Unit/SCF Inlet	Average Odor Strength Odor Unit/SCF Outlet	Effect of Incineration on Odor Strength
1	1,009°F	550	625	Unsatisfactory
2	1,250°F	380	53	Reduced to acceptable level
3	1,302°F	225	25	Reduced to acceptable level

Summary of Odor Panel Tests
Curing Oven for Hardboard Saturated with Tempering Oil
Lab Test—Portable Direct-Flame Fume Incinerator

Incineration Temperature	Average Odor Strength Odor Units/SCF Inlet	Average Odor Strength Odor Units/SCF Outlet	Effect of Incineration on Odor Strength
1,500°F	1,000	40	96% reduction
1,400°F	1,400	15	97.6% reduction

TABLE 6-3 (*Continued*)

Summary of Odor Panel Tests
Abrasive Wheel Curing Process
Lab Test—Portable Direct-Flame Fume Incinerator

Incineration Temperature	Average Odor Strength Odor Units/SCF Inlet	Average Odor Strength Odor Units/SCF Outlet	Effect of Incineration on Odor Strength
1,200°F	800	10	98% reduction
1,400°F	1,600	32	98% reduction

Summary of Odor Panel Tests
Auto Paint Bake Ovens
Field Test—Portable Direct-Flame Fume Incinerator

Incineration Temperature		Average Odor Strength Odor Units/SCF Inlet	Average Odor Strength Odor Units/SCF Outlet	Effect of Incineration on Odor Strength
Oven #1	1,450°F	170	10	94% reduction
	1,350°F	260	14	95% reduction
Oven #2	1,450°F	680	18	97% reduction
	1,350°F	650	10	98% reduction

[a] Reprinted with permission from Benforado *et al.*[8]

the amount of fuel required to reach oxidation temperatures. Efficiencies of this type of recuperative heat exchanger vary between 30 and 60%. However, the fuel requirements for high-temperature oxidation can be reduced by as much as 80% by use of a rotary heat exchanger, if applicable, if design conditions permit.

It is sometimes possible to recycle a portion of the clean, odorless hot gases directly into the process or plant heating equipment without using a heat exchanger. There has been a general reluctance to do this, because of the possibility of contaminating the material being manufactured with combustion products or with microscopic particles of insulation from the inner lining of the combustion chamber. The use of a heat exchanger eliminates this source of contamination.

In some applications, the incorporation of heat recovery helps in justifying high-temperature oxidation as a practical solution to the industrial odor problem by reducing operating costs to an acceptable value.[46,78]

COSTS

The costs of high-temperature oxidation equipment depend on a number of factors:

- The sizes of the equipment required, based on the volume and rate of gas flow to be processed.
- The nature and concentration of the pollutants in the odorous waste gases, the temperature of the gases, and the oxidation temperature required.
- The type of fuel (such as natural gas or fuel oil) used for the burner.
- The extent of heat recovery for preheating the odorous waste gases going to the oxidizer or for providing heat to other plant equipment.
- The hours of operation of the facility.

A recent study[76] by the Radian Corporation presented capital costs and annualized costs of direct-flame oxidizers and catalytic oxidizers for various solvent concentrations in waste gases. This report includes data originally presented in an EPA report.[24] hese costs are graphed in Figures 6-1 through 6-6 and are applicable to the control of odorous waste gases when qualified with the assumptions in Table 6-4. The basis used for annualizing the costs is presented in Table 6-5. It should be noted that these costs are for new installations or for existing installations that are easily retrofitted. Such fittings of existing installations may cost 2 or 3 times the values shown.

Add-on control equipment of this type is energy-intensive and costly. Even the application of heat-recovery equipment does not usually show a payout on the investment, i.e., a company does not save money by installing high-temperature oxidizers. Instead, it is an additional overhead cost that must be added to the cost of the product being manufactured. Heat-recovery equipment, if properly applied, enables a company to reduce this additional overhead cost.

Additional concerns in the use of high-temperature oxidizers are the availability and cost of natural gas or fuel oil. In some parts of the country, natural gas is not available, and fuel oil is scarce.

Many industrial odor problems involve odorous waste gases with very low organic-contaminant concentrations—up to a few hundred parts per million. In those cases, controlling the odors with high-temperature oxidation is extremely costly, because only negligible energy is recovered

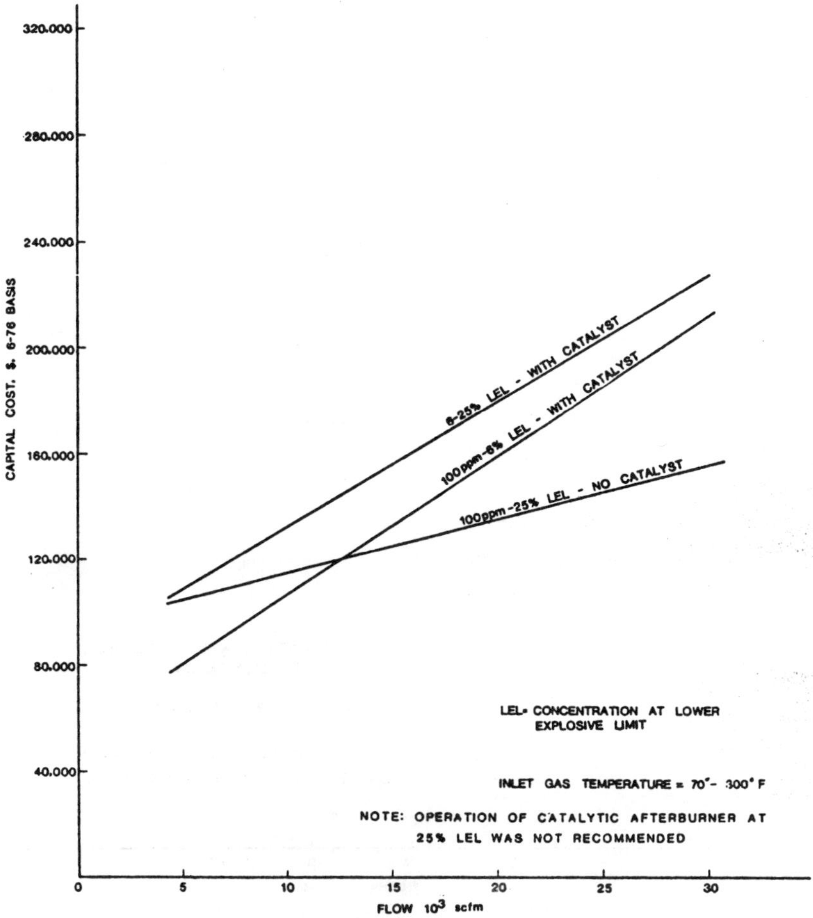

FIGURE 6-1 Capital costs of catalytic and thermal afterburners without primary heat recovery. Reprinted from Radian Corporation.[76]

from the oxidation of the pollutant. This can be seen by referring to Figure 6-5 or Figure 6-6, where the annualized costs may be compared for concentrations of 100 ppm and 25% of the LEL (about 2,500 ppm for an organic solvent with an LEL of 1%). For example, from Figure 6-6, for a 20,000-SCFM* noncatalyst system with primary and secondary heat recovery, the annualized control cost for a high pollutant concentration,

*SCFM is the volume rate of gas flow in standard cubic feet per minute at 70°F and a pressure of 1 atmosphere.

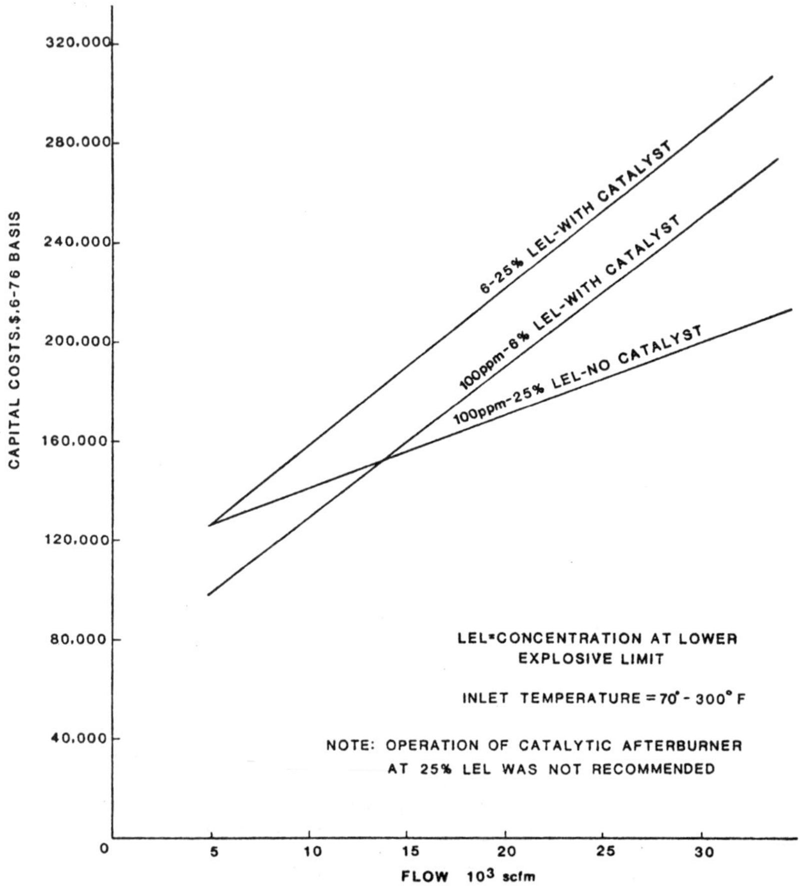

FIGURE 6-2 Capital costs of catalytic and thermal afterburners with heat recovery. Reprinted from Radian Corporation.[76]

25% of the LEL, is about $30,000/yr, compared with about $180,000/yr for a low pollutant concentration of 100 ppm.

This means that controlling odor sources with high-temperature oxidization can become prohibitively expensive. Companies are likely to install this type of control equipment only when it is absolutely necessary.

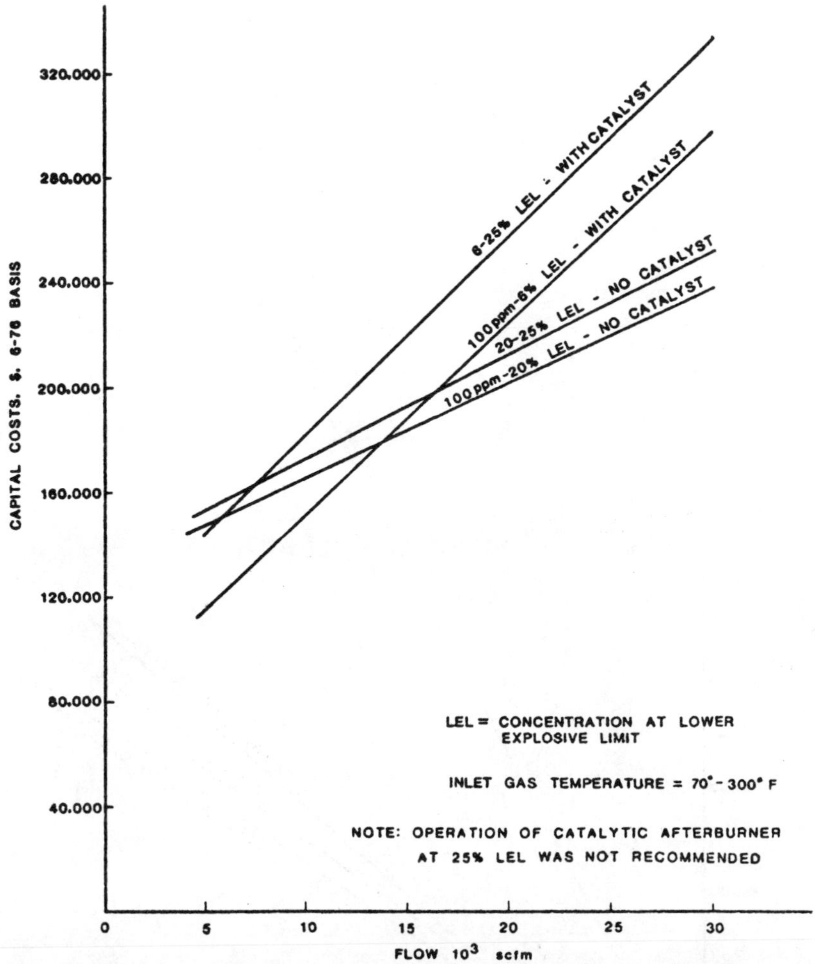

FIGURE 6-3 Capital costs of catalytic and thermal afterburners with primary and secondary heat recovery. Reprinted from Radian Corporation.[76]

SUMMARY AND RECOMMENDATIONS

High-temperature oxidation is a reliable proven method of destroying odorous waste gases. The incorporation of heat-recovery equipment to reduce fuel requirements makes it possible to minimize the high cost of this type of control.

There is a need to obtain sensory data on different categories of odor

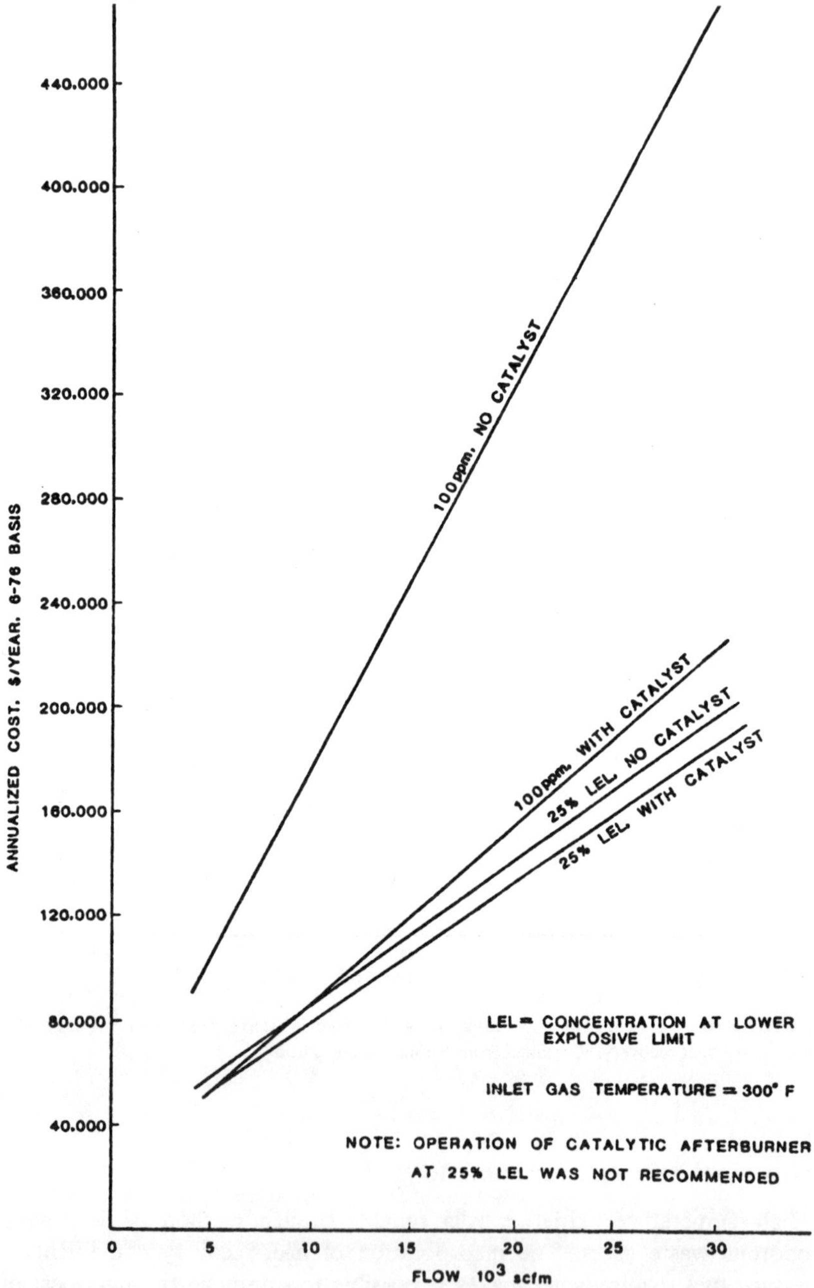

FIGURE 6-4 Annualized costs of afterburners without heat recovery. Reprinted from Radian Corporation.[76]

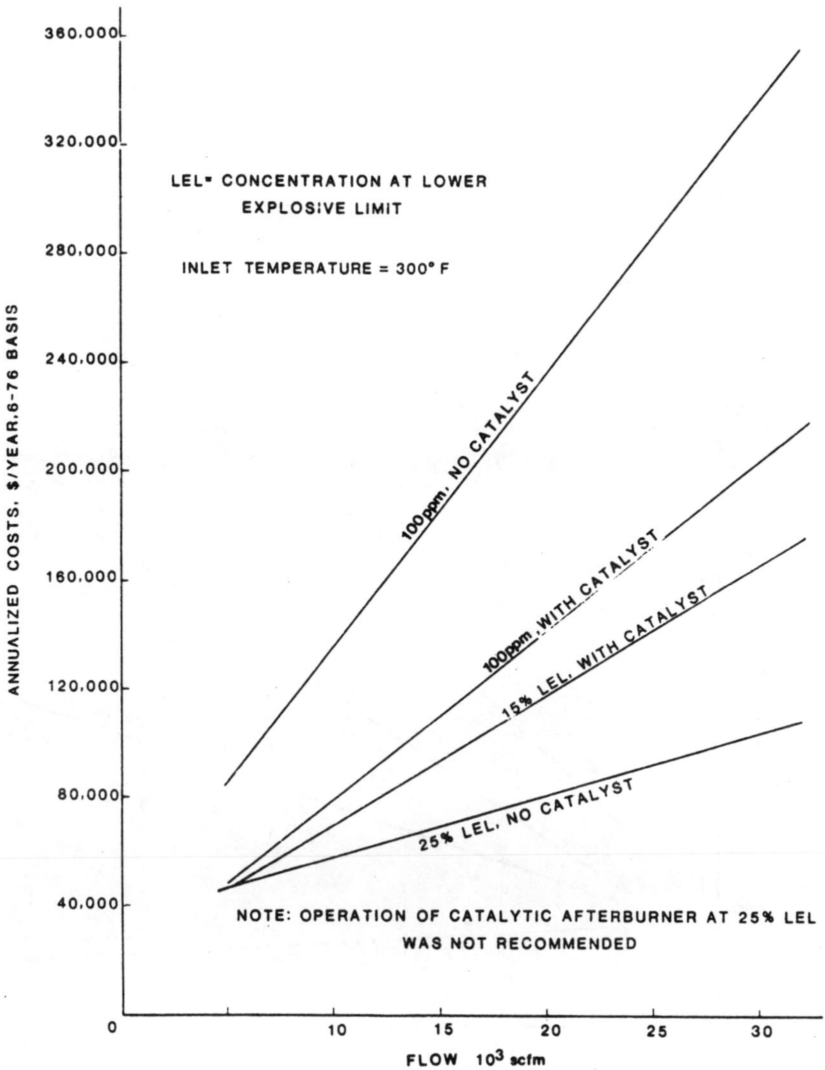

FIGURE 6-5 Annualized costs of afterburners with primary heat recovery. Reprinted from Radian Corporation.[76]

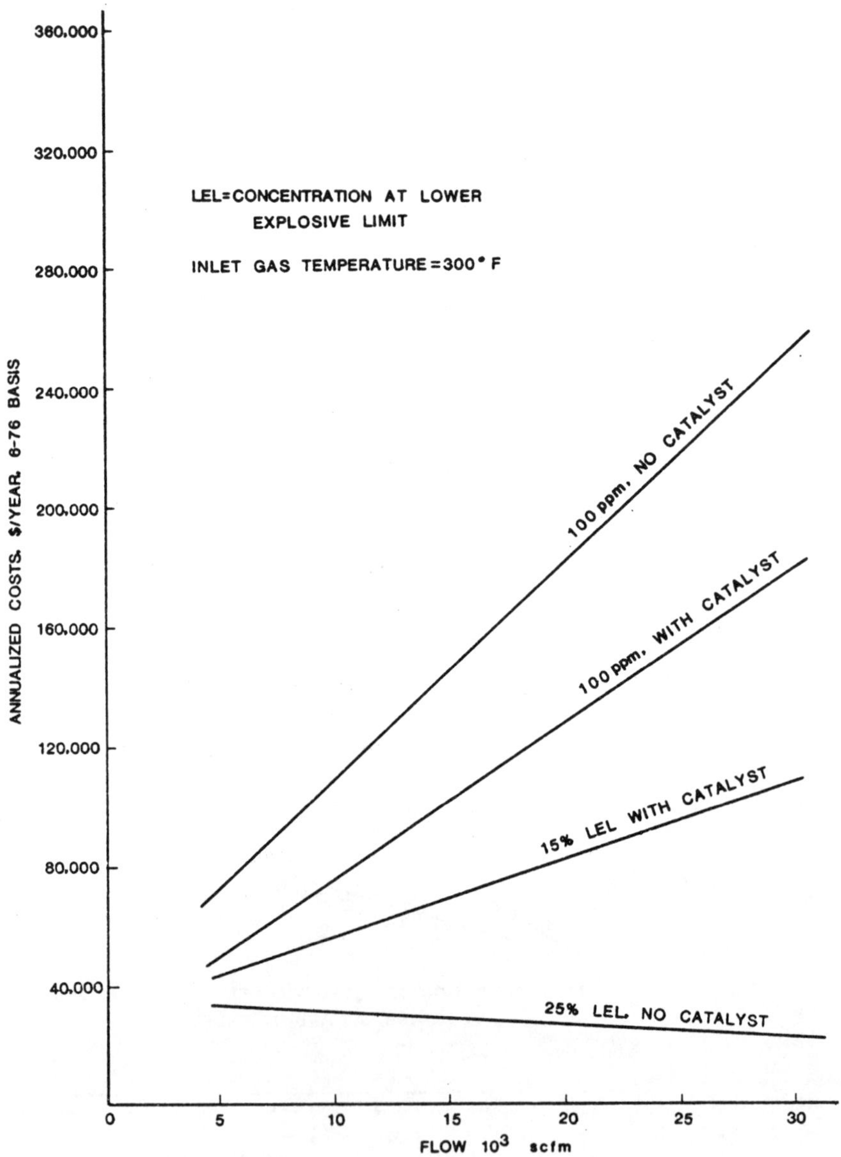

FIGURE 6-6 Annualized costs of afterburners with primary and secondary heat recovery. Reprinted from Radian Corporation.[76]

TABLE 6-4 Technical Assumptions Used in Developing Cost Estimates for Catalytic and Thermal Afterburners[a]

1. Thermal afterburners designed for both oil and natural-gas operation; catalytic afterburners designed for natural-gas and propane operation.
2. Catalytic afterburners capable of 800°F (427°C) operation at low pollutant concentrations, 1,200°F (649°C) at higher concentrations.
3. Equal weight percent hexane and benzene in air.
4. Afterburner operates 5,840 h/yr.
5. Catalyst lifetime is 3 yr.
6. 1,500°F (816°C) operation in thermal afterburner with a residence time of 0.5 s.
7. Primary heat-recovery efficiency of 35%, secondary heat-recovery efficiency of 55%.
8. Outdoor rooftop installation requires structural steel.
9. Gas inlet temperature of 300°F (149°C).
10. Materials to be burned and combustion products are noncorrosive.

[a] Data from U.S. National Air Pollution Control Administration[87] and R. C. Lunche (cited in U.S. Environmental Protection Agency[86]).

sources, to determine the minimal temperatures required for the destruction of odors with the various methods available and the various types of equipment being manufactured.

High-temperature oxidation is energy-intensive, so the impact of the expanded use of this type of equipment on national energy goals ought to be examined.

CONTROL OF ODORS BY ATMOSPHERIC DISPERSION AND LAND-USE PLANNING

Odor control by atmospheric dispersion is predicated on the assumption that dilution of odorants to below a sensory target value—e.g., the detection threshold—can be achieved and maintained by atmospheric dispersion. This assumption relies heavily both on the validity of reported odorant thresholds and on the accuracy of measurements of odorous emission. As mentioned in previous chapters, a published odorant threshold is more a function of the measurement method used than an absolute indicator. The same observation may be made with respect to measurement of odorant emission rates;[1] in this case, however, with careful selection of method and adaptation to the source, accurate and reliable measurement can be made, as described in Chapter 4.

Another assumption underlying the use of atmospheric dispersion for odor control is that odorant transport in the atmosphere is entirely a function of atmospheric turbulent diffusion processes. Dispersion models

TABLE 6-5 Typical Components of Annualized Costs of Catalytic Afterburners[a]

Gas-stream characteristics	
Flow	15,000 scfm (7 m^3/s)
Concentration	15% LEL
Inlet temperature	300°F (150°C)
Direct operating costs	
Utilities	\$20,000[b]
Direct labor	3,000[c]
Maintenance	7,800[d]
Annualized catalyst replacement	19,800[e]
Capital charges	41,000[f]
TOTAL	\$91,600

[a] Reprinted with permission from Hirt Combustion Engineers.[48]

[b] Fuel at \$1.56/GJ (\$1.65/10^6 Btu), electricity at \$9.17/GJ (\$0.033/kWh).

[c] Labor at \$8.25/man-hour.

[d] Maintenance as percentage of capital cost: 4%.

[e] Catalyst life of 3 yr.

[f] Capital charges include as percentages of capital cost: depreciation, 13%; and taxes, insurance, and administrative overhead, 4%.

can therefore be used with confidence in predicting downwind odor resulting from various stack and emission configurations at distances up to 20 km. At greater distances, however, other atmospheric transport methods, such as particulate adsorption or atmospheric reactions, may be more significant than turbulent dispersion.

Atmospheric dilution of odorous emission can be achieved in two main ways:[93]

- *Collection of all process and plant emission and discharge through a tall stack:* Contaminants emitted into the atmosphere are diluted by turbulence and diffusion. The dilution of a contaminant depends directly on the wind speed: the mass emitted in unit time is spread over the distance traveled in unit time by air blowing over the discharge point. In addition to this thinning of material in the direction of the mean wind, there is mixing along and across the mean wind horizontally and vertical mixing because of the natural turbulence resulting from the wind.
- *Relocation of the source at a greater distance from any receptor:*

Moving an odorous discharge farther from any receptor obviously results in a reduced odorant concentration at the receptor. Such a drastic step is currently limited to situations where the source may be relocated within the confines of property to take advantage of favorable atmospheric dispersion provided by local climate and topography. These situations arise most often during siting studies. Careful assessment of industrial-plant odor sources before the location of a plant is fixed can avoid odor problems. One automotive assembly plant has spent well over $1 million to solve a community odor problem that might have been prevented if the plant had been constructed at a different site on the property. This concept could be applied to determine property boundaries needed to prevent odor problems or to develop effective land-use policies.

Several factors must be carefully considered before reliance is placed on atmospheric dispersion for odor control. These include the characteristics of the odorants and their sources and local meteorologic characteristics. Point sources—such as ducts, ports, and vents—are suitable for dispersion. Fugitive emission—such as that from open doors, windows, or leaking flanges—should be prevented or captured and ducted to the stack. Terrain must also be considered. If the odor source is in a valley in rugged, complex terrain, odor control by dispersion is either impractical or impossible; in such cases, discharging odorants through a tall stack only transfers the location of complaints of malodors. Many factors contribute to the transport and dilution of odors, including temperature and humidity of vent gases and reaction with other materials in the atmosphere or photooxidation. Where these reactions are rapid—e.g., an hour or less, as in the oxidation of carbon disulfide, CS_2, to carbon oxysulfide, COS[71]—atmospheric dispersion is usually not feasible for odor control. As Turk *et al.*[84] pointed out, odors may be associated with particles. Odorants with very low thresholds—e.g., 1 ppb or less—are therefore usually not sufficiently controllable by straightforward dispersion. If the source is in a place where inversion conditions would be frequent, tall-stack dispersion for odor control may not be sufficient, so it is necessary to determine the frequency of such conditions for each specific site. Finally, there are many limitations on stack height: structural requirements, economic restrictions, and limitations related to air traffic in the area and other local regulations; these have been discussed in detail by Smith[80] and Cheremisinoff.[26]

Most current odor regulations are of the type that ban the release of objectionable odors but seldom define how objectionableness is to be determined.[56] To solve an odor problem, it is usually necessary to achieve essentially zero detectable odor in the community downwind from the odorant sources. The following steps are designed to achieve this objective when discharge through a tall stack is selected as the means of controlling odors:

- Measure the odor emission per source. Dilution ratio and standard flow rate are measured at each potentially odorous emission source.
- Perform modeling calculations from the present or design source configuration with a puff model, such as the TRC Odor Model,[65] which includes plume rise, building wake entrainment, and mixing-volume correction techniques designed to increase accuracy. The Briggs[18] plume-rise method is used in that model. The results obtained should be similar to those shown in Tables 6-6 and 6-7.
- Conduct an odor survey in the community. This involves the mapping of perceived odor qualities (or characteristic smell) and odor dilution ratios and/or intensities under different meteorologic conditions, usually on the basis of tests with expert odor judges.
- Correlate the model results with the survey results (i.e., calibrate the model). This will relate the odor intensities, dilution ratios, and types found in the community with those determined by modeling suspected sources. If there is no correlation, this could mean that the odor source configuration is more complex than anticipated or that a significant contribution to the odor problem has been neglected or incorrectly represented in the model. For example, building downwash effects could convert a stack source into a ground-level source, as shown in Figure 6-7.[49] The model should be adjusted accordingly.
- Determine required stack height. The adjusted model should be used for the determination of the stack height required to reduce the odor impact at receptors to less than the odor threshold or other allowable number. (This usually requires iterative calculations with output as shown in Figure 6-8.[25])

Some costs of two types of stacks used for odor control are given in Table 6-8. The practical limit for a guyed stack is approximately 150 ft (46 m), and this requires a large roof to provide adequate anchor sites for the guy wires. If a greater stack height is required—i.e., a self-supported stack—other means of odor control are usually more cost-effective.

ABSORPTION AND GAS-PHASE REACTIONS

Odor removal by scrubbing depends on the greater solubility of the odorant in a selected solvent (usually water) than in the odor-carrying gas (usually air). With intimate contact between gas and solvent, the odorant is transferred to the liquid phase from the gas and it then becomes acceptable to release the cleaned gas to the atmosphere. The contaminated solvent may be disposed of as waste or treated chemically to convert the dissolved odorant to another, nonvolatile chemical species and recirculated. Odor reduction by scrubbing is a mass-transfer process that takes place by a diffusional process known as absorption. It is an important unit process in chemical engineering practice, and the theoretical and practical operating principles are available in standard texts.[27,74,79]

Transfers of gaseous odorants to a scrubbing liquid are limited by the following physical factors:

- *Selection of a Suitable Solvent*: Although water is the preferred scrubbing liquid (or solvent) because of its ready availability, low cost, and ease of disposal, most unpleasant-smelling odorants are organic and have only limited solubility in water. Alternative solvents for organic odorants must be low in cost or easily repurified, nonvolatile, nonodorous, nonflammable, and easily disposed of, to be considered satisfactory for odor removal. Although a highly purified and deodorized mineral oil would meet most of these criteria (and other solvents could no doubt be hypothesized), water remains the only widely used scrubbing liquid. This has seriously limited the application of scrubbing for deodorizing effluent gases. Nevertheless, a number of stratagems have been used successfully to overcome the poor solubility of organic malodorants in water, including the use of gas-phase reactions that have the effect of improving the aqueous solubility of the reaction products and the addition of reacting chemicals to water to change the chemical form of the odorant and to lessen the tendency of these less easily absorbable compounds to revolatilize from the scrubbing liquid. These processes are discussed in more detail later. In spite of the prevalence of organic malodorants, many commonly encountered substances that produce odor complaints are inorganic (hydrogen sulfide is an outstanding example), and for these an aqueous scrubbing liquor is ideal.
- *Low Concentration of Malodorants in the Gas Phase*: The rate of absorption is directly proportional to the concentration difference between the malodorant in the gas phase and in the liquid phase, when expressed in consistent units. Inasmuch as offensive concentrations of most commonly encountered malodorants are in the parts-per-billion range, the concentration-difference driving force between gas and liquid phases will be vanishingly small under the best of circumstances, i.e., when a fresh solvent is used; but, as the solvent takes up the odorant, the concentration difference between phases becomes even smaller. This may be contrasted with the more usual application of absorption for the manufacture of chemicals. There the absorbable compound is measured in percent by

TABLE 6-6 Model Output from an Example Industrial Odor Study[a]

Odor model test; no control implemented. Distribution of odor levels at position 0.30, 0.00. Wind direction 270. Wind speed 1.00 m/s. Lateral stability class 1. Vertical stability class 5.

Class	Class Upper Limit Odor Level	Frequency of Occurrence	Cumulative Frequency of Occurrence	Cumulative Percentage Frequency of Occurrence	Comments
1	0.01	325	325	32.50	
2	0.05	130	455	45.50	1-h ave.
3	0.10	105	560	56.00	(at 50%)
4	0.20	109	669	66.90	
5	0.50	61	730	73.00	
6	0.80	19	749	74.90	
7	0.90	13	762	76.20	
8	1.00	7	769	76.90	Nondetectable
9	2.00	46	815	81.50	Detectable
10	3.00	47	862	86.20	
11	4.00	38	900	90.00	
12	5.00	33	933	93.30	
13	8.00	67	1000	100.00	Peak
14	10.00	0	1000	100.00	
15	20.00	0	1000	100.00	
16	50.00	0	1000	100.00	
17	100.00	0	1000	100.00	
18	200.00	0	1000	100.00	
19	500.00	0	1000	100.00	
20	1000.00	0	1000	100.00	

[a] Data from Murray *et al.*[65]

TABLE 6-7 Model Output for Industrial Source with 30-Meter Stack Installed[a]

Odor model test—30-m stack in place. Distribution of odor levels at position 0.30, 0.00. Wind direction 270. Wind speed 1.00 m/s. Lateral stability class 1. Vertical stability class 5.

Class	Class Upper Limit Odor Level	Frequency of Occurrence	Cumulative Frequency of Occurrence	Cumulative Percentage Frequency of Occurrence
1	0.01	523	523	52.30
2	0.05	176	699	69.90
3	0.10	129	828	82.80
4	0.20	135	963	96.30
5	0.50	24	987	98.70
6	0.80	5	992	99.20
7	0.90	3	995	99.50
8	1.00	1	996	99.60
9	2.00	4	1,000	100.00
10	3.00	0	1,000	100.00
11	4.00	0	1,000	100.00
12	5.00	0	1,000	100.00
13	8.00	0	1,000	100.00
14	10.00	0	1,000	100.00
15	20.00	0	1,000	100.00
16	50.00	0	1,000	100.00
17	100.00	0	1,000	100.00
18	200.00	0	1,000	100.00
19	500.00	0	1,000	100.00
20	1,000.00	0	1,000	100.00

[a] Data from Murray *et al.*[65]

volume rather than in parts per billion—a difference of seven orders of magnitude in concentration. When the liquid phase becomes saturated with the malodorant—i.e., when it is in equilibrium with the resulting concentration of the malodorant gas phase and is unable to absorb any more—the malodorant passes through the absorber unchanged. In spite of the extraordinarily low concentration differences commonly encountered in the deodorizing of gases by absorption, the rate of transfer can be maximized by the liberal use of reacting chemicals in the scrubbing water that serve to maintain the partial pressure of the absorbed odorants at zero. In addition, chemical treatment of the scrubbing water makes it possible to recirculate it through many cycles before it must be discarded or repurified.

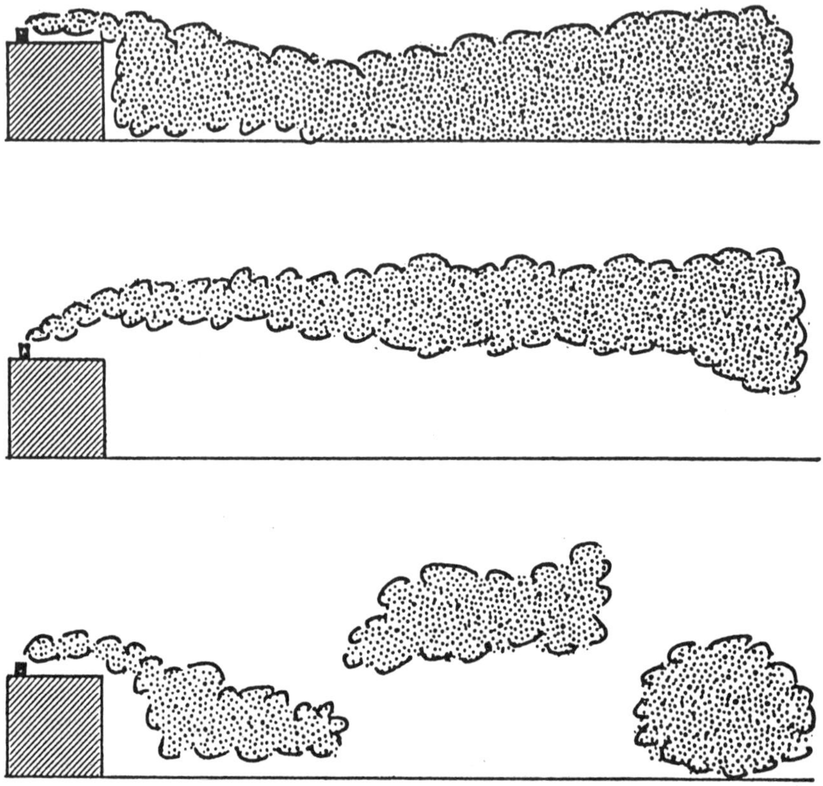

FIGURE 6-7 Plume behavior during full entrainment, nonentrainment, and partial entrainment. Reprinted with permission from Cheremisinoff.[26]

TABLE 6-8 Cost Estimates for Small-Diameter Tall Stacks for Odor Control

Height			
ft	m	Type	Approximate Cost, $
50	15	Guyed on structure	6,400
100	30	Guyed on structure	12,800
200	61	Self-supported	250,000
300	91	Self-supported	380,000

Gas- and liquid-phase reactions of odorous compounds have been studied extensively. It has been found that non-volatile oxidizing chemicals, such as potassium permanganate, and neutralizing chemicals, such as sodium hydroxide, are effective only after the odor-producing substance has been absorbed in the scrubbing liquid. It is often forgotten that the presence of a reacting or neutralizing chemical in the scrubbing liquid can have no influence on the rate of odorant transfer by physical absorption alone. However, a gaseous reactant—such as chlorine, chlorine dioxide, or

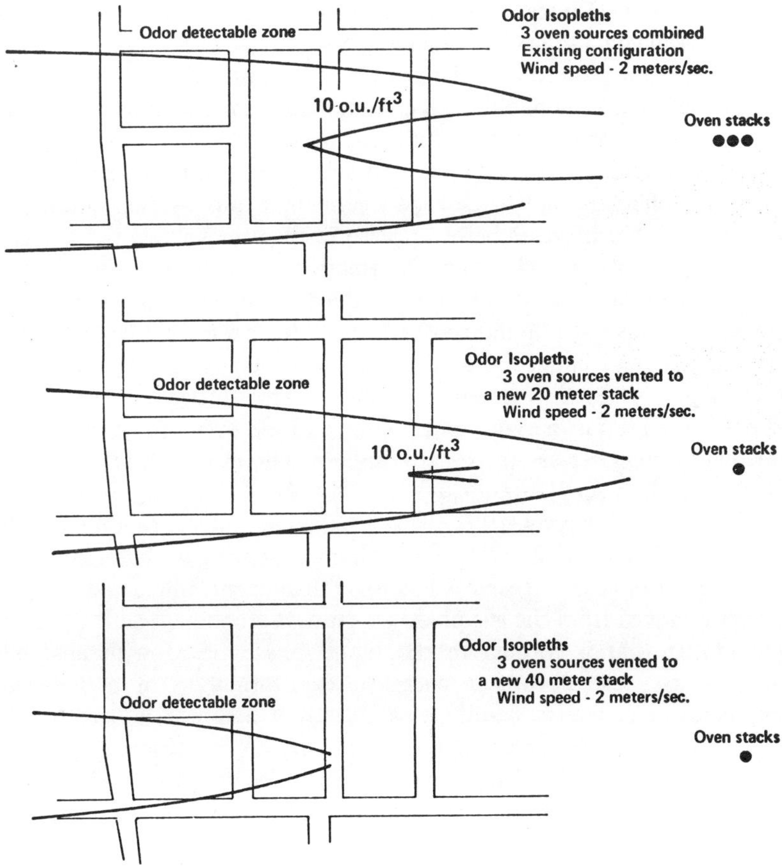

FIGURE 6-8 Odor isopleths for various plant stack heights. Top, area with odor diluton ratios greater than 10 from the existing plant. Middle, improvement achievable with a stack height of 20 m. Bottom, no odor dilution ratio above 10 with a 30-m stack height. Reprinted with permission from Cha.[25]

ozone—can change the chemical nature of odorants in the gas phase and make them less malodorous or prepare them for absorption by making them more soluble. In addition, these substances are soluble and, after absorption in the scrubbing liquid, act as powerful oxidizing agents for dissolved odorants by liquid-phase reactions.

When permanganate is used as an oxidizing agent in solution, it produces an insoluble manganese dioxide reaction product, which can clog the scrubber and plug the droplet carryover stage that follows it. When sodium hydroxide is used as a neutralizing agent for acid gases that have foul odors, such as hydrogen sulfide, it also traps carbon dioxide from the air or combustion gases that can combine with hardness chemicals in the scrubbing water to form insoluble calcium carbonate and magnesium carbonate scale on all the scrubber and mist-eliminator surfaces.

If ozone is introduced into the gas phase, it must be at 10 ppm, if not more, to effect a reasonable degree of oxidation of odorous compounds within an acceptable retention time (no more than a few seconds). A number of attempts have been made to ameliorate the rotten-egg odor of hydrogen sulfide emitted by sewage treatment plants by the addition of ozone to the effluent gases before release to the atmosphere, but the gas-phase reaction rate is so slow that this treatment is ineffective. In addition, a considerable excess of ozone must be added to the odorous gas stream to produce any effect at all, and most of the added ozone will be emitted to the atmosphere unchanged. This would represent an excessive atmospheric emission of ozone in most air-pollution control districts. Therefore, a scrubber of some sort is required to remove excess ozone from the treated gases before their release to the atmosphere. The use of chlorine dioxide has many of the same limitations.

Chlorine, another powerful gaseous oxidizing substance that can be added to malodorous effluent gases to induce useful gas-phase reactions, has a number of advantages: it is less toxic than ozone; unreacted chlorine is easily removed from the gas phase by caustic scrubbing, and the sodium hypochlorite that is formed in the liquid phase reacts with absorbed materials to cause a further chemical transformation of malodorous compounds; chlorine is readily available as a compressed gas (unlike ozone, which must be generated from clean, dry air or oxygen, as needed); automatic equipment for dispensing it in controlled amounts has been used universally for decades for the treatment of drinking water; and it is relatively inexpensive and very reliable.

Regardless of which chemicals may be added to the gas phase for the purpose of reacting with malodorants, experience has demonstrated the futility of depending on gas-phase reactions alone, because these reactions are always incomplete in practical applications (although they can

TABLE 6-9 Scrubber Reagents for Absorbing Frequently Encountered Odorants

Reagent	Odorant Class
Calcium bisulfite	Aldehydes
Hypochloric acid	Amines
Potassium permanganate	Aldehydes
Sodium bisulfite	Aldehydes
Sodium hydroxide	Mercaptans
Sulfamic acid	Amines
Sulfuric acid	Amines

sometimes be demonstrated in bench-scale laboratory apparatus) and a large excess of the injected gas-phase reactant must be used. Ultimately, this excess must be emitted to the atmosphere unchanged, and it will then behave as a pollutant in its own right. This makes it necessary to exercise extreme caution in the use of chemical additives (including reodorizing and masking agents) in effluent gases, unless adequate means are provided for their complete removal before emission to the atmosphere. For example, when the malodorous effluent from a spent grain dryer was treated with chlorine gas at 2 ppm in the gas phase, allowed to react for several seconds, and then passed through a multistage scrubber containing enough caustic to raise the pH to 11–12, there was a marked reduction in odor intensity of the cleaned effluent.[41] Increasing the gas-phase chlorine addition to 6 ppm produced an effluent that was almost odorless, with no hint of a chlorine or hypochlorite odor. Overall reduction in odorant concentration was 95% by gas chromatography; this indicated a high degree of collector efficiency by chemical measurement, as well as by sensory evaluation. An increase in chlorine injection to 9 ppm produced an equally good overall removal efficiency by chromatographic measurement, but the treated effluent now had a distinct hypochlorite odor that was not considered acceptable. This series of tests illustrated the usefulness of gas-phase reactions for altering the chemical nature of malodorants and the absolute need to remove reactants and reaction products by some step, such as absorption in a suitable solvent. Typical reagents for absorbing some frequently encountered classes of odorants are listed in Table 6-9. No reagent was found to be effective for all the odor classes considered.[66]

The effectiveness of odor removal by hypochlorite scrubbing solution alone—i.e., without preliminary gas-phase reactions—is shown in Figure 6-9.[4] The apparatus used was a countercurrent tower packed with 1/4-in. porcelain Raschig rings (hollow cylinders). The experiments were con-

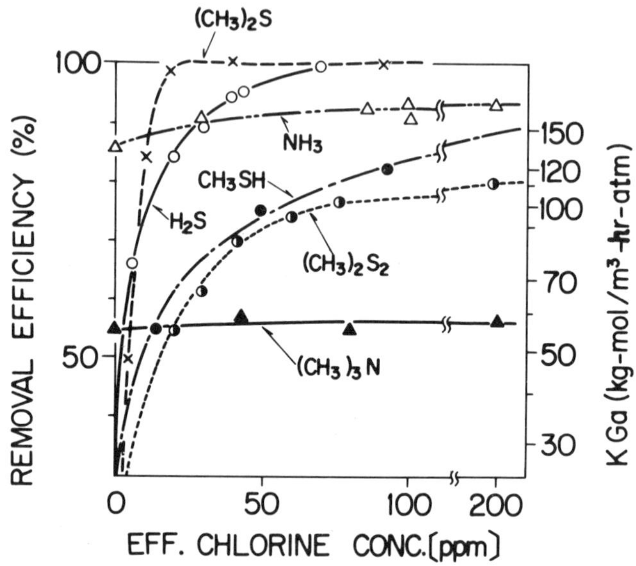

FIGURE 6-9 Dependence of malodorous-gas removal efficiency on effective chlorine concentration at top of packed tower. Reprinted with permission from Azuma *et al.*[4]

ducted at a superficial gas velocity of 0.25 m/s, a liquid-to-gas ratio of 2.5 kg/kg, and a gas retention time of 2 s. The molar ratio of sodium hypochlorite consumed by the odorants was 1.5 for methyl mercaptan, 1.8 for ammonia, 1.0 for dimethyl sulfide, and 2.8 for dimethyl disulfide. The odor quality and intensity of the effluent gases were not recorded. Additional examples of scrubbing for the removal of malodorous substances from process off-gases are cited later, in the discussion of specific industries. This control process is especially important for nonedible-fat rendering, fish- and poultry-offal meal production, and other industrial processes that emit substantial quantities of oils and fats that rapidly clog and inactivate adsorption carbons.

Careful scrubber design is essential to provide for long gas retention times, intimate contact between phases, and the absolutely necessary retention of strongly reactive chemicals wholly within the scrubbing apparatus. Often, the malodorous gases will be accompanied by solids that will clog the unit if special provisions are not made to remove them before entry into the contact sections of the absorber. This is a requirement of special urgency in the handling of off-gases that contain putrescible solids, lest they remain within the apparatus, putrify, and produce an additional source of intensely malodorous emission. Caustic scrubbing solutions that saponify grease and hypochlorite-containing scrubbing solutions that halt microbiologic decomposition can play a dual role when they are used in the gas-deodorizing scrubbers for off-gases that contain oil droplets and putrescible solid particles.

SCRUBBING EQUIPMENT

The gas scrubber most widely used by chemical engineers for chemical manufacture is the countercurrent packed tower in which the scrubbing liquid is introduced at the top and trickles down over and through the packing that is designed to expose an extended liquid surface in thin films to the rising gas. Packings include plastic, ceramic, metal, and graphite shapes ranging from rings, spheres, and saddles to porous beds of fibers (Figure 6-10).[22] The special advantage of countercurrent flow of liquid and gas phases is that it maximizes the concentration driving force throughout the apparatus, inasmuch as the fresh scrubbing liquid is in contact with the cleaned gases leaving the tower and the spent liquid at the bottom of the tower is in contact with the highest concentration of the entering gas. When gas and liquid flow rates are chosen with full consideration of the total amount of odorant that is to be transferred from gas to liquid, saturation of the liquid with respect to the gas concentration in contact

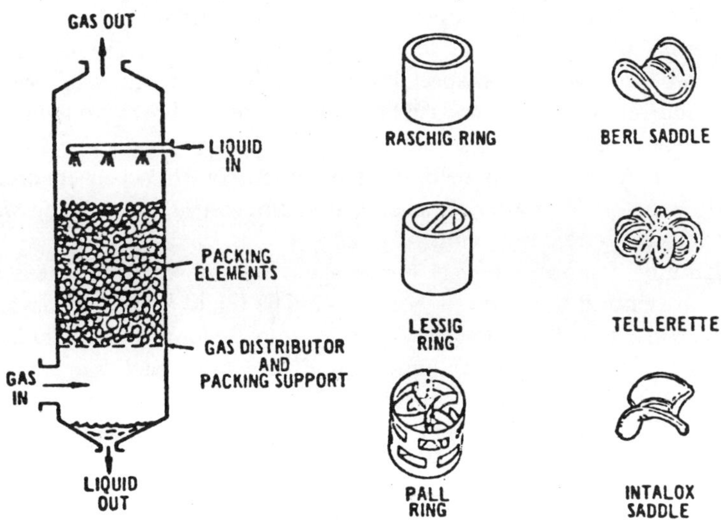

FIGURE 6-10 Countercurrent packed tower and packings. Reprinted from Calvert *et al.*[22]

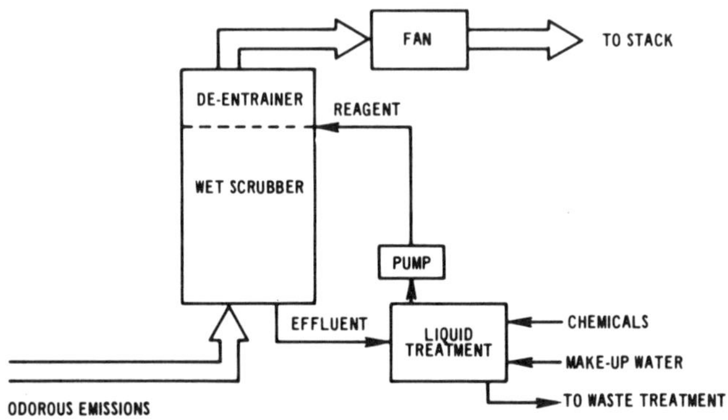

FIGURE 6-11 Odor-scrubbing system. Reprinted with permission from Dickerson *et al.*[33]

with it will not occur anywhere in the tower, and odorant transfer will occur smoothly throughout the tower height. A typical countercurrent odor-absorbing system is diagramed in Figure 6-11.[33] The system includes means to add chemicals to the scrubbing water. However, when a neutralizing or reacting chemical is added to the scrubbing liquid, so that there is no longer a tendency for the absorbed odorant to revolatilize into the gas phase, there is no special advantage in countercurrent scrubbers over cocurrent or crossflow scrubbers. Crossflow scrubbers are lighter (for roof installation), easier to service, and less expensive than packed towers. They have dominated the field of air-pollution control scrubbers and are widely used, with appropriate chemical additions to the scrubbing water, for removal of odors from effluent gases.

Figure 6-12 shows a typical multistage crossflow scrubber that is used for odor removal. Gas flow is horizontal. The liquid is sprayed on top of the packing and flows down by gravity into a sump, from which it is collected and resprayed. As shown in Figure 6-12, each stage may be sprayed with a different chemical to effect complete odor removal and retention of volatile treatment chemicals. For example, stage 1 might contain a reducing chemical for removal of aldehydes, stage 2 an oxidizing chemical for removal of amines, and stage 3 an alkali for removal of mercaptans and chlorine volatilized from stage 2 (Table 6-9). Figure 6-13 shows components of a single absorption stage, including packing, sprays, and mist eliminator.

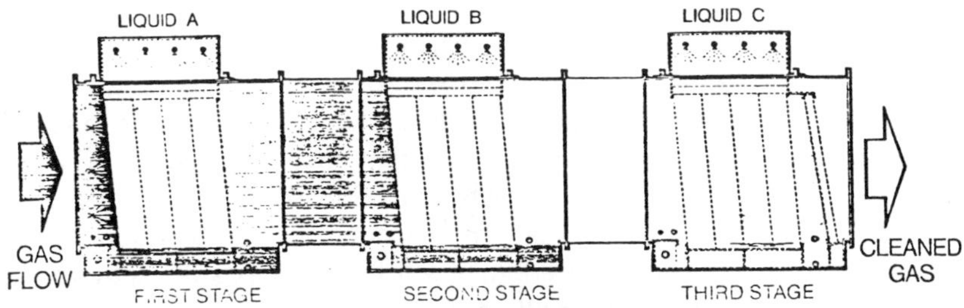

FIGURE 6-12 Multistage crossflow scrubber. Courtesy Buffalo Forge Co., Buffalo, N.Y.

COSTS OF GAS DEODORIZATON BY SCRUBBING

The cost of chemical scrubbing systems may be compared on the basis of the cost per pound of odorous material removed with the use of alternative chemical oxidants, but this "does not serve as a satisfactory basis for the selection of scrubbing over alternative forms of abatement."[45] For example,

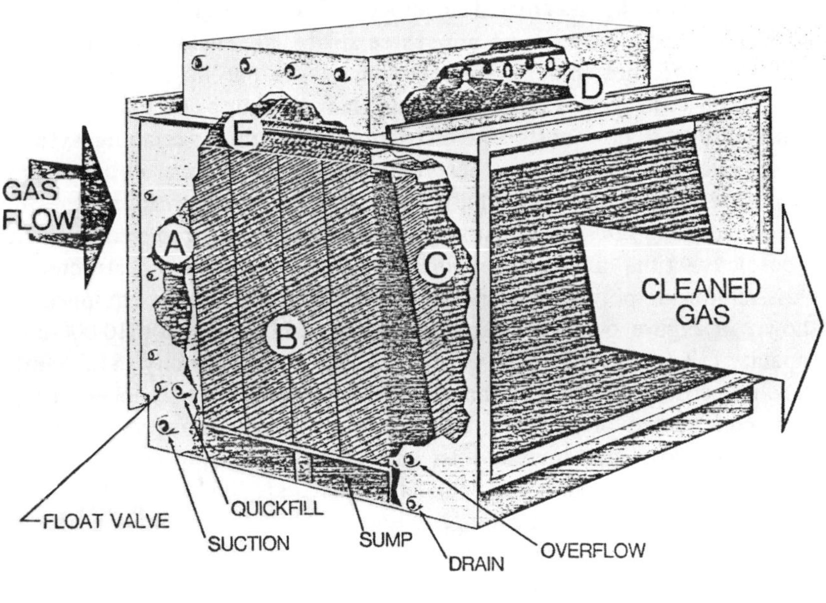

FIGURE 6-13 Components of a crossflow scrubber stage. A, inlet sprays. B, wetted packing. C, dry packing (mist eliminator). D, spray nozzles. E, liquid distribution packing. Courtesy Buffalo Forge Co., Buffalo, N.Y.

thermal destruction is likely to be much cheaper and easier than scrubbing for a small malodorous-gas stream, whereas the reverse is likely to be the case when gas streams substantially exceed an emission rate of 1,000 cfm (cubic feet per minute). This is because the cost of fuel is directly proportional to the volumetric gas rate that must be treated, whereas substantial economies of scale are obtained as the scrubber size increases. Although chemical costs increase in proportion to the amount of odorant removed, the concentration of odorant will in most cases be measured in parts per billion or parts per million, by volume, and the chemicals will seldom represent a major fraction of total scrubber costs when the scrubbing liquid is recirculated and only the odorant reacts with the chemical. If the odorant concentration increased to the point where it provided all the fuel required to sustain the thermal reaction with the aid of heat regeneration and catalytic burners, the economics would reverse; but such high concentrations are likely to represent air-pollution threats of a more serious nature than odor and are likely to be controlled for other reasons. Similar considerations apply in a comparison of scrubber costs with those associated with adsorption. As the concentration of odorants decreases and the gas volume to be deodorized increases, the cost advantage of scrubbing over gas adsorption widens.

Average cost figures for all manner of large, complex air-cleaning systems are hard to obtain, because these devices are custom-designed and constructed to meet precise specifications of the purchaser and because installation costs vary severalfold from plant to plant, depending on site availability and how costs are distributed between air-cleaning system construction and process or structural improvement. Furthermore, each of these several costs changes rapidly in response to widespread changes in prices and wage rates, so a cost survey can scarcely be completed before it is obsolete. At the time of this report, a three-stage scrubber constructed of glass-reinforced polyester resin for chemical resistance and equipped as shown in Figure 6-12 costs approximately $23,000 FOB for 10,000-cfm capacity ($2.3/cfm) and $66,000 *fob* for 50,000-cfm capacity ($1.3/cfm). Purchase cost would be less than half these figures if the scrubber casing were constructed of carbon steel, instead of glass-reinforced polyester. Installation costs that include inlet and outlet piping, blower and motor, pumps, electric and plumbing connections, etc., are likely to be 2–5 times the purchase cost, with an average of about 2.5 times. Operating costs are for chemicals, electric power, water, wastewater treatment, and maintenance and repair of equipment. These are highly variable. If corrosion-resistant materials of construction are selected, repair costs are negligible for this type of equipment. Maintenance may involve a substantial number of weekly man-hours whenever scale formation occurs, because the scrubbing liquid contains substantial hardness and is maintained at high alkalinity, or when insoluble particles accumulate in the scrubber packing.

Spent scrubbing solutions may require neutralization before disposal, to prevent rapid corrosion of waste pipes and waste-treatment plant equipment. In some cases, bactericidal chemicals in the spent scrubbing water have to be destroyed, to avoid disruption of microbiologic sewage-treatment processes. Scrubbing solutions heavily contaminated with fats, oils, and soluble and insoluble putrescible substances can usually be treated satisfactorily by existing wastewater-treatment processes that are an integral part of the process equipment of the industries involved, e.g., stickwater-treatment tanks in fishmeal and nonedible-fat rendering plants. This is especially feasible when the quantity of scrubber wastewater can be severely limited by chemical treatments that result in extensive liquid recirculation before discharge.

With meticulous scrubber design, rigorous constraints on overdosage of scrubbing-water chemicals, and careful restrictions on wastewater volumes, unit costs for gas deodorization by scrubbing compare favorably with those associated with adsorption and incineration whenever choices may be made among effective treatment systems. In consideration of the probable future trend of fuel prices, the cost advantage of scrubbing (in cases where it is an effective means of gas deodorization) is likely to increase substantially. Typical applications of scrubbing for control of odorous emission from a number of industries are shown in Table 6-10.[31]

ADSORPTION*

GENERAL PRINCIPLES

Any gas or vapor will adhere to some degree to any solid surface. This phenomenon is called "adsorption." Practical adsorbents have extensive areas of inner surface, by reason of extensive honeycombing. When adsorbed matter condenses in the submicroscopic pores of an adsorbent, the phenomenon is called "capillary condensation." Adsorption is useful in odor control, because it is a means of concentrating gaseous odorants from an airstream, thus facilitating their disposal, their recovery, or their conversion to innocuous or valuable products. When an odorous airstream is passed through a fresh adsorbent bed, almost all the odorant molecules that reach the surface are adsorbed, and desorption is very slow. Furthermore, if the bed consists of closely packed granules, the distance

*Some of the material in this section is summarized from Turk.[82] Information on costs has been obtained from a Radian Corporation report.[76]

TABLE 6-10 Odor Control by Liquid Scrubbing for Selected Industrial Processes[a]

Industry	Odorous Gases	Process Sources	Scrubbing Solutions
Petroleum-refining	Hydrogen sulfide, mercaptans, ammonia, amines	Air and steam stripped of sour waters, process vessels, product-treating, wastewater, tank vents	Ethanolamine, sodium hydroxide
Natural gas	Hydrogen sulfide, mercaptans	Gas wells	Ethanolamine, sodium hydroxide, metal salts
Coke ovens	Hydrogen sulfide	Coking vent	Ethanolamine, sodium hydroxide
Viscose rayon	Hydrogen sulfide	Acid-treating	Sodium hydroxide, metal salts
Kraft pulp mill	Hydrogen sulfide, methyl mercaptan, organic sulfides, terpenes	Digesters, evaporators, recovery furnaces, lime kilns	Sodium hydroxide, sodium hypochlorite, chlorine, water, black liquor
Chlorine chemicals	Chlorine, hydrochloric acid, carbonyl chloride	Beach plants, metal-plating, chemical operations	Water, ammonium hydroxide
Meat-packing	Amines, organic acids, organic gases	Cooking, packaging, washing	Water, chlorine
Animal-rendering	Amines, organic acids, ammonia, hydrogen sulfide, mercaptans	Cooking, grinding	Water, chlorine dioxide
Fish meal reduction	Amines, skatoles, acids	Cooking, dryers	Chlorine, sodium hydroxide
Soap-making	Organic esters, ammonia, amines, hydrogen sulfide	Storage tanks	Sulfuric acid, sodium hydroxide
Varnish-cooking	Maleic anhydride, alkyd resins, amines	Cooking	Water acids, sodium hydroxide

[a] Data from Danielson.[31]

the molecules must travel to reach some point on the surface is small, and the transfer rate is therefore high. In practice, the half-life of airborne molecules streaming through a packed adsorbent bed is around 0.01 s, and a 95% removal occurs in about 4 half-lives, or around 0.04 s.[83] Thus, the very high efficiencies required to deodorize a highly odorous airstream may be achieved with a bed of moderate depth at reasonable airflow rates.

The quantity of material that can be adsorbed by a given weight of adsorbent depends on the following factors: the concentration of the material in the space around the adsorbent, the total surface area of the adsorbent, the total volume of pores in the adsorbent with diameters small enough to facilitate condensation of adsorbed gases, the temperature, the presence of other gases in the environment that may compete for a place on the adsorbent, the characteristics of the molecules to be adsorbed (especially their weight, electric polarity, size, and shape), and the electric polarity of the adsorbent surface. Maximal capacity for adsorption of a given substance is favored by a high concentration of the substance in the space adjoining the adsorbent, a large adsorbing surface, freedom from competing substances, low temperature, and aggregation of the substance in large molecules that fit and are strongly attracted to the receiving shapes of the adsorbent.

Adsorbed odorants may be disposed of in any of the following ways: the adsorbent with its adsorbate may be discarded; the adsorbate may be desorbed and recovered, if it is valuable, or discarded (the adsorbent is recovered in either case); or the adsorbate may be chemically converted to a more easily disposable product, preferably with preservation and recovery of the adsorbent.

ACTIVATED CARBON

Adsorbents are most significantly characterized by their chemical natures, by their surface areas, and by the volume and diameter of their pores. The most important chemical differences among adsorbents are related to electric polarity.

Activated carbon, consisting largely of neutral atoms of a single element, presents a surface with a relatively homogeneous distribution of electric charge. As a result, it has less preference for highly polar molecules, such as water, than for most organic substances; it is therefore suitable for the decontamination of an airstream that contains odorous organic matter.

Table 6-11 shows ranges of surface areas and pore volumes of several adsorbents. Among these, activated carbon is generally highest in surface

TABLE 6-11 Surface Areas and Pore Sizes of Adsorbents[a]

	Activated Carbon	Activated Alumina	Silica Gel	Molecular Sieve
Surface area, m^2/g	1100-1600	210-360	750	—
Surface area, m^2/cm^3	300-560	210-320	520	—
Pore volume, cm^3/g	0.80-1.20	0.29-0.37	0.40	0.27-0.38
Pore volume, cm^3/cm^3	0.40-0.42	0.29-0.33	0.28	0.22-0.30
Mean pore diameter, Å	15-20[b]	18-20	22	3-9

[a] Reprinted with permission from Turk.[82]

[b] Refers to micropore volume (<25 Å diameter); macropores (>25 Å) not included.

area and pore volume, and these are the properties that primarily determine overall adsorptive capacity.

Activated carbon (also called "active carbon" or "activated charcoal") consists of particles of moderately to highly pure carbon that have a large surface area per unit weight and unit volume of solid. For use in a fixed bed for air or gas purification, the particles must be so sized that they impose little resistance to flow for a given sorption efficiency; the range of 4–20 mesh (U.S. Sieve Series) encompasses the predominant portion of carbon for such use. To minimize mechanical attrition during transportation and use, the activated carbon should be hard. Hardness is determined in part by the nature of the raw material used for manufacture and in part by the manufacturing process. Raw materials include coconut and other nut shells, fruit pits, bituminous coal, hard woods, and petroleum residues.

As stated above, activated carbon is effective in adsorbing molecules of organic substances, even from a humid gas stream, with less selectivity than is exhibited by other, more polar sorbents. Water molecules, being highly polar, exhibit stronger attraction for each other than the nonpolar carbon surface; consequently, larger, less polar organic molecules are selectively adsorbed.

The total adsorptive capacity of a sample of activated carbon may be measured by its activity or retentivity for a standard vapor. The activity is the maximal amount of a vapor that can be adsorbed by a given weight of carbon under specified conditions of temperature, concentration of the vapor in question, and concentration of other vapors (usually water). The retentivity is the maximal amount of adsorbed vapor that can be retained by the carbon after the vapor concentration in the ambient air or gas stream passing through it has been reduced to zero. Because an adsorbent may be required to retain its adsorbate even in pure air, the retentivity represents the practical capacity of the carbon in service. Typical specifications for activated carbon to be used for air purification are given in Table 6-12.

The pore sizes of activated carbons are important determinants of their adsorptive properties. Pores less than about 25 Å in diameter are generally designated as micropores, larger ones as macropores. The distinction is important, because the molecules of most odorants range in diameter from about 4.0 to about 8.5 or 9.0 Å. If the pores are not much larger than twice the molecular diameter, opposite-wall effects play an important role in the adsorption process by facilitating capillary condensation. Maximal adsorption capacity is determined by the liquid packing that can occur in such small pores.

A plot of adsorption capacity against partial pressure of the adsorbate at a given temperature is called the adsorption isotherm. Figure 6-14 shows how adsorption capacity increases with increasing partial pressure, as well as with increasing molecular weight in a series of compounds of related chemical structure.

Figure 6-15, taken from a study of the saturation of activated carbon in an apple-storage atmosphere at 85% relative humidity and 35°F (1.7°C), shows how the initially adsorbed moisture is gradually displaced by the adsorbed organic vapors.[85]

When odorant concentrations are low, thin-bed (about 2-cm) adsorbers often provide a useful service life while offering the advantage of low resistance to airflow. The carbon is retained between perforated metal in flat, cylindric, or pleated shapes (Figures 6-16, 6-17, and 6-18). Commercially available cylindric canisters are designed for air at about 25 cfm; the

TABLE 6-12 Typical Specifications for Activated Carbon Used for Air Purification[a]

Activity for CCl_4[b]	At least 50%
Retentivity for CCl_4[c]	At least 30%
Apparent density	At least 0.4 g/ml
Hardness (ball abrasion)[d]	At least 80%
Mesh distribution	6-14 range (Tyler Sieve Series)

[a] Reprinted with permission from Turk.[82]

[b] Maximal saturation of carbon, at 20°C and 760 torr in an airstream equilibrated with CCl_4 at 0°C.

[c] Maximal weight of adsorbed CCl_4 retained by carbon exposure to pure air at 20°C and 760 torr.

[d] Percent of 6-8 mesh carbon that remains on a 14-mesh screen after shaking with 30 steel balls of 0.25-0.37 in. (0.635-0.940 cm) per 50 g of carbon, for 30 min in a vibrating or tapping machine.

larger pleated cells handle 750–1,000 cfm, and cells comprising aggregates of flat bed components handle 2,000 cfm.

Thick-bed adsorbers are used when large adsorbing capacity is needed and when on-site regeneration is used. Bed depths are in the range of 1–6 ft (0.3–1.8 m). Design airflow capacities are up to 40,000 cfm (67,960 m^3/h). The ratio of weight of carbon to design airflow capacity is typically about 0.5 lb/cfm (0.27 $kg/m^3 \cdot h$). Typical thick-bed adsorbers, such as are used in solvent recovery systems, are shown in Figure 6-19. Other systems include fluidized-, rotating-, and falling-bed adsorbers.

The period between regenerations of the adsorbent is limited by its capacity and by the contaminating load. Provisions must therefore be made for determining when the adsorbent is saturated and for renewing or reactivating it. The weight of an adsorbent is not a valid measure of its saturation, because its moisture content, which depends on the relative

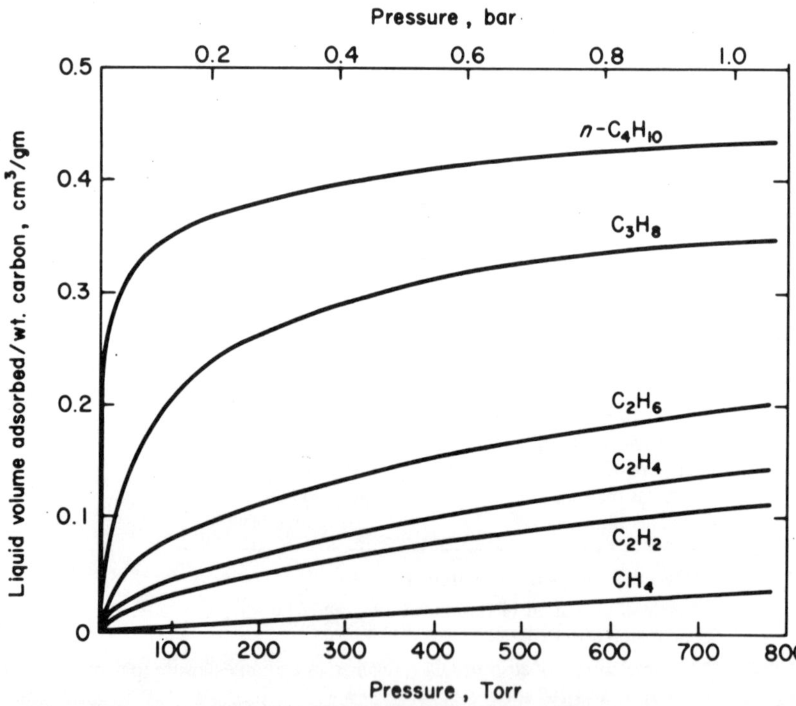

FIGURE 6-14 Adsorption isotherms of hydrocarbon vapors at 100°F (38°C) on air-purification activated carbon. Liquid volumes measured at boiling points of the hydrocarbons. Reprinted with permission from Turk.[82]

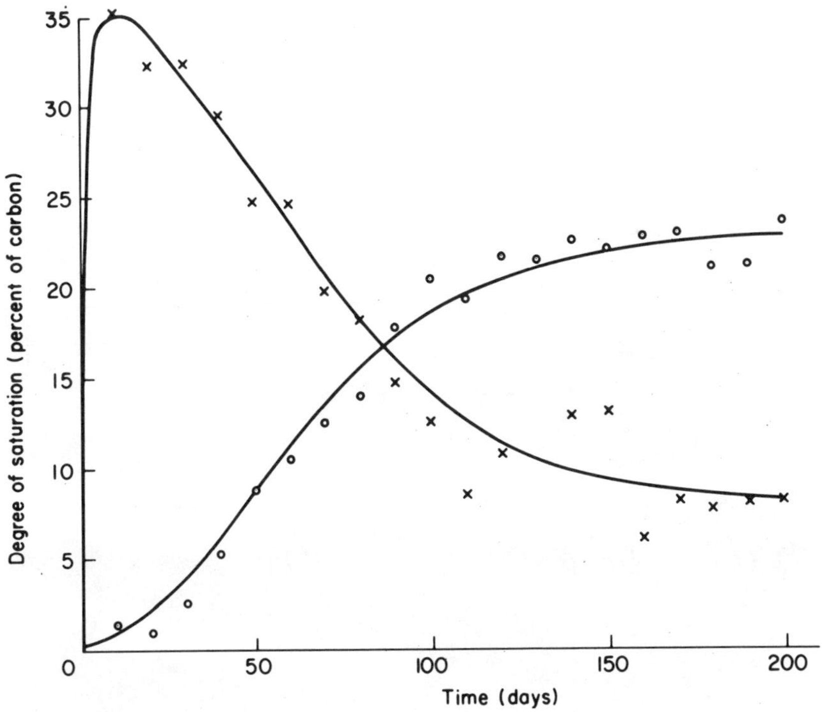

FIGURE 6-15 Saturation of coconut-shell activated carbon in commercial apple storage (Entiat, Washington). X, adsorbed water. O, adsorbed organic vapors. Reprinted with permission from Turk and Van Doren.[85]

humidity of the gas streaming through it, is likely to be variable. If it is mechanically feasible, a representative element or portion of the adsorbent bed may be removed and chemically analyzed to determine the degree of saturation of the entire bed.[83] In many cases, a schedule for renewal of adsorbent is determined by actual deterioration of performance (odor breakthrough); or it may be based on a schedule calculated from previous performance history.

Thin-bed adsorbers, which are used for light odorant loads, are expected to have service lives of several months, and are normally replaced when they are exhausted. For thick-bed adsorbers and heavy contaminant loads, it is generally economical to regenerate the adsorbent by on-site stripping with superheated steam. The adsorbate is thereby also removed and may be recovered if it is valuable. When the adsorbate is not worth recovering, either because its intrinsic value is low or because the recovery procedure

FIGURE 6-16 Aggregated flat-cell thin-bed adsorber. The small test element located on the upstream side of the cell contains carbon that is to be analyzed after some period of service for degree of saturation, to predict the remaining capacity of the cell. Courtesy of Connor Engineering and Manufacturing, Inc., Danbury, Conn.

is too difficult or expensive, it may nonetheless pay to regenerate the adsorbent at the site. The desorbed matter is then disposed of or destroyed. The desorbate may then be removed from the effluent stream by incineration or scrubbing. In effect, the adsorber serves as a vapor-concentrating medium. For example, benzene at a concentration of 150 ppm can be effectively stripped from air by a carbon bed and returned to a regenerating airstream at concentrations up to about 3%, or 30,000 ppm.[37] This represents a 200-fold magnification, which greatly reduces the cost of later treatment.

The oxidation of the adsorbate by air may also occur on the adsorbent surface, preferably in the presence of a catalyst. It has been shown[67-69] that various oxide and noble-metal catalysts are effective for such applications, that hydrocarbons and oxygenates can be completely oxidized before the carbon bed itself starts to oxidize, and that repeated cycles of adsorption and catalytic oxidation can be carried out without impairing the function of the carbon.

There are several possible adsorption modes for odor control other than physical adsorption by activated carbon. These alternatives are all more selective in their action and are therefore suitable only for special cases. They are summarized in Table 6-13.

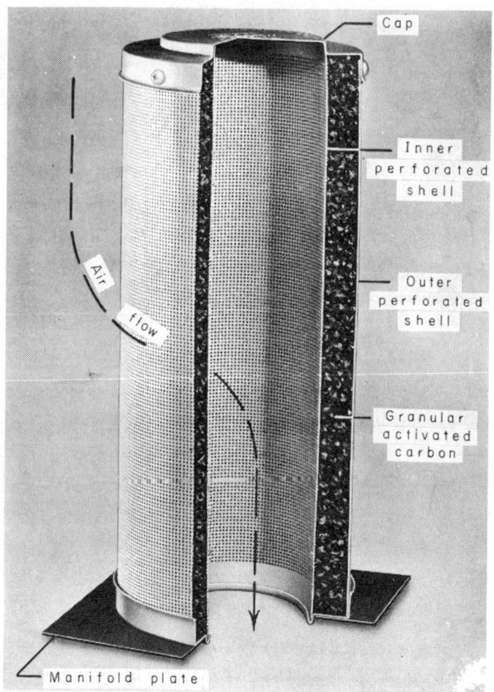

FIGURE 6-17 Cylindrical thin-bed canister adsorber. Courtesy of Connor Engineering and Manufacturing, Inc., Danbury, Conn.

FIGURE 6-18 Pleated-cell thin-bed adsorber. Courtesy of Barnebey-Cheney Co., Columbus, Ohio.

COSTS

The costs of adsorption systems vary with the difficulty of fitting existing equipment, the nature of contaminants in the waste gas, the concentrations of organics in the gas, the temperature at which gases enter the adsorber, the adsorbent, the regeneration technique, the value of recovered solvent, the type of adsorber, and the gas volume flow rate. Tables 6-14 and 6-15 summarize the assumptions on which the following cost estimates are based.

Adsorption capital costs include costs of the basic equipment, costs of auxiliary equipment, costs of equipment installation, and interest charges on investment during construction. The capital costs for a dual fixed-bed adsorber with recovery of desorbed vapors are shown in Figure 6-20. Costs for moving- and fluidized-bed adsorbers are slightly lower than those for fixed-bed adsorbers. The cost advantages arise from a reduction in the size requirement for moving- and fluidized-bed adsorbers. For this cost analysis, the pollutant vapors are considered to be easily adsorbed and desorbed from activated carbon. In addition, the adsorbed organics are considered to be capable of being stripped by steam and to be insoluble in water. If the organic vapors were soluble in water, additional equipment would be required to separate the steam from the desorbed vapors. The organics could be separated from the steam by distillation followed by water treatment, or the steam-vapor mixture could be incinerated. The first alternative increases capital requirements. If the vapor is not combustible, hot air can be used as the regenerant. Hot-air regeneration has no significant effect on capital costs. Costs for a regenerative adsorption system with incineration of the desorbed vapors are shown in Figure 6-21. In general, capital costs for an adsorption-incineration system are lower than those for adsorption recovery systems, except in the treatment of high-flow-rate gas streams.

The capital costs reported in Figures 6-20 and 6-21 are representative of

FIGURE 6-19 Thick-bed adsorbers used in a solvent recovery system. Courtesy of Union Carbide Corp., New York.

TABLE 6-13 Adsorbent Impregnations[a]

Adsorbent	Impregnant	Pollutant	Action
Activated carbon	Bromine	Ethylene; other alkenes	Conversion to dibromide, which remains on carbon
	Lead acetate	H_2S	Conversion to PbS
	Phosphoric acid	NH_3; amines	Neutralization
	Sodium silicate	HF	Conversion to fluorosilicates
	Iodine	Mercury	Conversion to HgI_2
	Sulfur	Mercury	Conversion to HgS
	Sodium sulfite	Formaldehyde	Conversion to addition product
	Sodium carbonate or bicarbonate	Acidic vapors	Neutralization
	Oxides of Cu, Cr, V, etc.; noble metals (Pd, Pt)	Oxidizable gases, including reduced sulfur compounds, such as H_2S, COS, and mercaptans	Catalysis of air oxidation
Activated alumina	Potassium permanganate	Easily oxidizable gases, especially formaldehyde	Oxidation
	Sodium carbonate or bicarbonate	Acidic gases	Neutralization

[a] Reprinted with permission from Turk.[82]

TABLE 6-14 Technical Assumptions Used in Developing Cost Estimates for Regenerative Carbon Adsorption Systems with Recovery of Desorbed Vapors[a]

1. Organic vapors of 50 wt% benzene, 50 wt% hexane in air
2. Activated carbon with useful life of 5 yr
3. Dual fixed-bed adsorber operating at 40°C (104°F)
4. Inlet gas at 77°C (171°F)
5. No water-soluble compounds in the airstream
6. No particles in the entering airstream
7. Steam regeneration and solvent recovery by condenser/decanter
8. Adsorber operating 5,840 h/yr

[a] Derived from Radian Corporation.[76]

TABLE 6-15 Technical Assumptions Used in Developing Cost Estimates for Regenerative Carbon Adsorption Systems with Incineration of Desorbed Vapors[a]

1. Organic vapors of 50 wt% benzene, 50 wt% hexane in air
2. Activated carbon with useful life of 5 yr
3. Dual fixed-bed adsorber operating at 40°C (104°F)
4. Inlet gas at 190°C (374°F)
5. No particles in the entering airstream
6. Steam regeneration and incineration of desorbed vapors
7. Adsorber operating 5,840 h/yr

[a] Derived from Radian Corporation.[76]

"typical" cases in which existing installations are fitted with adsorption equipment. Costs for new installations are 50–70% of the costs shown in the figures. Difficult fitting of existing installations may cost 2–2.5 times the reported values.

As expected, the higher the gas flow rate and the higher the concentration of organic vapors, the greater the required capital investment.

Annualized costs include labor and maintenance costs, utility and material costs, capital-related charges, and credits for chemical (solvent) recovery. The annualized adsorption costs presented in Figure 6-22 include recovery credits at fuel value ($1.65/10^6 Btu) and at market value (benzene at $0.90/gal, hexane at $0.50/gal). Other economic assumptions are summarized in the footnotes in Table 6-16. When recovered organics are credited at their market value, the adsorption operation shows a substantial profit. Reuse of the recovered organics, however, is not always economically desirable when more than one solvent is recovered. Product separation may be too costly to warrant the organic compounds' reuse in the process.

If it is not economically desirable to recover the organic vapors, the desorbed vapors can be incinerated. Annualized costs for an adsorption-incineration system are presented in Figure 6-23. Annualized costs for the adsorption-incineration system are comparable with those for the adsorption-recovery system only when no credit is allowed for recovery of organics. Allowances for chemical recovery at fuel or market values give a significant advantage to the recovery system.

Annualized-cost components for two carbon adsorption systems are reported in Table 6-16. About 75% of the utility costs for the organic recovery system is related to cooling-water and steam requirements. Steam

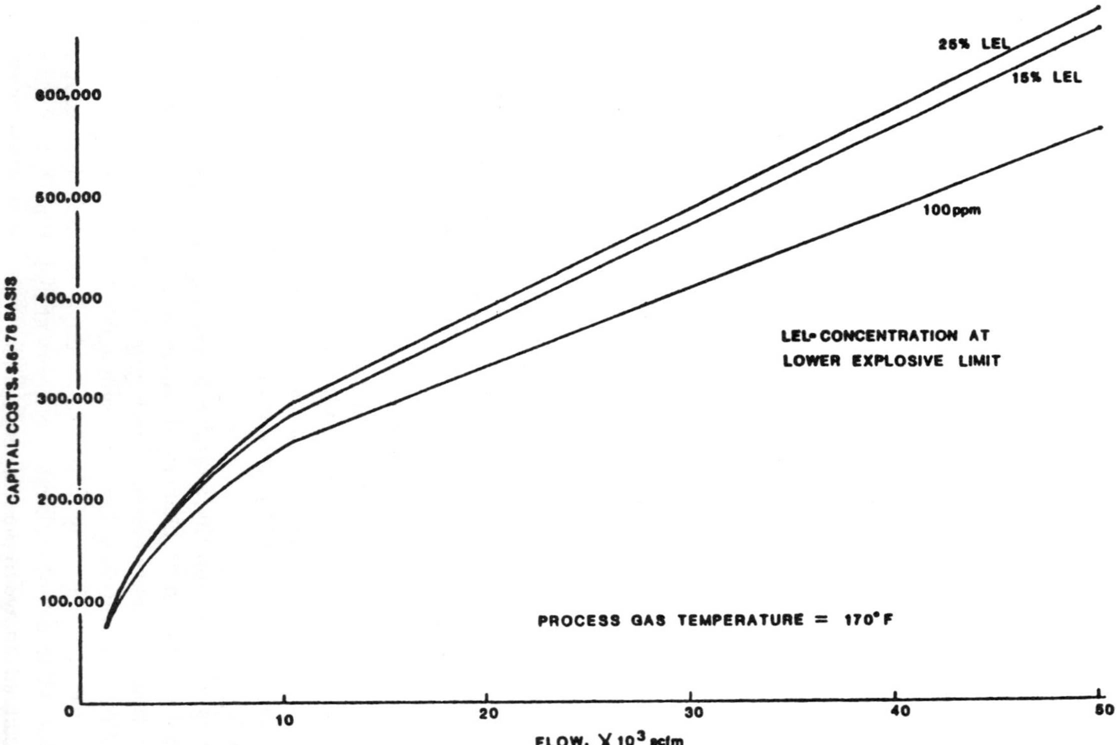

FIGURE 6-20 Capital costs for adsorption with dual fixed-bed adsorber with recovery of desorbed vapors. Reprinted from Radian Corporation.[76]

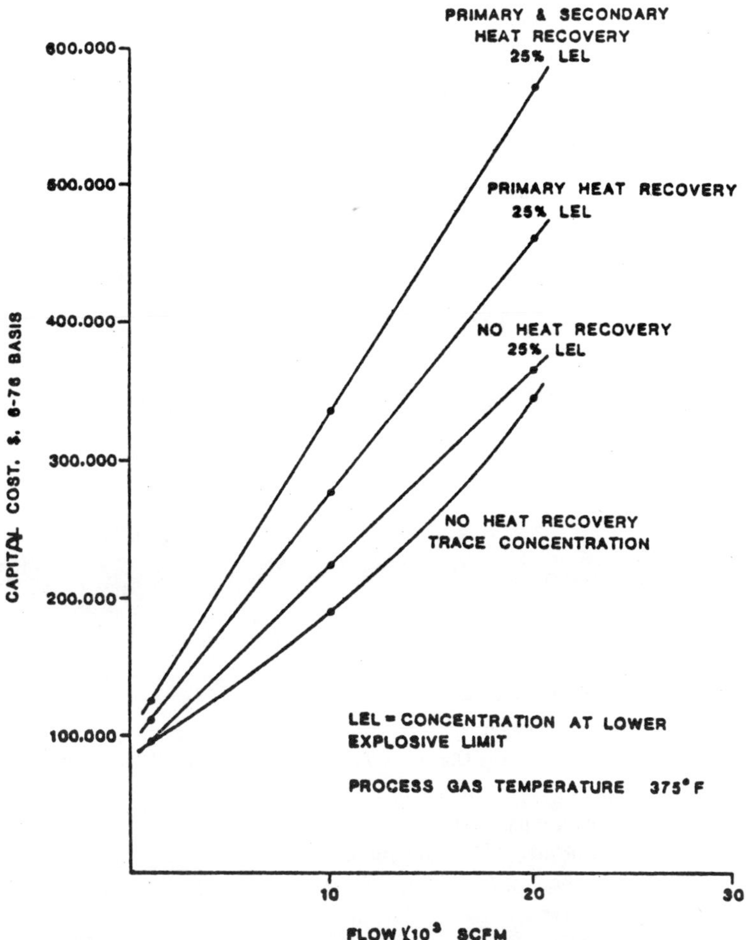

FIGURE 6-21 Capital costs for adsorption with incineration. Reprinted from Radian Corporation.[76]

requirements are significantly larger for organics that are difficult to desorb from the adsorbent. Cooling-water requirements depend on the temperature of the incoming gas stream and on the condensation temperature of the organic vapors. Power costs for moving- and fluidized-bed adsorbers are smaller than those for fixed-bed adsorbers. Total utility costs amount to about one-third of all cost elements (direct costs and capital charges) in a carbon adsorption-recovery system. Utility costs

TABLE 6-16 Components of Annualized Costs (Savings) for Adsorption and Adsorption-Incineration Systems[a]

Configuration	1. Dual fixed-bed adsorber operating at 100°F (38°C) 2. Solvent recovery with condenser and decanter	1. Dual fixed-bed adsorber operating at 100°F (38°C) 2. Thermal incineration with primary heat recovery
Gas-stream characteristics		
Flow	20,000 scfm	20,000 scfm
Concentration	25% LEL	25% LEL
Process-gas temperature	170°F (77°C)	375°F (191°C)
Direct operating costs		
Utilities	$48,700[b]	$70,200[b]
Direct labor	3,000[c]	3,000[c]
Maintenance	15,400[d]	18,400[d]
Carbon replacement	11,500[e]	11,500[e]
Capital charges	80,850[f]	96,500[f]
Recovery (credits)	(333,400)[g]	—
Total net annualized costs (credits)	(173,950)[h]	199,600

[a] Derived from Radian Corporation.[76]

[b] Cooling water at $0.045/1,000 gal, steam at $2/1,000 lb, electricity at $0.033/kWh.

[c] Labor at $8.25/h.

[d] Maintenance as 4% of the capital cost.

[e] Carbon at $0.72/lb, with 20% of carbon replenished each year.

[f] Capital charges included as percent of capital cost: depreciation, 12%; taxes, insurance, and overhead, 4%; interest, 5%.

[g] Benzene credited at $0.90/gal, hexane at $0.50 gal.

[h] Net costs calculated as capital charges + direct operating costs — recovery credits.

amount to about one-third of the costs for an adsorption-incineration system.

Adsorbent-replacement costs account for only 7% of all cost items in the carbon adsorption systems of Table 6-16. Other adsorbents may have significantly higher costs than carbon, but the total impact of adsorbent-replacement cost is minor.

Capital charges amount to over half the total operating and capital-related costs. As stated above, adsorption capital costs are representative of "typical" cases in which existing installations are fitted with adsorption

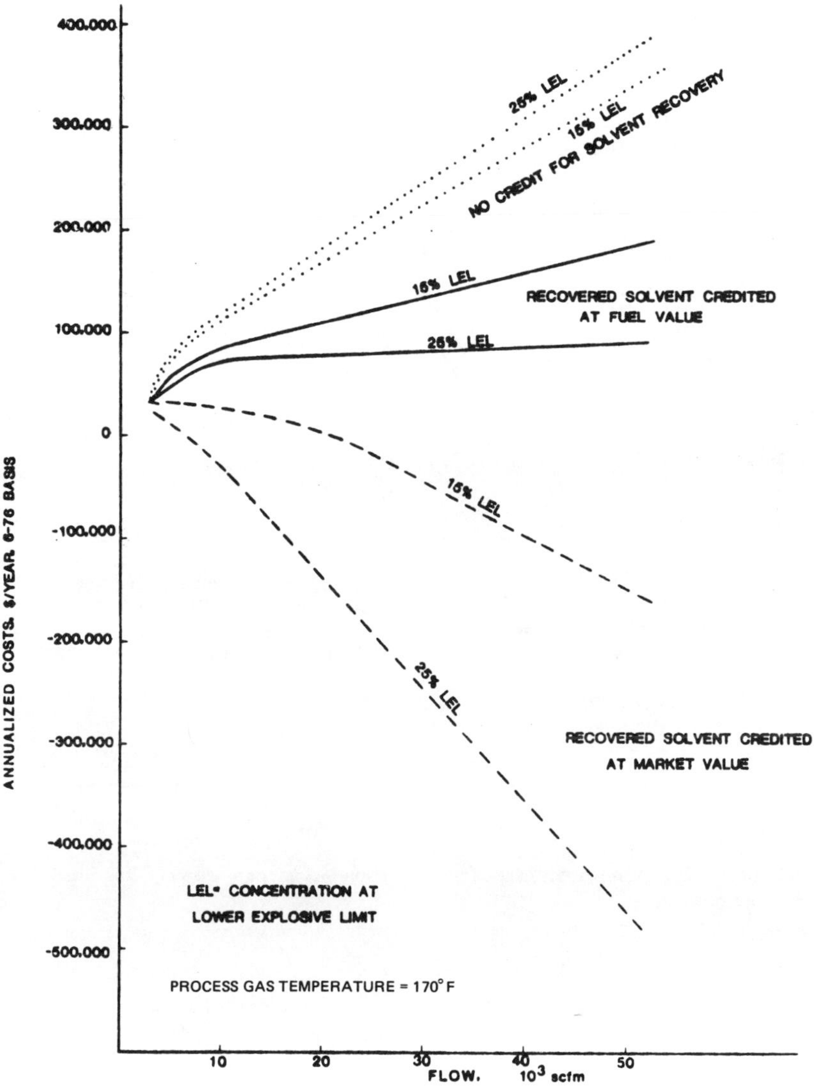

FIGURE 6-22 Annualized costs for adsorption system. Reprinted from Radian Corporation.[76]

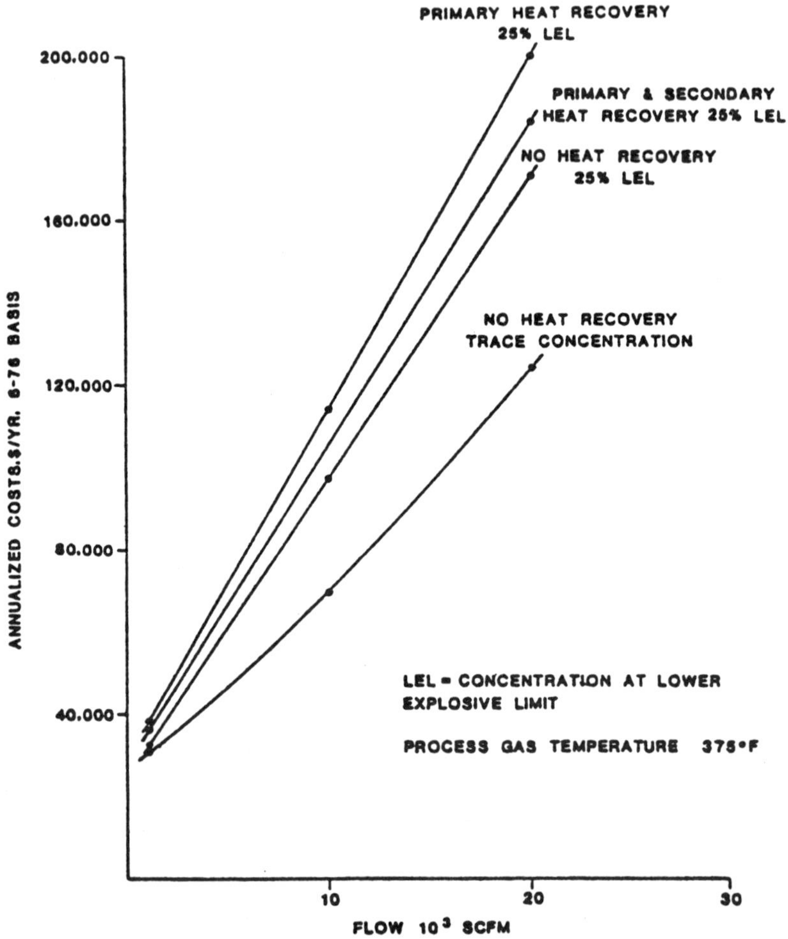

FIGURE 6-23 Annualized costs for adsorption-incineration system. Reprinted from Radian Corporation.[76]

equipment. Capital costs (and charges) for new installations are 50–70% of the costs reported here. Difficult fitting may cost 2–2.5 times the values reported.

The total net annualized costs are most significantly affected by values credited for recovered organics. When the recovered chemicals cannot be reused in the process, they may be credited with a significant fuel value (as shown in Figure 6-22). The main advantage accruing to adsorption-recovery systems is the credit associated with recovered chemicals.

SUMMARY

In general, activated carbon adsorption is the method of choice for deodorizing at ambient temperature an odorous airstream whose vapor concentrations are low (in the parts-per-million range or lower). At higher temperatures and concentrations, other methods become progressively more attractive, and the choice of activated carbon usually must be justified by some additional benefit, such as recovery of a valuable solvent. When a less efficient but cheaper method can remove the bulk of contaminant organic matter from an airstream, an activated-carbon adsorbent may be used as a final stage to advance the cleanup to a condition of complete deodorization. Various selective adsorption systems are suitable for special cases.

RECOMMENDATIONS FOR FUTURE RESEARCH

The further improvement of activated carbons for physical adsorption and the search for alternatives to activated carbon probably offer only limited opportunity for progress. It is possible to manufacture activated carbon with a considerably higher adsorptive capacity than that now commercially available, but such carbon is softer and less able to survive the normal stresses of handling. Substitutes for activated carbon would have to be nonpolar solids that, like carbon, could function in moist airstreams. Although this objective can be realized with various synthetic polymers, all such materials available now are much more expensive than carbon. In any event, regardless of cost, there seems to be little prospect of improving the physical capacity of an adsorbent bed, operating under practical conditions, by more than a factor of perhaps 1.5 or 2 over the capacity of existing carbon systems.

One approach that offers greater promise—at least it operates under a considerably higher theoretical ceiling—is a further improvement in methods of catalytic reactivation of saturated carbon, along the lines suggested by Nwanko and Turk.[69] If such systems could be developed to a state of reliable operation on a commercial scale, the range of conditions in which activated-carbon systems are economically attractive could be considerably expanded.

ODOR MODIFICATION BY ADDITIVES

When a mixture of odorants is smelled, the odor qualities of the components may be perceived separately or blend into one quality so that the individual components cannot be recognized. The odor intensity of the mixture is generally less than the sum of the odor intensities of the components. Likewise, the odor of any component of such a mixture is usually less intense than the odor of that component in its pure state.

Interaction effects on odor intensity have been studied for some two-component mixtures.[21] For example, the perceived intensity of vapor-phase mixtures of various concentrations of pyridine and a second component—such as linalyl acetate, linalool, or lavandin oil—is less than the sum of the perceived intensities of the two components smelled alone. The addition of the second component to a relatively weak stimulus of pyridine causes an increase in overall odor intensity, but the addition of the same amount of the second component to a relatively intense stimulus of pyridine causes a reduction in overall odor intensity. Mixtures of 1-propanol and *n*-amylbutyrate have been reported to interact similarly.[20] These data suggest the existence of complex interactions in the perceived intensity of odorous mixtures.

A vector-summation model of odor interaction has been suggested as a means of predicting the odor intensity of mixtures of malodorants, such as dimethyl disulfide, dimethyl monosulfide, hydrogen sulfide, methyl mercaptan, and pyridine.[11-13] For components equal in perceived intensity when smelled alone, a direct proportionality has been reported between odor intensity of the mixtures and the arithmetic sum of the odor intensities of the components.

The interpretation of the application of these phenomena to practical odor-control objectives presents difficulties, and the common industrial terminology does not make matters easier. "Counteraction" has been used to connote reduction of intensity, although it is not always clear whether it is the odor of the blend or of the malodorant alone that is reduced. "Cancellation" means reduction to zero intensity—a phenomenon that has never been convincingly documented. "Masking" refers to a change in odor quality that makes the malodorant unrecognizable; the connotation of concealment has made the term unpopular. "Odor modification" is a more noncommittal expression. Finally, "reodorization" is sometimes used, but is generally applied to the treatment of products (for example, making used cars smell like new), rather than to airstreams.

In spite of this variety of terminology, the odor-control practices to which the words refer are operationally indistinguishable. The materials used are selected from industrially available high-intensity odorants, often from byproduct sources. They may be applied in undiluted form or as aqueous emulsions. They may be incorporated into the process or product that constitutes the source of malodors, sprayed into a stack or over a stack exit, or vaporized over a large outdoor area. Pumps, metering devices, spray nozzles, compressors, evaporators, and assorted ducts and piping constitute the types of equipment usually needed.

The general method has the important practical advantages of low initial equipment costs, negligible space requirements, and greater freedom from the necessity of confining the atmosphere into a closed space for treatment. Examples of unconfined odor sources that are sometimes treated by these methods are exposed sites for disposal of undigested sewage sludge, drainage ditches contaminated by odorous spills, trickling filters in sewage-treatment plants, and holding ponds for industrial liquid wastes.

Clearly, it is very difficult to estimate the effectiveness of this category of odor-control methods.[40] Not the least of the problems is the choice of criteria for evaluation. Furthermore, industrial and commercial odor-control installations are not designed for controlled experiments. Instead, they are generally combined with other beneficial actions, such as improvements in sanitation and general housekeeping, to maximize the opportunities for odor reduction. As a result, information concerning the performance of such systems consists entirely of descriptions of actual operations, anecdotal reports, and undocumented claims.[43,51,53,54,64,70,72]

Lauren and Young[54] estimated capital investment for vaporizing an odor modifier around the perimeter of a malodor source at $1.00 per linear foot of the perimeter, and operating costs at $0.07–0.25/h per 100 linear feet. Anonymous reports from a rendering plant that used such a system cited operating costs, in 1973, of $0.40/h per 100 linear feet.

CONCLUSIONS

Odor modifiers are not air cleaners. Instead, they constitute an additional discharge of vapors into the atmosphere.

Odor modifiers change the perceived character or intensity of odors. They should never be allowed to interfere with the perception of the odors of toxic gases or of gases used as warning agents (such as the odorants in natural gas).

There is no documented evidence that odor modifiers eliminate community malodors in practical situations. Therefore, it cannot be assumed that a "modified" odor, even if originally acceptable in a community, will continue to be acceptable with the passage of time. It may be necessary to change the odor character of the modifier, or to abandon it in favor of an air-cleaning alternative.

RECOMMENDATIONS

Because the compositions of odor modifiers are proprietary, investigations that use such materials cannot meet the criteria of independently verifiable research. Therefore, we make no recommendations for further study.

REFERENCES

1. Air Pollution Control Association, TT-4 Odor Committee. Dynamic Olfactomer Evaluation Round-Robin Test—Preliminary Results. 1977. (unpublished)

2. American Industrial Hygiene Association. Air Pollution Manual Part II, pp. 112–124. Detroit: AIHA, 1968.

3. American Society for Testing and Materials. ASTM D 1391-57, Measurement of Odor in Atmospheres (Dilution Method). Philadelphia: ASTM, 1957.

4. Azuma, K., K. Doi, A. Ikeda, S. Suzuki, S. Yamauchi, and H. Takahara. Exhaust gas deodorization with chemical scrubbing and ozonation, pp. 743–746. In Proceedings, 4th International Clean Air Congress. Tokyo: Japanese Union of Air Pollution Prevention Associations, 1977.

5. Benforado, D. Control by incineration, pp. 99–109. In R. A. Fox, Ed. Proceedings—MECAR Symposium. New Developments in Air Pollution Control. New York: Metropolitan Engineers Council on Air Resources, 1967.

6. Benforado, D. M., and G. Cooper. The Application of Direct Flame Incineration as an Odor Control Process in Kraft Pulp Mills. Paper No. 2-4, presented at TAPPI 22nd Engineering Conference, Atlanta, September 19, 1967. 19 pp.

7. Benforado, D. M., C. E. Pauletta, and N. D. Hazzard. Economics of heat recovery in direct-flame fume incineration. Air Eng. 9(3):29–32, 1967.

8. Benforado, D. M., W. J. Rotella, and D. L. Horton. Development of an odor panel for evaluation of odor control equipment. J. Air Poll. Cont. Assoc. 19:101–105, 1969.

9. Benforado, D. M., and J. Waitkus. Exploring the applicability of direct-flame incineration to wire enameling fume control. Wire Prod. 42:1971–1978, 1967.

10. Benforado, D. M., and J. Waitkus. Fume control in wire enameling by direct-flame incineration. J. Air Poll. Cont. Assoc. 18:24–26, 1968.

11. Berglund, B. Quantitative and qualitative analysis of industrial odors with human observers. Ann. N.Y. Acad. Sci. 237:35–51, 1974.

12. Berglund, B., U. Berglund, and T. Lindvall. On the principle of odor interaction. Acta Psychol. 35:255–268, 1971.

13. Berglund, B., U. Berglund, T. Lindvall, and L. T. Svensson. A quantitative principle of perceived intensity summation in odor mixtures. J. Exp. Psychol. 100:29–38, 1973.

14. Bethune, W. J., and L. J. Foord. Fume abaters clean exhaust from wire enamel curing ovens. Wire Prod. 44(7):50–51, 94, 1969.

15. Blosser, R. O., and H. B. H. Cooper. Current Practices in Thermal Oxidation of Noncondensable Gases in the Kraft Industry. Atmospheric Quality Improvement Technical Bulletin No. 34. New York: National Council for Air and Stream Improvement, October 1967. 24 pp.

16. Bolduc, M. J., and R. K. Severs. A Modified Total Combustion Analyzer for Use in Air Pollution Source Testing, VII. Presented at 149th Meeting of the American Chemical Society, Detroit, 1965. (abstract)

17. Brewer, G. L. Fume incineration. Chem. Eng. 75(22):160–165, 1968.

18. Briggs, G. A. A plume rise model compared with observation. J. Air Poll. Cont. Assoc. 15:433–438, 1965.

19. Byrd, J. F., H. A. Mills, C. H. Schellhase, and H. E. Stokes. Solving a major odor problem in a chemical process. J. Air Poll. Cont. Assoc. 14:509–516, 1964.

20. Cain, W. S. Odor intensity: Mixtures and masking. Chem. Sens. Flav. 1:339–352, 1975.

21. Cain, W. S., and M. Drexler. Scope and evaluation of odor counteraction and masking. Ann. N.Y. Acad. Sci. 237:427–439, 1974.

22. Calvert, S., J. Goldshmid, D. Leith, and D. Mehta. Wet Scrubber System Study. Vol. 1, Scrubber Handbook. Publ. No. EPA-R2-72-118a. Cincinnati: U.S. Environmental Protection Agency, 1972. 828 pp.

23. Catalytic system removes odors from acrylate monomers. Chem. Proc. 32(9):77–78, 1969.

24. C-E Air Preheaters, Wellsville, N.Y. Report of Fuel Requirements, Capital Cost and Operating Expense for Catalytic After-Burners. Publ. No. EPA-450/3-76-031. Contract No. 68-02-1473. Research Triangle Park, N.C.: U.S. Environmental Protection Agency, Office of Air Quality Planning and Standards, 1976. 242 pp.

25. Cha, S. S. Odor control from paint and coating operations through process simulation and dispersion modeling, pp. 89–99. In Proceedings of the State-of-the-Art of Odor Control Technology II—Specialty Conference. Pittsburgh: Air Pollution Control Association, 1977.

26. Cheremisinoff, P. Atmospheric transport, dispersion, and stack design considerations, pp. 443–456. In P. Cheremisinoff and R. Young. Industrial Odor Technology Assessment. Ann Arbor: Ann Arbor Science Publishers, Inc., 1975.

27. Christian, J. G., and J. E. Johnson. Catalytic Combustion of Nuclear Submarine Atmospheric Contaminants. NRL Report No. 6040. AD-437-177. Washington, D.C.: U.S. Naval Research Laboratory, Mar. 20, 1964. 15 pp.

28. Crouse, L. F., and D. E. Waid. Direct-flame industrial-fume incineration. Ind. Gas 49:7–13, 1969.

29. Crouse, L. F., and D. E. Waid. Efficient design of afterburners for incineration of many industrial fumes. Air Eng. 9(8):20–29, 1967.

30. Crouse, L. F., and D. E. Waid. Incineration of Industrial Fumes by Direct Gas Flame. Paper Presented at the American Gas Association Industrial/Commercial Sales Conference, Chicago, April 26, 1967. 8 pp.

31. Danielson, J. A., Ed. Air Pollution Engineering Manual. Los Angeles County Air Pollution Control District. Public Health Service Publication No. 999-AP-40. Cincinnati: U.S. Department of Health, Education, and Welfare, Public Health Service, National Center for Air Pollution Control, 1967. 919 pp.

32. Decker, L. D. Odor Control by Incineration. Paper No. 5-048. Greenwich: Universal Oil Products, Air Correction Division, 1966. 15 pp.

33. Dickerson, R. C., and B. N. Murthy. Scope of wet scrubbers for odor control. In W. S. Cain, Ed. Odors: Evaluation, Utilization, and Control. Ann. N.Y. Acad. Sci. 237:374–388, 1974.

34. Dravnieks, A. Development of a Dynamic Forced-Choice Method for the Measurement of Emission and Ambient Odors. IIT Research Institute Project No. C8217 for Fats and Proteins Research Foundation, Inc. Des Plaines, Ill.: Fats and Proteins Research Foundation, Inc., 1976. 96 pp.

35. Dravnieks, A. Instructions for Dynamic Triangle Olfactometer. A Manual of the Illinois Institute of Technology Research Institute. Chicago: IIT Research Institute, 1977. 18 pp.

36. Dravnieks, A., and W. H. Prokop. Source emission odor measurement by a dynamic forced-choice triangle olfactometer. J. Air Poll. Cont. Assoc. 25:28–35, 1975.

37. Faulkner, W. D., W. G. Schuliger, and J. E. Urbanic. Odor Control Methods Using Activated Carbon. Paper 36C. Presented at the 74th National Meeting of the American Institute of Chemical Engineers, New York, 1973. 31 pp.

38. Fazzalari, F. A., Ed. Compilation of Odor and Taste Threshold Values Data. Sponsored by ASTM Committee E-18 on Sensory Evaluation of Materials and Products. ASTM Data Series 48A. Philadelphia: American Society for Testing and Materials, 1978. 497 pp.

39. First, M. W. Control of haze and odours from curing of plastics, pp. 188–191. In Proceedings of the International Clean Air Congress. London: National Society for Clean Air, 1967.

40. First, M. Effectiveness of odor counteractants for the treatment of stack gases. J. Air Poll. Cont. Assoc. 9:28–31, 1959.

41. First, M. W., W. Shilling, J. H. Govan, and A. H. Quinby. Control of odors and aerosols from spent grain dryers. J. Air Poll. Cont. Assoc. 24:653–659, 1974.

42. Gamble, B. L. Control of Organic Solvent Emissions in Industry. Paper No. 68-48. Presented at the 62nd Annual Meeting, Air Pollution Control Association, St. Paul, June 1968. 31 pp.

43. Goldman, A. The odor phase of air pollution. Ind. Wastes 1:217–220, 1956.

44. Harding, C. I., and J. E. Landry. Future trends in air pollution control in Kraft pulping industry. TAPPI 49(8):61A–67A, 1966.

45. Hardison, L. C., and L. R. Steenberg. Criteria for chemical odor control systems. In W. S. Cain, Ed. Odors: Evaluation, Utilization, and Control. Ann. N.Y. Acad. Sci. 237:363–373, 1974.

46. Heilman, W., and J. Welling. Odor Control in Rendering by Direct-Flame Incineration. Paper No. 69-43. Presented at the 62nd Annual Meeting, Air Pollution Control Association, New York, June 1969. 8 pp.

47. Hein, G. M. Odor control by catalytic and high-temperature oxidation. Ann. N.Y. Acad. Sci. 116:656–662, 1964.

48. Hirt Combustion Engineers. A Discussion of Organic Vapor Disposal by Direct-Flame Incineration. Bulletin No. S-209-2. Montebello, Calif.: Hirt Combustion Engineers. 4 pp.

49. Johnson, W. B., H. B. Singh, L. A. Cavanagh, and J. H. Smith. Stanford Research Institute Project No. 3400. AIF Publ. No. AIF/NESP-007a. Washington, D.C.: Atomic Industrial Forum, Inc., December 1975.

50. Kilby, J. L. Flare system explosions. Chem. Eng. Progr. 64(6):49–52, 1968.

51. Lamoreaux, C. Stack effluent "reodorized"; wind direction, velocity analyzed for control. Plant Eng. 9:96–97, July 1955.

52. Lauber, J. D. The Control of Solvent Vapor Emissions. Paper No. 69-42. Presented at the 62nd Annual Meeting, Air Pollution Control Association, New York, June 1969. 23 pp.

53. Lauren, O. B. Odor modification. Poll. Eng. 5:27–29, July 1973.

54. Lauren, O. B., and R. A. Young. Odor modification and masking, pp. 190–198. In Proceedings of the State-of-the-Art of Odor Control Technology II—Specialty Conference. Pittsburgh: Air Pollution Control Association, 1977.

55. Leonard, J., R. D. MacPhee, and R. J. Bryan. Analysis and Testing of Organic Solvents. Los Angeles: Los Angeles County Air Pollution Control District, Technical Services Division, November 1966.

56. Leonardos, G. A critical review of regulations for the control of odors. J. Air Poll. Cont. Assoc. 24:456–458, 1974.

57. Los Angeles County Air Pollution Control District. Control equipment for gases and vapors, pp. 171–232. In J. A. Danielson, Ed. Air Pollution Engineering Manual. Public Health Service Publication No. 999-AP-40. Cincinnati: U.S. Department of Health, Education, and Welfare, Public Health Service, National Center for Air Pollution Control, 1967. 919 pp.

58. Los Angeles County Air Pollution Control District. Recommended Test Methods for Organic Solvents and Vapors (Rule 66). Los Angeles: L.A. County APCD, April 1968. 16 pp.

59. Los Angeles County Air Pollution Control District. Technical Progress Report, Control of Stationary Sources, Vol. 1, Chapters 7, 8, and 10. Los Angeles: L.A. County APCD, April 1960.

60. Magill, P. L., F. R. Holden, and C. Ackley, Eds. Air Pollution Handbook, pp. 13–25 to 13–28. New York: McGraw-Hill Book Co., 1956.

61. Meyers, F. D., and J. Waitkus. Fume incineration with combustion air at elevated temperature. J. Air Poll. Cont. Assoc. 16:378–382, 1966.

62. Miller, M. R., and H. J. Wilhoyte. A study of catalyst support systems for fume-abatement of hydrocarbon solvents. J. Air Poll. Cont. Assoc. 17:791–795, 1967.

63. Mills, J. L., R. T. Walsh, and K. D. Luedtke. Quantitative odor measurement. J. Air Poll. Cont. Assoc. 13:467–475, 1963.

64. Moncrieff, R. W. How air fresheners work. Perf. Essent. Oil Rec. (Lond.) 46:189–196, 227–231, 1955.

65. Murray, D. R., S. S. Cha, and N. E. Bowne. Use of a Fluctuating Plume Puff Model for Prediction of the Impact of Odorous Emissions. Preprint No. 78-68.6. Paper presented at the 71st Annual Meeting, Air Pollution Control Association, Houston, Texas, June 1978.

66. Murthy, B. N., and J. S. Bowen. Research and development for improving odor control technology, pp. E108–E111. In Proceedings of the Third International Clean Air Congress. Düsseldorf: Verein Deutscher Ingenieure, 1973.

67. Nwankwo, J., and A. Turk. Catalytic oxidation of adsorbed hydrocarbons. Ann. N.Y. Acad. Sci. 237:397–408, 1974.

68. Nwankwo, J. N., and A. Turk. Catalytic oxidation of vapors adsorbed on activated carbon. Environ. Sci. Technol. 9:846–849, 1975.

69. Nwanko, J., and A. Turk. Rapid evaluation of oxidation catalysts supported on activated carbon. Carbon 13:495–499, 1975.

70. Olson, L. E. Reodorization for Ventura. Mayor Manag. Mag. 3(5):20–21, 1960.

71. Olszyna, K. J., and J. Heicklen. The reaction of ozone with carbon disulfide. J. Phys. Chem. 74:4188–4198, 1970.

72. Pantaleoni, R. Odors and fragrances. Chem. Eng. News 31:1730–1734, 1953.

73. Parsons, W. Thermal oxidation system actually saves energy for coil coater. Poll. Eng. 9(5):46–48, 1977.

74. Perry, R. H., C.H. Chilton, and S. D. Kirkpatrick, Eds. Perry's Chemical Engineers Handbook, 5th ed. New York: McGraw-Hill Book Co., 1973.

75. Prass, H. J., and J. O. Ross. Waste gases can be cleaned at low cost. Power 107(4):74–77, 1963.

76. Radian Corporation. Control Techniques for Volatile Organic Emissions from Stationary Sources. Final Report to the U.S. Environmental Protection Agency. EPA Contract No. 68-02-2608, Task 12. Austin, Tex.: Radian Corporation, February 1978.

77. Reed, R. D. Design and operation of flare systems. Chem. Eng. Progr. 64(6):53–57, 1968.

78. Sandomirsky, A. G., D. M. Benforado, L. D. Grames, and C. E. Pauletta. Fume control in rubber processing by direct-flame incineration. J. Air Poll. Cont. Assoc. 16:673–676, 1966.

79. Sherwood, T. K., and R. L. Pigford. Adsorption and Extraction, 2nd ed. New York: McGraw-Hill Book Co., 1952. 478 pp.

80. Smith, H., Ed. Recommended Guide for the Prediction of the Dispersion of Airborne Effluents. New York: American Society of Mechanical Engineers, 1968. 85 pp.

81. Sullivan, J. L., F. L. Kafka, and L. M. Ferrari. An evaluation of catalytic and direct fired afterburners for coffee and chicory roasting odors. J. Air Poll. Cont. Assoc. 15:583–586, 1965.

82. Turk, A. Adsorption, pp. 329–363. In A. C. Stern, Ed. Air Pollution, 3rd ed. Vol. IV, Engineering Control of Air Pollution. New York: Academic Press, 1977.

83. Turk, A., H. Mark, and S. Mehlman. Tracer gas nondestructive testing of activated carbon cells. Mater. Res. Stand. 9(11):24–26, 1969.

84. Turk, A., L. R. Reckner, and G. D. Robinson. The Mechanism of Odor Transport. Paper No. 69-123. Presented at the 62nd Annual Meeting, Air Pollution Control Association, New York, June 1969.

85. Turk, A., and A. Van Doren. Saturation of activated carbon used for air purification. J. Agr. Food Chem. 1:145–151, 1953.

86. U.S. Environmental Protection Agency. Air Pollution Aspects of Odors. 2 vols. February 1971. (unpublished)

87. U.S. National Air Pollution Control Administration. Control Techniques for Hydrocarbon and Organic Solvent Emissions from Stationary Sources, pp. xx,1-1–7-3,I-1–I-3. Publication No. AP 68. Washington, D.C.: U. S. Department of Health, Education, and Welfare, Public Health Service, Environmental Health Service, National Air Pollution Control Administration, March 1970.

88. Waid, D. E. Direct Gas Flame Oxidation of Hydrocarbons for Smoke, Gaseous Hydrocarbons and Odor Control by Thermal Incineration. Paper No. 69-45. Presented at the 62nd Annual Meeting, Air Pollution Control Association, New York, June 1969. 8 pp.

89. Waid, D. E. Incineration of Organic Materials by Direct Gas Flame for Air Pollution Control. Paper No. 118. Presented at American Industrial Hygiene Conference, St. Louis, May 13–17, 1968. 6 pp.

90. Wallach, A. Some data and observations on combustion of gaseous effluents from baked lithograph coatings. J. Air Poll. Cont. Assoc. 12:109–110, 1962.

91. Walther, J. E., and H. R. Amberg. A positive air quality control program at a new Kraft mill. J. Air Poll. Cont. Assoc. 20:9–18, 1970.

92. Walther, J. E., and H. R. Amberg. Odor control in the Kraft pulp industry. Chem. Eng. Progr. 66(3):73–80, 1970.

93. Yocom, J. E., and R. A. Duffee. Controlling industrial odors. Chem. Eng. 77(13):160–168, 1970.

7 Control of Agricultural Odors

This chapter discusses the application of odor-control technology to various agricultural enterprises. These examples were chosen because they are representative of the agricultural applications of this technology, and their inclusion does not imply any ranking as sources of odor pollution.

LIVESTOCK AND POULTRY PRODUCTION

Livestock and poultry production is a major agricultural activity in the United States. It takes place in all 50 states and includes dairy farms; swine-, beef-, chicken-, turkey-, and sheep-feeding operations; and egg production. The vast majority of these enterprises are sufficiently small and dispersed that there has been little effort expended or needed to control their odors.

Financial pressures and technologic development since 1950 have prompted the concentration of livestock and poultry production into an ever-decreasing number of larger operations that can benefit from mechanization and labor specialization. This concentration generates large quantities of manure in relatively small areas and requires skilled management to avoid degradation of water and air quality. The scope, magnitude, and distribution of the major livestock and poultry production enterprises in the United States are indicated in Tables 7-1 through 7-5.

Livestock and poultry production generates about 2 billion tons (about 1.8×10^9 metric tons, or t) of manure a year. Manure production for the various species is summarized in Table 7-6. The collection, storage,

TABLE 7-1 Regional Distribution of Farm Animals in the United States[a]

Region	Distribution, %				
	Hogs	All Cattle	Dairy Cattle	Chickens, Including Broilers	Broilers
North Atlantic[b]	2	5	17	14	6
East North Central[c]	30	13	26	13	2
West North Central[d]	48	32	23	15	2
South Atlantic[e]	8	7	8	19	42
South Central[f]	10	24	15	21	44
Western[g]	2	19	11	18	4
TOTAL	100	100	100	100	100

[a] Data from U.S. Department of Agriculture.[46]

[b] Maine, Vermont, New Hampshire, New York, Massachusetts, Rhode Island, Connecticut, and Pennsylvania.

[c] Wisconsin, Illinois, Indiana, Michigan, and Ohio.

[d] North Dakota, South Dakota, Nebraska, Kansas, Minnesota, Iowa, and Missouri.

[e] West Virginia, Maryland, Delaware, New Jersey, Virginia, North Carolina, South Carolina, Georgia, and Florida.

[f] Oklahoma, Texas, Arkansas, Louisiana, Kentucky, Tennessee, Mississippi, and Alabama.

[g] Washington, Oregon, California, Nevada, Idaho, Montana, Wyoming, Utah, Colorado, Arizona, and New Mexico.

TABLE 7-2 Beef-Cattle Feedlots in the United States, 1977[a]

Size (Head Capacity)	No. Feedlots	No. Head Marketed, millions
<100	101,070	3.0
100-499	28,600	5.7
500-999	10,300	6.1
1,000-9,999	2,100	8.5
$\geq 10,000$	430	15.0
TOTAL	142,500	38.3

[a] Data from Development Planning and Research Associates.[11]

TABLE 7-3 Dairy Farms in the United States, 1977[a]

Size (No. Head)	No. Farms	Milk Sold, 10^9 lb
<30	115,800	18.9
30-49	60,800	23.5
50-99	55,700	39.1
100-199	10,300	15.6
200-699	3,000	11.2
≥700	410	3.4
TOTAL	246,010	111.7

[a] Data from Development Planning and Research Associates.[11]

TABLE 7-4 Hog-Feeding Operations in the United States, 1977[a]

Size (Head Capacity)	No. Operations	Hogs Marketed, millions
<200	252,000	17.6
200-999	111,700	57.2
1,000-2,499	4,700	12.2
≥2,500	1,600	9.2
TOTAL	370,000	96.2

[a] Data from Development Planning and Research Associates.[11]

TABLE 7-5 Poultry and Egg Production in the United States, 1977[a]

Item	Quantity
Eggs	5.35×10^9 dozen
Layers	272×10^6
Broiler production, live weight	12.85×10^9 lb
Turkey production, live weight	2.55×10^9 lb

[a] Data from U.S. Department of Agriculture.[47]

TABLE 7-6 Manure Production and Characteristics per 454 kg (1,000 lb) of Live Weight[a, b]

Item	Units	Dairy		Beef		Swine			Poultry		
		Cow	Heifer	Yearling 182-318 kg (400-700 lb)	Feeder >318 kg (>700 lb)	Feeder	Breeder	Sheep	Layer	Broiler	Horse
Raw Waste (RW)	kg/day	37.2	38.6	40.8	27.2	29.5	22.7	18.1	24.0	32.2	20.4
	lb/day	82.0	85.0	90.0	60.0	65.0	50.0	40.0	53.0	71.0	45.0
Feces/Urine Ratio		2.2	1.2	1.8	2.4	1.2		1.0			4.0
Density	kg/m^3	1,005.0	1,005.0	1,010.0	1,010.0	1,010.0	1,010.0		1,050.0	1,050.0	
	lb/cu ft	62.7	62.7	63.0	63.0	63.0	63.0		65.5	65.5	
Total Solids (TS)	kg/day	4.7	4.2	5.2	3.1	2.7	1.9	4.5	6.1	7.7	4.3
	lb/day	10.4	9.2	11.5	6.9	6.0	4.3	10.0	13.4	17.1	9.4
	% of RW	12.7	10.8	12.8	11.6	9.2	8.6	25.0	25.2	25.2	20.5
Volatile Solids	kg/day	3.8			2.7	2.2	1.4	3.8	4.3	5.4	3.4
	lb/day	8.6			5.9	4.8	3.2	8.5	9.4	12.0	7.5
	% of TS	82.5			85.0	80.0	75.0	85.0	70.0	70.0	80.0
BOD_5[c]	% of TS	16.5			23.0	33.0	30.0	9.0	27.0		
COD[d]	% of TS	88.1			95.0	95.0	90.0	118.0	90.0		
TKN[e]	% of TS	3.9	3.4	3.5	4.9	7.5		4.5	5.4	6.8	2.9
P[f]	% of TS	0.7	3.9		1.6	2.5		0.66	2.1	1.5	0.49
K[g]	% of TS	2.6			3.6	4.9		3.2	2.3	2.1	1.8

[a] Reprinted with permission from American Society of Agricultural Engineers.[1 (p. 465)]

[b] Numerical values for kg/day/1,000 kg live weight are the same as those for lb/day/1,000 lb live weight.

[c] Five-day biochemical oxygen demand.

[d] Chemical oxygen demand.

[e] Total Kjeldahl nitrogen.

[f] Phosphorus as P.

[g] Potassium as K.

transport, treatment, and disposal of manure account for major odor problems. Most manure is currently applied to cropland for economic recovery of plant nutrients. However, alternative uses, including refeeding and energy production, are of increasing importance. Less than 2% of manure produced in the United States is refed or used for energy production.

Odor control is a significant problem for livestock producers throughout the country. The problem consists most often of neighbors' complaints with occasional legal actions seeking monetary damages or court-imposed injunctions. To operate compatibly within the community, the livestock producer must be aware of some basic odor controls and practice the techniques appropriate to his location.

Livestock-production enterprises have led to odor complaints from nearby residents, commercial operations, and recreational interests. More recently, odor concern has been supplemented by allegations that ammonia and other water-soluble pollutants discharged into the air from livestock operations are being transported and later absorbed by nearby surface waters. Increased ammonia in the atmosphere near livestock-feeding operations has been documented by Hutchinson and Viets[22] and by Luebs, Laag, and Davis.[27]

Conflicts between livestock producers and the public concerning odor complaints have been documented.[50] State and local rules and regulations are now being designed to reduce the malodors generated by livestock production, and additional restrictions are being implemented on the location, design, and operation of commercial livestock- and poultry-production enterprises.[41]

Livestock-production odor problems involve a complex release mechanism, transport system, and receptor reaction. Research[29] has identified more than 40 compounds in the air near manure storage or treatment devices (Table 7-7). Many of these compounds are known to be odorous in trace concentrations and hazardous to human health at higher concentrations.

Most quantitative measurements of odorant concentrations suggest, however, that perceived odors from livestock production are results of mixtures of odorous compounds, inasmuch as all the measured compounds are present in concentrations below their thresholds.[31]

Release rates of specific odorants from the livestock industry have not been documented to the same extent as those from other sources. There are two reasons for that: the emission is diffuse and from a variety of physical situations, and there is no established, predominant odorant to serve as an indicator of odor release. Ammonia has been measured most frequently in relation to animal odors. Table 7-8 summarizes ammonia evolution rates that have been measured under a variety of livestock-

TABLE 7-7 Compounds Identified in the Air from the Anaerobic Decomposition of Livestock and Poultry Manure[a]

Alcohols	*Acids*	*Amines*	*Fixed Gases*
Methanol	Butyric	Methylamine	Carbon dioxide
Ethanol	Acetic	Ethylamine	Methane
2-Propanol	Propionic	Trimethylamine	Ammonia
n-Propanol	Isobutyric	Triethylamine	Hydrogen sulfide
n-Butanol	Isovaleric		
Isobutanol			
Isopentanol			
Carbonyls	*Esters*	*Sulfides*	*Mercaptans*
Acetaldehyde	Methyl formate	Dimethyl sulfide	Methyl mercaptan
Propionaldehyde	Methyl acetate	Diethyl sulfide	Nitrogen
Isobutyraldehyde	Isopropyl acetate		
Hexanal	Isobutyl acetate	*Disulfides*	*Heterocycles*
Acetone	Isopropyl propionate		Indole
3-Pentanone	Propyl acetate		Skatole
Formaldehyde	*n*-Butyl acetate		
Heptaldehyde			
Valeraldehyde			
Octaldehyde			
Decaldehyde			

[a] Data from Miner.[30]

production situations. They are characterized by extensive variability, but reflect the conditions under which the phenomenon occurs.

Emission for the cattle industry in 1972 was estimated[36] to include 20,500 t of total suspended particles and 3,480 t of ammonia. Total amine and sulfur compound emission was calculated to be 139 and 522 t, respectively.

MEASUREMENT OF ODORS FROM LIVESTOCK PRODUCTION

Various schemes have been proposed for the measurement of odors. The concentrations of individual odorous compounds have been used by some investigators as an indication of odor transport and odor pollution. Ammonia and hydrogen sulfide have been widely used for this application, because of their ease of measurement and known odorous characteristics. In the concentrations measured, however, they would not be detectable by the human nose if they were not accompanied by other odorous

TABLE 7-8 Ammonia Volatilization from Various Surfaces Associated with Livestock Production

Situation	Rate, kg/ha-day	Reference
Sheep pasture, late summer	0.26	10
Beef feedlot, Nebraska	0.41	14
Pasture, Nebraska	0.044	14
Beef feedlot, Idaho summer, dry	3.48 (average)	33
	0.31-12.2 (range)	33
Manure, free pasture	0.01-0.02	32
Pasture after recent application of liquid manure	0.05-0.2	32
Manure-covered aisle in dairy barn	0.5-1.0	32
Anaerobic lagoon for swine manure, Iowa	17-98	24

compounds. Studies of Hill and Barth[20] substantiated the synergistic nature of the combinations of compounds typical of manure odors.

The measurement of odor intensity based on the number of dilutions required to reduce the concentration to a barely detectable point has been the most generally accepted method for evaluating odor concentrations. The Scentometer, as distributed by Barnebey-Cheney and described by Rowe,[38] has been used by researchers[43] and, more recently, regulatory agencies,[42] to evaluate odor-control techniques and odor transport. Similar measurements have been made in the laboratory with more sophisticated devices[34] and in a mobile unit described by Lindvall *et al.*[26]

The frequency of odor detection, either measured or estimated, has been widely used as a means for evaluating odor problems. This approach attempts to determine the percentage of time that an odor can be detected at the site where the receiver is. For example, if a home is near a cattle feedlot, it might well be important to be able to estimate the percentage of time that that odor would be detectable at that site. By consulting published data on wind direction, velocity, temperature, and relative humidity, one can calculate an estimated odor detection or frequency. This calculation is helpful in assessing the severity of an odor problem.[28]

ODOR SOURCES

To avoid odor complaints and to minimize the escape of potential water pollutants to the air, livestock producers are faced with the need to control the evolution of odorous compounds. These gases may arise from various sources, which may be categorized as feed materials, fresh manure, and stored or decomposing manure. The greatest research efforts have been directed toward the control of volatile compounds released by the storage and treatment of manures. Feed odors are most commonly associated with the feeding of waste materials or fermented products that have objectionable odors. Among the materials most commonly related to feed odors are potato and other food-processing wastes and fermented feeds.

The odor of fresh manure is generally described as less objectionable than that of anaerobically decomposing manure. Fresh manure evolves large quantities of ammonia, but this ammonia is generally not accompanied by other decomposition products, which contribute the most objectionable characteristics. The odorous compounds evolved from manure-covered surfaces or treatment facilities are a function of the biologic reactions taking place in the material, the nature of the material as excreted, and the configuration of the storage or treatment surface.

Roofed confinement facilities—common for poultry, swine, and to a lesser extent dairy and beef production—have high odor-production potential, owing to the high animal density involved, the large inventory of manure frequently in storage, and the limited rate of air exchange. Of particular importance are manure-covered floor and animal surfaces, manure storage tanks beneath slotted floors, and anaerobic lagoons often used for manure storage and treatment.

Feedlots, nonroofed intensive confinement facilities most commonly used for beef-cattle production, pose special odor-control difficulties. Feedlots involve extensive manure-covered areas subject to all the climatic extremes of the locale. When surfaces are excessively dry, there is a potential for dust problems. These problems have been most severe in the arid cattle-producing areas of Arizona, California, and Texas.[36] When feedlot surfaces are wet, particularly in warm weather, they support widespread anaerobic decomposition with an associated large surface area for the evolution of odorous gases. Feedlot odor problems are most frequent in warm humid areas or in feedlots constructed in areas of inadequate drainage or poor drying conditions.

ODOR-CONTROL PRINCIPLES

Techniques for the control of odor release are based on a limited number of rather specific control principles. Some volatile compounds present in the feed, manure, or manure slurry can be converted to a less volatile or less odorous form by pH control or by biologic conversion. An example is the addition of lime as a base to control the release of hydrogen sulfide.[9] The dissociation of hydrogen sulfide is a strong function of pH; if the pH is raised above 9.5, the escape of hydrogen sulfide is insignificant. Sulfides are oxidized to sulfates under aerobic waste treatment, as is practiced in oxidation ditches and aerated lagoons. Paraformaldehyde has been added to manure as a means of converting ammonia to nonvolatile hexamethylenetetramine.[40]

Another approach to the control of odor is to inhibit the anaerobic decomposition of manure. This is most frequently accomplished by keeping feedlots dry enough to allow oxygen permeation of the surface. In systems that use bedding, the bedding serves a similar role. Prompt removal of manure from indoor pens or outside lots may also be a management technique for odor reduction. In liquid systems, odorant-producing decomposition is most often inhibited by maintaining the slurry in an aerobic condition. An example of this practice is the use of oxidation ditches and aerated lagoons. Bacterial decomposition of manure is also inhibited by low temperatures, but, except in site selection, little use has been made of this fact.

Physical confinement of the odorants offers a third potential for odor control. Covers on manure storage tanks and anaerobic manure-treatment devices are effective in controlling the escape of odorants. When the air exchange over manure slurries is reduced, the volatilization of odorous gases is reduced. Gases escaping from enclosed tanks may be further managed by incineration, liquid scrubbing, or soil-column absorption. Although no entirely satisfactory cover material has been developed, covering of anaerobic lagoons and venting of the released gases to a burner or a soil absorption field offer potential for odor control.

ODOR-CONTROL CHEMICALS

The addition of odor-control chemicals to manure storage tanks or animal feeding areas has attracted widespread interest. Materials having potential for odor control, when used in this manner, are those which can prevent the release of odorous compounds, inhibit their formation, or mask their odor.

Oxidizing agents have potential as odor-control chemicals. Faith[15] proposed the use of potassium permanganate on cattle feedlots in Arizona at a rate of 22 kg/ha (20 lb/acre). Ford and Ulich[19] conducted a series of tests and concluded that potassium permanganate at 28 g/kg (56 lb/ton) of manure totally suppressed the release of odorous gases. It was also judged effective by their panel in reducing malodors at lower application rates. They reported that potassium permanganate was effective in reducing the odor of well-managed feedlot surfaces when applied at 22 kg/ha (20 lb/acre) as a 1% water solution. Potassium nitrate, paraformaldehyde, hydrogen peroxide, and Ozene (a commercial formulation of *o*-

dichlorobenzene) were also evaluated by Ford and Ulich,[19] but judged less effective than potassium permanganate. Hill and Barth[21] later evaluated the use of ozone for the control of ammonia and methylamine odors from animal and poultry manure. They reported less success than had been anticipated on the basis of the response of other odor sources to ozone.

Enzymes and other digestive aids have been proposed for the control of livestock-production odors. These products have been difficult to evaluate under controlled conditions, because of the unwillingness of manufacturers to disclose their compositions. Ford and Ulich[19] subjected "Formula 2" and a digestive deodorant to their testing program. Their odor panel concluded that neither was effective in reducing the strength or offensiveness of manure odors. This conclusion was similar to the earlier results of Burnett and Dondero.[5]

Masking agents have been applied to manure as a means to improve its acceptability. Burnett and Dondero[5] evaluated several commercially available agents and found them neither satisfactory nor economically feasible for long-term use. Nine commercially available products were individually applied to one or more cattle feedlot pens at an eastern Idaho site to determine their effectiveness in reducing odor release from this source.[33] Only sodium bentonite, Odor Control Plus, and two natural zeolites were found to reduce the rate of ammonia release consistently when the treated areas were compared with untreated control areas. Odor-intensity measurements confirmed the effectiveness of sodium bentonite. The pen treated with Odor Control Plus had a measurably less intense odor 5 days after treatment, but not 10 days after treatment. Only one of two odor observers was able to distinguish the zeolite-treated pens from the control. The cost of the effective materials ranged from $300 to $600 per acre for treatment during the odor-production season.

Feed additives for the modification of manure odors remain of interest to researchers. Although it is known that ration ingredients influence the odor of fresh feces and urine, no usable feed additives have gained widespread application. Matsuhima[39] used sagebrush as an additive in cattle rations and reported an odor reduction. R. J. Kellems (Oregon State University, unpublished data) fed sagebrush, up to 2% of the ration, to beef animals and was unable to detect any influence on the odor of decomposing manure. He did, however, report a significant carryover of mint-oil odor from ration to urine. Sagebrush, dry lactobacillus acidophilus, lyophilized yeast culture, and activated charcoal, all at 5%, were fed to swine by Ingram *et al.*,[23] with no significant change in fecal odor, according to an odor panel.

DESIGN AND MANAGEMENT PRACTICES TO CONTROL ODORS

Application of odor-control techniques requires specific attention to the operation under discussion. Perhaps the most critical and effective opportunities for reducing odor complaints occur in the initial site selection. Although it is difficult to establish definitive perimeters beyond which odor complaints will not be problems, a livestock producer must seriously consider odor control in the selection of a site. A site may be ideally suited for livestock production with respect to transportation, feed supply, and zoning regulations, but be inappropriate because of existing or proposed development in the area.

Although wind direction is important in evaluating a site with regard to odors, most locations have winds from several directions during the year. The simple location "downwind" of development is not sufficient to ensure acceptability. By referring to published data, one can estimate the percentage of time that the wind will blow from the odor source to the point in question and thereby make a more rational decision concerning the site suitability. Where distance is used as the only criterion, it must be expected that odors can be transported further than a mile downwind under appropriate climatic conditions.

The second opportunity for reducing odor problems occurs during the design and construction of a facility. By application of odor-control principles, one can minimize the probability of odor production. Designing outdoor lots that are well drained, watering systems that do not flow onto the lot surface, and runoff control facilities that are remote from areas of odor sensitivity will achieve some odor reduction. In confinement facilities, the methods of manure removal from the pens, manure transport, and the handling approach are most important for odor control. Also, the animals must be kept clean and dry. Among approaches used for accomplishing this are slotted floors, flushing gutters, and frequent pen-scraping. Covered manure storage tanks control odor release from stored manure. Where treatment is required and odor control is important, aerobic systems, such as oxidation ditches and floating surface aerators, although more expensive, can be effectively used to maintain low odor intensities.

The operation and management of a livestock-production facility also offer considerable opportunity for exercising odor control. Maintaining the operating system in functional order is probably the most important. Overflowing manure storage tanks, broken scrapers, leaking waterers, and ruptured retention ponds and dikes are among the most common causes of odor complaints.

Anaerobic lagoons for swine-waste treatment are of special concern in odor control. Properly designed and managed lagoons are not free of odors, but are much less likely to cause serious odor problems than overloaded or shock-loaded lagoons. Where multiple-celled lagoons are used, it is important that the cell or cells receiving fresh manure not be loaded in excess of the recommendations for a particular area. Anaerobic-lagoon odors are most common in the late spring and early summer, when the water temperature increases and manure accumulated during the winter undergoes rapid decomposition. Where odor control is critical, it has been found helpful to remove as much of the lagoon contents as possible and to refill to the normal operating level with clean water. Another helpful recommendation to reduce lagoon odors is to lower the loading rate, because overloading is a significant contributor to increased odor intensity. Loading rates are decreased by building larger lagoons or reducing the number of animals served. A final alternative is to add a surface aerator. Where practical, it is desirable to locate lagoons as far as possible from residences, roads, and other odor-sensitive areas. Recommendations range from 300 ft to 0.5 mile (about 0.8 km), or more. Shielding lagoons from view is also helpful.

Manure-disposal techniques and timing are also very important for odor control. When manure is to be applied to cropland, selection of a field downwind of residences on the day in question is important. Morning application of manure is more desirable than late afternoon application, which limits potential drying time. Neighbors are generally most sensitive to odor problems in early evening, when they are using outdoor recreational facilities.

When manure disposal is necessary and odor control is critical, immediate covering of the manure with soil can effectively reduce odor. Where the soil is suitable and neighbors are particularly close, direct manure injection beneath the soil surface is a valuable technique.

INDUSTRIAL RESPONSE

In response to public pressure, local regulatory agencies, and widely publicized legal actions, livestock and poultry producers have initiated a variety of odor-control measures. Problem systems have generally been those in sensitive areas, such as those near residential, recreational, or commercial areas; unusually large operations; or those feeding highly odorous materials.

Over the last 20 yr, swine production has moved progressively toward confinement in environmentally controlled buildings. Manure in such buildings may be managed by storage beneath the floor pits, piping to an exterior anaerobic lagoon or earthen storage pit, or transfer to an alternative storage or treatment device. Final disposal is to cropland with either a mobile tank system or irrigation equipment. Early systems, designed without particular attention to odor control, were characterized by dirty animals, exposed manure on floors, and overloaded storage-treatment systems. More recent designs have included slotted or flushed floors for cleaner animals, prompt manure removal from the buildings, improved ventilation, and waste-treatment or -storage devices designed to minimize odor release. Land application systems can be a less severe odor source through the use of judicious scheduling and appropriate equipment selection. Because all these modifications are applied to the basic production system, it is difficult to estimate specific odor-control costs. Equipping of older structures is frequently either too expensive or physically infeasible.

Beef-cattle production generally involves a finishing phase in nonroofed feedlots. The larger and more numerous feedlots are in central and western states, as indicated in Table 7-1. Odor complaints have been voiced relative to larger feedlots and those in particularly sensitive locations. Odorous emission is primarily related to volatilization from manure-covered surfaces. Warm, wet weather promotes decomposition, which increases the quantity of volatile material available. Thus, feedlot odor is a function of local weather, as well as design and management. Odor control is difficult, owing to the large land area covered and the continual renewal of the surface. Odor-control costs are largely buried in system construction costs, except where the experimental use of surface-applied chemicals is practiced. A small percentage of beef animals are in roofed confinement; the number is sufficiently small to make the attendant odor problem much less frequent. The same technology as is applied to swine in confinement is generally used. Most beef animals are on pasture at any particular time, and there the odor problem is handled by dispersion.

Dairy production is more dispersed than beef feedlots, and herds are typically smaller. Dairy cows are managed in combinations of pasture, dry lot, and confinement. Pastures are of little odor concern. Dry lots are managed like beef feedlots and have similar odor problems. Confinement facilities require manure-handling systems and have odor problems simliar to those associated with other confinement units.

Poultry production—including that of broilers, layers, and turkeys—has moved almost exclusively to confinement. Odor control is related to the handling of manure, which may be liquid-based, dry, or mixed with litter. Liquid systems generally involve either collection beneath the cages in pits or daily flushing of a collection channel. Frequent removal is necessary, to control ammonia concentration in the building. Liquid from these systems may be transferred to lagoons, stored, or otherwise treated before land application. The difficulties involved in controlling odors from these systems has contributed to their unpopularity and general nonuse. Dry poultry systems are designed to promote rapid drying and keep the manure dry enough to promote air permeation and to restrict decomposition. Specific designs vary in pit depth and ventilation, but all have the same goal. Litter systems are used to absorb moisture and maintain aerobic conditions. Sufficient dry litter is added to keep the moisture content below that required for active composting.

The production of other animals—including horses, dogs, sheep, cats, mink, ducks, and goats—is associated with manure odors, but their numbers are smaller, their production more localized, and their problems less severe. Control technology, when applied, is essentially like that described for other species.

SUMMARY

Livestock production is a major agricultural pursuit involving over 700,000 individual farms. Recent trends have indicated a shift to larger but less numerous livestock and poultry production enterprises. As the sizes have increased, greater confinement of animals has been adopted to minimize labor requirements. Increased size and intensive confinement result in extensive manure accumulations in small areas. The decomposition of this manure is the predominant source of odors. The public has responded to livestock production odors by complaining to owners, complaining to regulatory agencies, promoting more rigid regulations, and, in a few isolated instances, initiating legal actions to seek property damages or injunctive relief.

Anaerobically decomposing manure is particularly prone to odor release, whether on a feedlot surface, in a manure-storage tank, or in an anaerobic lagoon. Other potential odor sources are fermenting feeds and improperly handled dead animals.

Current odor-control techniques include careful site selection to avoid locations near residential or other sensitive areas, inhibition of anaerobic decomposition by aeration or moisture control, and confinement of manure in covered storage tanks. Air scrubbing and filtration have not generally proved feasible with existing animal-raising facilities. Many odor-control chemicals for inclusion in feed or direct application to manure have been proposed, but none has received widespread acceptance.

CONCLUSIONS

- Livestock-production odors are a major concern to residents in the proximity of large enterprises.

- There are over 700,000 livestock and poultry producers in the United States operating under a wide variety of climatic conditions, with varied feed supplies, and with varied facility designs.
- Although considerable research has been devoted to the problem, the complexity of odor generation and transport has not yielded a straightforward technical solution having widespread application.
- The greatest success in controlling livestock odors has been achieved by development of an overall odor-control strategy that incorporates judicious site selection, appropriate facility design, and responsible management, with due recognition of odor control as an operating objective.
- Odor-control chemicals have achieved little success or acceptance in the control of livestock-production odors.

RECOMMENDATIONS

- Owing to the widespread nature of livestock and poultry production and the regional variability in design, feed-supply, and management techniques, a regulatory program for odor control must be responsive to diverse needs.
- To ensure a continuing abundance of low-cost high-quality meat, milk, and eggs, a research program directed to identifying appropriate odor-control technology for the various livestock-production systems should be initiated.
- Avoidance of livestock-odor conflicts is a function of an adequate cost-effective technology and an appropriate regulatory and land-use-planning process. This combination is worthy of pursuit.

FOOD-PROCESSING[48]

Food-processing operations are distributed throughout the United States. Typical processes include coffee-roasters, canneries, smokehouses, and fermentation processes.

COFFEE-ROASTING

Coffee is roasted from coast to coast, but the bulk of it is processed in and around New York, Los Angeles, Chicago, and New Orleans. Coffee-processing consists of the following: precleaning green coffee beans to remove foreign matter, blending green beans of various quality for flavor, roasting to change the beans chemically, cleaning and cooling the beans to remove heavy contaminants, and grinding and packaging for consumption.

TABLE 7-9 Emission Inventory Factors for Coffee-Processing[a]

Process	Solid Emission, lb/1,000 lb of green beans	
	With No Control	With Usual Control
Roaster		
Direct-fired	3.8	1.1[b]
Indirect-fired	2.1	0.6[b]
Stoner and cooler	0.7	0.2[b]
Instant-coffee spray dryer	(Control always used)	0.7[c]

[a] Data from U.S. Environmental Protection Agency.[48]

[b] Cyclone.

[c] Cyclone and wet scrubber.

Coffee-processing produces four types of emission: dust, chaff, odor, and smoke. The odor and smoke are combinations of organic constituents volatilized at roasting temperatures and steam produced when the roast is quenched with water. Further processing to produce instant coffee causes an additional emission in the form of powdered coffee, which escapes during the drying process. During decaffeination, odors can be produced by trichloroethylene, the solvent used in extracting caffeine from the green coffee beans.

Currently, the best method of smoke elimination involves use of a separate afterburner. Powdered-coffee particles, being highly soluble, can be effectively controlled with a simple water scrubber. Emission inventory factors are presented in Table 7-9. The most effective means of control now available are centrifugal collectors for dust and chaff, afterburners for smoke and odor, and water scrubbers for instant-coffee particles.

Roasting is the most important step in coffee-making, because it develops the flavor. For the average roast, about 370 Btu of heat energy is required per pound of green coffee. More than 95% of modern roasters are gas-fired; the remainder are oil-fired.

A roaster consists of a perforated, horizontal cylinder with internal helical flanges, enclosed in a metal jacket. Of the three types of roasters in use, two are batch-fed; they roast up to 1,500 lb/h in 500-lb batches that require 15–20 min of roasting time each. The older, direct-fired roasters use a gas jet inside the cylinder, which operates at an air temperature of about 2,000°F (about 1,100°C) and heats the roaster by radiation. In the newer, indirect-fired roasters, the gas burner is in a separate chamber behind the cylinder. Hot combustion gases are recirculated, for more efficient heat transfer. This reduces the operating temperature range to 850–900°F (455–480°C) and produces a more uniform, higher-quality roast. The third type, and the most modern, is a continuous roaster with a rated capacity of up to 10,000 lb/h and a roasting time of only 5 min; it is operated at 450–500°F (230–260°C).

During the first 10 min of batch-fed roasting, the charge heats at a fairly uniform rate and moisture is driven off. In the last 5–10 min, the temperature rises rapidly, and the chemical degradation that produces the familiar odor and flavor occurs. The beans swell and turn brown. While the beans are roasting, the operator constantly compares samples from the roaster with beans of a standard prescribed by cup testers as having the desired flavor. At the moment the color of the roasting beans matches that of the standard beans, the operator applies a water quench to stop the roasting action. Because of the high temperature in the roaster, the water flashes off and passes out of the stack as steam, carrying coffee odorants with it.

Because of the size and nature of the dust and chaff particles, cyclone collectors provide a simple, economical means of emission control. A simple cyclone will handle, with great efficiency, emitted particles greater than 20–40 μm in diameter. It can tolerate temperatures to 750°F (400°C) and accommodate flow rates in excess of 25,000 cfm. Care must be exercised in removing the collected solids so that another air-pollution problem is not generated. In some locations, the chaff is burned, with the inherent possibility of creating more smoke and odor.

Even where cyclones are used, the submicrometer particles in the smoke and odor leaving the roaster are not controlled. There is, however, a so-called smokeless roaster. Its manufacturers claim that this roaster eliminates smoke and odor completely; however, not everyone accepts this claim as entirely true. A damper system recirculates the combustion gases that are ordinarily vented directly into the atmosphere through the gas flame of the roaster. The additional heat required increases fuel consumption by about 40%. Better smoke control is experienced with an afterburner in the roaster stack, but fuel requirements are increased by 100–150% over those for a conventional roaster.

Other methods of emission control—such as catalytic oxidation, scrubbers, and ultrasonic agglomeration of submicrometer particles to the point that they can be handled by conventional collection equipment—have been attempted. They have failed for various reasons, usually poor economics or poor performance.

In modern coffee plants, cyclones are included as an integral part of the roaster design. Particularly in installations of small capacity, however, the effluent airstreams from the cleaning, cooling, and stoning processes may have no such individual collection system. In general, these can be connected in manifold to a common exhaust stack and serviced by a single cyclone collector. Efficiency ratings as high as 97% may be expected from such devices. In some communities, however, the remaining 3% may constitute a nuisance problem. In such situations, water scrubbers have been used to eliminate the remaining emission.

CANNERIES

The most important processes for the preservation of fruits and vegetables are canning, dehydration, and quick-freezing. The only important air-pollution problem is related to the disposal of hulls, leaves, rinds, pods, cuttings, etc. If held too long, these materials decay and produce revolting odors.

Water-pollution regulations prohibit sending the waste to streams unless it has been processed to reduce biologic oxygen demand (BOD). One method commonly used to reduce BOD is digestion of the cannery waste in either anaerobic or aerobic lagoons. Even the best-operated lagoon has upsets during which odorous gases (hydrogen sulfide, mercaptans, fatty acids, amines, and other nitrogenous gases) are evolved. If the upsets are only occasional, odor-counteracting or -masking agents are sometimes released on the downwind side of the lagoon. If the condition persists, addition of a nutrient, precipitation of excess sulfide, reestablishment of the active organism, or other treatment is required. Some food-processors have replaced lagoons with spray irrigation plots to advantage. The National Canners Association is studying methods of composting.

Leaves, stalks, and cuttings can sometimes be disposed of odorlessly in incinerators. Sanitary landfill is an alternative method.

SMOKEHOUSES

Smoking has been used for centuries to preserve meat and fish products. Modern smoking operations do not differ greatly from those used by our forefathers, although the prime purposes of smoking today appear to be the imparting of flavor, color, and "customer appeal" to food products. Curing and storage processes have been improved to the point where preservation is no longer the principal objective.

Most smoked products are meats of porcine and bovine origin. Some fish and poultry and, in rare instances, vegetable products are also smoked as gourmet items.

Smoke from ovens and smokehouses used for curing meat is sometimes objectionable because of both smoke and odor. One installation required a water scrubber, a low-voltage electrostatic precipitator, and an afterburner. In this case, the air-pollution control equipment cost $42,000, whereas the basic oven cost only $18,000. In many operations, a water scrubber followed by an afterburner is satisfactory.

FERMENTATION PROCESSES

The manufacture of beer, wine, whiskey, and other fermented beverages results in the emission of organic compounds—largely alcohols. Such odorous emission occurs from distilling operations, cooking or brewing kettles, vacuum systems, fermenters, and aging or storage processes. Air-pollution controls for organic emission, other than condensers (which are an integral part of the process), are not used. In the manufacture of distilled liquors (whiskey, gin, etc.), one of the larger sources of organic emission is the aging warehouse, where the liquor is kept for 1–5 yr in wooden barrels. During this time, evaporation occurs through the barrel staves. This evaporation is largely water, but obviously (from the odor) some alcohol is also lost. The drying of the "slop" or still bottoms, which are used for animal feed, also results in odorous organic emission. This drying is accomplished in either rotary or vacuum drum dryers.

Incineration of exhaust gases from dryers is most commonly used for the control of odors from fermentation processes. Ozonation has been used recently in the control of fermentation odor. The data reported on this system showed capital costs of $55,000 for the ozonator and $215,000 for the stainless-steel ductwork. The system handled the fermentation odor from an average-size facility, discharging 140,000–150,000 cfm. Half this is sent to the boiler house for incineration, and the other half is sent to the ozonation unit, where air is pulled into the ozonation unit at 75–150 cfm. It is compressed and dried with a dual-column system using activated alumina, and this dry ozonated air is sent back to the main stream, held in contact for about 10 s, and sent to a stack. Ozone is kept at 1–2 mg/L. There have been no failures over the last 2 yr. The operating cost for steam, air, electricity, and maintenance is roughly $0.30/1,000 cfm.

RENDERING

The United States is the world's largest producer, consumer, and exporter of tallow and greases. This country is responsible for 55–60% of the world's production of tallow and grease.[25] The 1971 production of inedible

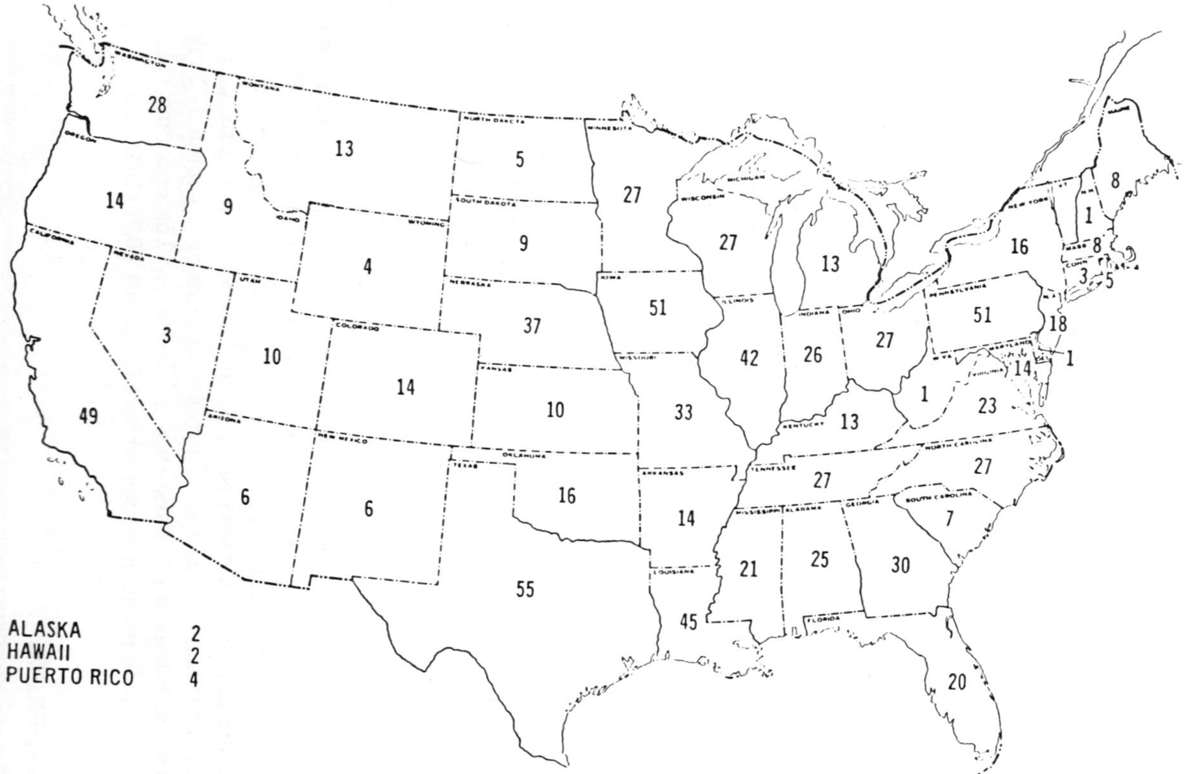

FIGURE 7-1 Rendering and fish-reduction plants, 1968. Reprinted from Osag and Crane.[135 (p. 2-3)]

tallow and grease was an estimated 5.4 billion pounds (2.4×10^6 t), worth approximately $430 million.

In 1968, there were approximately 900 rendering facilities, which were operated by 770 firms. The number of facilities had declined to about 750 by 1972.[35] The 1968 distribution of facilities by states is shown in Figure 7-1.

PROCESS

Animal matter not suitable as food for human consumption is converted into salable byproducts through various reduction processes. Cows, horses, sheep, and poultry that have died through natural or accidental causes and the byproducts from slaughterhouses, butcher shops, and poultry dressers are processed into proteinaceous meal and tallow. Rendering operations that constitute part of a meat-packing or poultry-processing plant are designed to process blood, meat, offal, and feathers produced on the premises and are referred to as captive plants. Off-site, or independent, rendering plants are operated independently and normally rely on a number of local sources for raw material.

Although the rendering process involves mainly the heated reduction of fat-containing materials into tallow and proteinaceous solids, it can also include such operations as blood-drying, feather-drying, and grease-reclaiming.

Batch Process

The process raw material is placed in a dump pit and conveyed to a hogger, where the meat and bones are ground to facilitate mechanical handling and heat transfer. The ground material is then conveyed to the cookers for processing.

In the cooking process, heat breaks down the flesh and bone structure, causing tallow to separate from solids and water. In the batch process, cookers are charged with 3,000–12,000 lb of animal matter and heated for 1–4 h. Batch cookers may be operated either at pressures greater than 50 psig to digest bones, hooves, hides, and hair, or under a vacuum to produce high-quality tallow. The cookers are equipped with paddles to mix the charge during processing.

Dry rendering is used almost exclusively for inedible materials and is carried out at atmospheric pressure or under partial vacuum. Moderate-sized agitating vessels are used for batch operation. The material is cooked until all the free moisture in the tissue is driven off. The separated fat is then screened to remove the solid proteinaceous residue.

Wet rendering is used primarily for edible materials and is seldom used to process inedible material. It involves cooking under pressure with the direct addition of live steam. The fat and water are separated after cooking. The solids are screened out of the water, and the water is evaporated to a thick, protein-rich material that can be added to animal feeds.

Continuous Process

Continuous rendering is a highly mechanized dry rendering process. In the continuous system, raw material is screw-conveyed to the hogger, where it is ground and fed into the end of a multicompartment cooker as the raw material passes through the cooker compartments. On the other end of the cooker, processed material is removed by the control wheel and placed in a drainer to separate tallow and solids. The entrainment trap prevents solids from escaping from the cooker and fouling the air-pollution control system.

The scrubber-condenser handles vapors and noncondensables from the cooker. Condensation takes place in a fully enclosed tubular condensing section that is cooled with a water spray. Condensed vapors flow off as waste, and noncondensables go to a small wet scrubber and incinerator. The same water spray also scrubs plant ventilating air, which is passed axially upward outside the tubes, also cooling the condenser waste spray. The condensate and spray water are kept apart within the unit; the only mixing occurs on sewering. Because of evaporation and to avoid scaling, some spray water is bled off and some makeup is added.

Refining of Rendering Products

Tallow and solids separated during cooking require further processing to obtain finished products. The solids, or cracklings, are pressed to remove residual tallow and usually are ground to a meal before marketing. Tallow is maintained at 200°F (93°C) or above and is then processed in settling tanks, centrifuges, or filters to remove the solids. Moisture is removed from the tallow by flash drying, either at atmospheric pressure or under vacuum, and by blowing air through the tallow. The tallow is further refined by adding caustic soda to neutralize the free fatty acids. The saponified free fatty acid settles out and is known as foots or soap stock. The tallow is usually filtered to remove all traces of the foots and other solids.

EMISSION

Malodors are the principal air contaminants from rendering and companion processes. Cookers are a primary source of malodors in rendering plants.[35] When animal matter is heated, the cells break down, liberating gases and vapors. Further heating causes chemical decomposition, and the resulting products are often highly odorous.

Cooker streams contain 95% or more water by volume. The remainder, however, includes compounds that are highly malodorous. Rates of emission of odorous contaminants are functions of the rate of moisture evaporation. The maximal emission from atmospheric cookers occurs in the initial portion of the cook, whereas in pressure cookers the moisture evaporation rate and emission proceed as the temperature builds up.

Processing tanks (in which tallow is dehydrated by boiling or air blowing), feather-dryers, tallow presses, and blood spray dryers are smaller but significant sources of malodors. Dryers can be a large source of malodors, particularly if feedstocks are putrified or not completely cooked beforehand, or if meal is overheated in dryers.

Odor concentrations from air blowing of tallow may be significant. Percolator pans are also a source of significant odors for short periods. At the end of a cooking cycle, when tallow and solids are discharged into percolator pans, substantial quantities of steam and odors are released. Percolator-pan emission is especially difficult to control, because of the necessity to gather the vapors in suitable hoods, but at least one western plant has recently demonstrated control by this method.[35]

Storage areas, dump pits, and hoggers are a significant source of malodors if raw materials are not fresh. Ideally, raw materials should not be over 24 h old when processed.

Rendering-plant malodors have been attributed to a variety of organic compounds belonging to such classes as aldehydes, fatty acids, amines, mercaptans, and sulfides.[4,16] Aldehydes and fatty acids are the principal odorous breakdown products from fats; putrescine and cadaverine are two extremely malodorous organic nitrogen compounds associated with decaying flesh. Keratins, the primary constituents of horny material (skin, hair, nails, feathers, etc.), are the principal source of sulfides and mercaptans.

Some specific compounds that have been identified in rendering-plant odors are trimethylamine, quinoline, dimethyl pyrazine, skatole, ammonia, and hydrogen sulfide.[35] Recent studies have identified such compounds as methyl and dimethyl sulfides; butylamine and trimethylamine; the methyl pyrazines; aldehydes, ketones, and alcohols; and organic acids, including butyric acid.[12] Odor-threshold concentrations are extremely low for some

TABLE 7-10 Odor Concentrations and Emission Rates from Inedible-Material Rendering Processes[a]

	Typical Average Odor Concentration, o.u./scf	Odor Emission Rate, o.u./ton of feed	Exhaust Products, scf/ton of feed	Typical Moisture Content of Feeding Stock, %
Rendering cooker, dry-batch type[b]	50,000	$1{,}000 \times 10^6$	20,000	50
Blood-cooker, dry-batch type[b]	100,000	$3{,}800 \times 10^6$	38,000	90
Feather-dryer steam tube[c]	2,000	153×10^6	77,000	50
Blood spray dryer[c,d]	800	80×10^6	—	60

[a] Reprinted from Danielson.[6] (pp. 776-778)

[b] Noncondensable gases are neglected in determining emission rates.

[c] Exhaust gases are assumed to contain 25% moisture.

[d] Blood is handled in spray dryer before any appreciable decomposition occurs.

of the malodorous compounds, and they can be detected in concentrations as low as 0.2 ppb. Many odorous compounds have not been identified, nor have their detectability limits been established.

An odor unit is defined as the quantity of any odorous substance or combination thereof that, when completely dispersed in 1 ft^3 of odor-free air, is detectable by a median number of observers in a panel of at least eight persons.[35]

Table 7-10 lists odor concentrations and emission rates from various inedible-material rendering processes. These odor concentrations were determined by the Mills modification of the ASTM syringe method. Of the facilities listed in Table 7-10, the cookers are the predominant source of malodor emission. Typical batch cookers release 250–750 scfm of exhaust gases over a 2- to 4-h cooking cycle, and continuous cookers release 3,000–4,000 scfm.[35] At an average emission concentration of 50,000 odor units (o.u.) per scf, a batch cooker would release up to 37.5 million odor units per minute. However, a 2,000-lb/h feather-dryer (1,600-scfm exhaust volume) with an average emission concentration of 2,000 o.u./scf would release 3.3 million odor units per minute.

STATE AND LOCAL ODOR REGULATIONS

Several state and local jurisdictions have adopted odor-control regulations that apply to rendering plants. Many are general prohibitions that apply to all odor sources; others apply more specifically to rendering operations.

The most common type of statute is a general-nuisance regulation. The enforcement of such regulations requires that a great deal of evidence be assembled to show that a given source causes nuisance or annoyance; endangers comfort, repose, health, or welfare of persons; or causes injury or damage to property or to business.

Fenceline regulations stipulate maximal allowable odor limits at the rendering-plant property line. These limits are not necessarily indicative of the extent of emission from the plant. The Scentometer is normally used to enforce fenceline regulations. The allowable dilution limits generally range from 7 to 15.[35]

The most common type of rendering-plant odor regulation is an equipment standard that directs the operator to use incinerators with specific characteristics. Two such characteristics are listed in Table 7-11. They usually apply to discrete processes within the plant—e.g., cookers, dryers, and heated reduction processes—rather than to the entire facility. As noted, required temperatures and residence times vary from 1,200 to 1,600°F (about 650 to 870°C) and from 0.3 to 0.5 s. In jurisdictions where such equipment standards are in use, the operator usually has the option of

TABLE 7-11 State and Local Regulations for the Incineration of Rendering-Plant Odorants[a]

State or Locality	Incineration Temperature, °F	Incineration Time, s
Pennsylvania	1,200	0.3
Minnesota	1,500	0.3
St. Louis, Mo.	1,200	0.3
Los Angeles County	1,200	0.3
Pittsburgh, Allegheny County	1,600	0.5
Arizona	1,200	0.3
Montana	1,200	0.3

[a] Reprinted from Osag and Crane.[35]

using any other control system if it can be shown that it provides equivalent odor abatement.

DISPERSION OF ODORS

There is no evidence that rendering-plant odors are harmful to health when dilute.[35] Therefore, a reasonable objective is the prevention of detectable odors at ground level outside the plant. Regulations attempt to provide this assurance by limiting odor emission or concentration at the stack or fenceline or by requiring air-pollution control equipment that will achieve these limits. Some typical installations—uncontrolled and controlled—are examined here, to consider the effect of atmospheric dispersion.

Odors from Uncontrolled Plants

Calculations have been made to estimate ground-level odor concentrations for an uncontrolled rendering plant. The plant had a stack-gas flow rate of 6,000 cfm. This plant uses a shell-and-tube condenser for cooker off-gases; for simplicity, all plant odors are assumed to exit from the stack. Because it emits about 670 cfm from the condenser, there must be a large amount of dilution air to make up 6,000 cfm. The stack temperature is assumed to be 90°F (32°C), and the stack velocity, 10 ft/s (3 m/s). The stack-gas odor concentration is taken as 100,000 o.u./ft^3, which is derived from the dilution and the usual range of exit concentrations from rendering-plant shell-and-tube condensers. The stack is assumed to be 75 ft (23 m) high, with a windspeed of 1 m/s.

Under the preceding conditions, the maximal ground-level concentration was estimated to lie 1,600 ft (490 m) downwind from the stack and to be about 1,400 o.u./ft^3.[45] There would still be 170 o.u./ft^3 at ground level 2.5 miles (4 km) from the stack, and a distance of about 30 miles (48 km) would be required for complete dissipation of odor.

Residual Odors from Chemical Scrubber

The maximal ground-level odor concentration was estimated[35] for a rendering plant controlled by a hypochlorite water scrubber. The plant was assumed to be well kept, with negligible odor sources other than the scrubber exit gas. A stack-gas flow rate of 6,000 cfm at 90°F (32°C) was assumed, with an odor concentration of 200 o.u./ft^3. The plume rise was found to be negligible, because the gas temperature was so close to ambient. With such a small plume rise, the maximal ground-level concentration will occur at low windspeeds. Assuming a windspeed of 1 m/s, it was initially determined that a stack height of 30 ft (9 m) would be required to prevent the ground-level concentration from exceeding 1.0 o.u./ft^3. The actual odor source, however, will have a building under it and may have other buildings near it. When air flows past a building at any velocity, the airstream lines are bent. This bending is small at 1 m/s, but increases with velocity, finally resulting in turbulent eddies just downwind of the building. Therefore, odorous air passing over the roof may be deflected groundward, or may be pushed downward in turbulent eddies. This phenomenon is called downwash. With a 30-ft (9-m) stack, downwash is likely and will occasionally cause ground-level concentrations to exceed 1 o.u./ft^3 up to several hundred feet downwind. A rule of thumb is that the stack height—or the height of stack plus building if the stack is atop the building—should be 2.5 times the building height. Downwash effects from a building will extend 5–10 building heights downwind.

For the average rendering-plant building 30 ft (9 m) high, the stack should extend to 75 ft (23 m) above ground. Under the described conditions, the maximal ground-level concentration will then be 0.06 o.u./ft^3 when a chemical scrubber is used. This determination was made by calculating the critical windspeed at which ground-level odor concentration is maximal.[35]

Residual Odors from Afterburner

The maximal ground-level concentration was estimated for an afterburner on the basis of the same assumption as for the chemical scrubber, except for a stack temperature of 750°F (400°C). The resulting plume rise was about 40 ft (12 m). The maximal ground-level odor concentration from the 75-ft (23-m) stack was only 0.01 o.u./ft^3, because of the improvement in plume rise.

Odors from Rendering Buildings

The preceding discussion assumes that all odor in a rendering plant is directed up the stack. If some odors escape from the building itself, ground-level concentrations will be greater. Ground-level concentration at a given point is a linear function of emission rate in odor units per second if *all* other assumptions remain unchanged. Although building emission rates are not available, a typical dry-rendering cooker vented through a surface condenser with condensate temperature of 80°F (27°C) might release 12,500,000 o.u./min. In the absence of any other control or a stack, the ground-level concentration would be about 19 o.u./ft^3 200 m downwind from the rendering building.

Actual emission rates—and therefore ground-level concentrations—are potentially much greater when odors escape from the building itself than when they are directed up a stack and then subjected to downwash. This is primarily because odor-control devices are assumed to be operating when odors are directed up the stack, whereas escaping odors are not subject to control.

In view of the preceding discussion, it is imperative that odors be directed through control devices and a sufficiently tall stack and that no strong odor source be permitted to vent directly into a building, unless the building itself is vented through an effective odor-control system.

CONTROL TECHNIQUES FOR ODORS

There are several suitable control techniques that the rendering industry can use to comply with typical state and local regulations. These techniques involve treatment of odorous streams by condensation, incineration, combinations of condensation and incineration, and chemical scrubbing.

TABLE 7-12 Odor-Removal Efficiencies[a]

Inlet Concentration, o.u./min	Condenser Type	Condensate Temperature, °F	Outlet Concentration, o.u./min	Odor-Removal Efficiency, %
25,000,000	Surface	80	12,500,000	50
25,000,000	Direct contact	80	250,000	99

[a] Reprinted from Danielson.[6]

Condensers

Although significant control of many kinds of high-moisture emission can be accomplished by condensation alone, this technique is not effective enough to be used independently as a control for rendering-plant malodors. Condensation is useful, however, when applied in conjunction with incineration or chemical scrubbing. Under these conditions, a condenser reduces the load and energy requirement of secondary control equipment. For example, condensation of steam from high-moisture gas streams (rendering-cooker or blood-cooker exhaust) reduces the gas volume by a factor of 10 or more.

Most condensers are designed to provide subcooling of the gas stream and condensate to approximately 120–140°F (50–60°C). The major purpose of a condenser is to reduce the volume and moisture content of the gas stream before additional treatment; however, some malodors condense or dissolve in the condensate. Table 7-12 presents measurements of odor-removal efficiency for a direct-contact condenser and a surface condenser.

Contact Condensers A rendering-plant process stream can be cooled through the use of either a direct-contact condenser or a surface condenser. Either type will result in some odor reduction, because of the condensation of malodorous material. The use of a direct-contact condenser results in more efficient odor removal because of the scrubbing action associated with direct-contact cooling and because of greater liquid dilution.

Contact condensers are relatively uncomplicated pieces of equipment in which coolant, vapor, and condensate are brought into intimate contact. Water is the usual coolant. The direct-contact gas-liquid heat exchange can be accomplished in baffle-tray columns, spray chambers, barometric condensers, packed columns, or high-velocity jets. The emphasis in this discussion is on spray chambers and barometric condensers, because it is thought that these pieces of equipment are the most applicable and the most commonly used.

Spray chambers are among the simplest types of direct-contact coolers. The liquid coolant is introduced into the chamber by either spray nozzles or atomizers and brought into contact with a countercurrent gas stream. The moisture in the gas stream is condensed and exits with the cooling water. In the case of rendering plants, this contaminated water must be sent to a sewage-treatment facility. The noncondensables leave the spray chamber and are either sent to additional odor-control equipment or exhausted to the atmosphere.

In comparison with surface condensers, contact condensers are more flexible, simpler to operate, and less difficult to maintain. Although the initial equipment cost for a direct-contact condenser is less than that for a surface condenser designed to provide an equivalent amount of cooling, the operating costs for the former are higher. Because cooling water is not recycled, contact condensers require far more water than surface condensers and produce 10–20 times more wastewater. This large volume of water can create a disposal problem.[35]

The amount of treatment required by a particular sewage system is usually measured on one of two bases: the amount of suspended solids or the biologic oxygen demand (BOD), which is a measure of the amount of impurities on the basis of the amount of oxygen required to oxidize them. Liquid effluent from rendering plants (condenser condensate, washwater, etc.) is often high in both suspended solids and BOD.

Grease traps are usually installed before any condensate- or sewage-treatment facility. Where there is significant grease in the water, treatment with alum or similar chemicals could be required.

Rendering plants in rural areas may find it necessary to construct the required sewage-treatment plant. Three types of treatment are currently in favor:

- A preliminary catch basin involving gravity, screening, or other simple mechanical separation, followed by air flotation to remove grease. This is one of the cheapest and least effective treatments.
- A simple holding pond with no outlet. The waste is taken care of by ground seepage or evaporation from the pond surface or is used in irrigation. The odors near the pond may be objectionable.
- An anaerobic lagoon followed by natural aeration with a long ditch or another lagoon.

The estimation of control costs for the rendering industry is difficult,

TABLE 7-13 Batch-Type Model Rendering Plants[a]

Model	Annual Sales, $	No. Cookers	Raw Material Process Rate, lb/h	Peak Gas Volume to Condenser, acfm at 250°F
A	184,000	1	1,560	700
B	915,000	3	5,200	2,300
C	2,620,000	6	15,000	6,700
D	3,980,000	9	22,500	10,000

[a] Reprinted from Osag and Crane.[35 (p. 6-5)]

because of the wide range of plant sizes. Costs are presented in Table 7-13 for four model plants that are assumed to provide a representative view of the industry.[35]

Approximately 80% of the existing rendering plants are thought to be batch operations. Models A through D are representative of this portion of the industry. The remaining 20% are newer continuous operations that are often adequately controlled.

The capital investment, direct operating cost, and total annual cost of a direct-contact condenser for model plants A and B are listed in Table 7-14. The use of direct-contact condensers for plants similar to models C and D was not considered because of the large cooling-water requirements and the high sewage-treatment cost involved. Captive rendering plants and plants near large bodies of water, however, may be able to use kill-floor water or local water if they already have adequate BOD treatment for their wastewater. Operating costs for these plants might be lower than the figures given in Table 7-14, and contact condensers might be economically attractive for most of the model plants.

The costs of contact condensers for model plants A and B were obtained from direct communication with several equipment vendors and manufacturers. Estimates include the cost of a temperature-flow control on the cooling-water stream. All equipment costs reflect a January 1973 basis.

Water-Cooled Surface Condensers Rendering-plant vapors can also be condensed through the use of a surface condenser. In a surface condenser, a heat-transfer surface separates the coolant from the vapor stream. The advantage of this type of condenser is that it produces a much smaller quantity of condensate, which results in reduced sewage-treatment costs. Surface condensers also have lower operating costs, compared with those of contact condensers, because the coolant can be recycled.

TABLE 7-14 Capital Investment, Direct Operating Cost, and Total Annual Cost of a Direct-Contact Condenser[a]

	Model A	Model B
Raw material process rate, lb/h	1,560	5,200
Capital investment, $	2,900	5,000
Operation, h/yr	2,140	3,200
Operating costs, $/yr		
Water	720	3,500
Sewage	2,280	11,400
Electricity	20	70
Total direct cost,[b] $	3,000	15,000
Depreciation (10-yr straight line), $/yr	290	500
Interest, taxes, insurance (10%), $/yr	290	500
Maintenance (3%), $/yr	90	150
Total annual cost,[b] $/yr	3,700	16,000

[a] Reprinted from Osag and Crane.[35 (p. 6-6)]

[b] Total has been rounded to two significant figures.

The most commonly used surface condensers are shell-and-tube condensers with water as the coolant and extended-surface condensers with ambient air as the coolant. This section discusses the use of shell-and-tube (water-cooled) units. Most water-cooled surface condensers are of the shell-and-tube type with the coolant on the tube side and the condensing vapors on the shell side.

The vapor stream enters the condenser on the shell side and is condensed by contact with the cool tube surface. The condensate is then drained from the bottom of the condenser and sent to a sewage-treatment facility. Noncondensable gases are vented from the unit and sent to additional control equipment. The cooling water leaves the condenser and is usually sent to a cooling tower, where it is chilled and recycled to a condenser for reuse.

A surface condenser designed to provide subcooling of the condensate will reduce odor emission. In one case, a surface condenser that subcooled the condensate to 80°F (27°C) reduced odor emission by approximately 50%.[6] Although the 50% reduction was based on measured odor concentrations, the flow rate from the condenser was estimated because of its low velocity. This reduction is usually impractical, because cooling-tower water cannot be cooled enough to cool the condensate to 80°F at

TABLE 7-15 Capital Investment, Direct Operating Cost, and Total Annual Cost of a Water-Cooled Shell-and-Tube Condenser[a]

	Model B	Model C	Model D
Raw material process rate, lb/h	5,200	15,000	22,500
Capital investment, $	22,000	49,000	74,000
Operation, h/yr	3,200	3,200	3,200
Operating costs, $/yr	1,300	3,800	5,700
Depreciation (10-yr straight line), $/yr	2,200	4,900	7,400
Interest, taxes, insurance (10%), $/yr	2,200	4,900	7,400
Maintenance (3%), $/yr	600	1,500	2,200
Total annual cost,[b] $/yr	6,400	15,000	23,000

[a] Derived from Osag and Crane.[35 (p. 6-8)]

[b] Total has been rounded to two significant figures.

usual wet-bulb temperatures. These facts necessitate the venting of noncondensable gases to additional control equipment.

The cost of a water-cooled shell-and-tube condenser is presented for model plants B, C, and D in Table 7-15. It is thought that a plant similar in size to model A would elect to install a direct-contact condenser under most circumstances. Table 7-15 lists the capital investment, direct operating cost, and total annual cost associated with the use of a shell-and-tube condenser. All costs reflect a January 1973 basis.

Air-Cooled Surface Condensers Air-cooled surface condensers are used extensively where heat rejection from a process is possible with the use of ambient air as the coolant. Air-cooled condensers are usually constructed with either fin tubes or some other form of extended surface to increase the heat-transfer area. Air is placed in the fin side to take advantage of the large heat-transfer area. Condensation occurs inside the tubes.

Air-cooled condensers offer advantages over water-cooled units in requiring no water connections, cooling towers, or cooling-water treatment and in being simpler to install. Operating costs may be higher for an air-cooled condenser than for an equivalent water-cooled condenser, however, because of the larger power consumption of the fan.

The initial cost of air-cooled condensers for model plants B, C, and D is relatively high, making their use impractical in small plants. Table 7-16 presents estimates of the capital investment, direct operating cost, and total annual cost.

TABLE 7-16 Capital Investment, Direct Operating Cost, and Total Annual Cost of an Air-Cooled Surface Condenser[a]

	Model B	Model C	Model D
Raw material process rate, lb/h	5,200	15,000	22,500
Capital investment, $	16,000	44,000	58,000
Operation, h/yr	3,200	3,200	3,200
Operating costs, $/yr	1,100	3,300	4,700
Depreciation (10-yr straight line), $/yr	1,600	4,400	5,800
Interest, taxes, insurance (10%), $/yr	1,600	4,400	5,800
Maintenance (3%), $/yr	480	1,400	1,700
Total annual cost,[b] $/yr	4,800	13,000	18,000

[a] Derived from Osag and Crane.[35 (p. 6-11)]
[b] Total has been rounded to two significant figures.

Incinerators

Flame incineration is an effective control method for rendering-plant odors, provided that the incineration time and temperature are sufficient for complete oxidation of odorous vapors. In fact, many state and local control agencies have established odor-control regulations based on flame-incineration time and temperature.

The presence of the flame appears to have a very important effect on the efficiency of odor removal. It has been suggested that in the absence of a flame—using electric heat alone—much higher temperatures would be required to obtain the same efficiency achieved with a direct-flame oxidation system.

Incinerators have been used alone and in combination with other control equipment, principally condensers. Total incineration is used to control low-volume, low-moisture streams, such as cooker noncondensables. In some instances, a dust collector (centrifugal collector, baghouse, or precipitator) must be used ahead of an incinerator to remove particulate matter.

A rendering-plant burner has a mixing plate or other suitable design so that all the air used for full combustion comes from an odorous airstream. If the combustion air were taken from outside, a greater amount of fuel would be required to raise to 1,200°F (650°C) the additional air introduced into the system. The total gas stream is raised to 1,200°F to destroy odor, and about half the heat is recovered by exchange against the feed stream. From a practical standpoint, 65% heat recovery is about the maximum attainable.[35]

The usual incinerator has a steel outer shell lined with a refractory material. The purposes of the refractory are to protect the steel shell from direct exposure to high temperatures and corrosive materials and to improve thermal efficiency by limiting heat losses. The refractory may have any of a number of chemical compositions and physical forms. Most refractories used in incinerators are made up of heavy-duty fire clay and are in the form of bricks and castables. In some instances, high-temperature alloys are used as liners.

The fuel shortage and the rising costs of fuel should draw much attention to the improvement of afterburners and heat recovery, as well as furnishing added incentive for the improvement of scrubbers and of odor-testing to demonstrate their performance.

To minimize incineration costs, some rendering plants are using boiler fireboxes to incinerate cooker noncondensables and other low-volume, low-moisture reduction streams. Firebox incineration is attractive, because it lowers initial capital investment and operating costs by eliminating the need for an incinerator and its associated fuel, operation, and maintenance.

The Los Angeles County Air Pollution Control District tested four rendering plants that use incineration to control cookers, blood-dryers, and feather-meal dryers. Incineration units at these plants were operated at 1,200°F (650°C) with a stream retention time of 0.3 s. Outlet emission ranged from 70 to 140 o.u./scf. These results represented an odor-removal efficiency of better than 99%.[49]

Incinerator costs were determined for low-moisture, low-volume streams from cookers, presses, process tanks, and dryers. Condenser costs must be added to these figures to obtain the total cost of control. Total costs are listed in Table 7-17. The capital investment of flame incineration is proportionately higher for smaller plants. Annual costs are more in line with incinerator sizes, but are still proportionately higher for smaller plants, because of depreciation, taxes, and maintenance costs.

Condenser-incinerator combinations are usually more practical than incinerators, as well as more efficient, especially for controlling cooker streams. When a condenser is used to remove steam from cooker streams before incineration, the volume of the stream is reduced and a considerable portion of malodors is removed with the condensate. Other streams that contain 15–40% moisture may also warrant the use of a condenser. Such factors as volumes, exit temperatures, fuel costs, water availability, and equipment cost determine condenser feasibility for these streams.

Tests by the Los Angeles County Air Pollution Control District showed

TABLE 7-17 Capital Investment, Direct Operating Cost, and Total Annual Cost of a Condenser-Incinerator System[a]

	Model A	Model B	Model C	Model D
Raw material process rate, lb/h	1,560	5,200	15,000	22,500
Capital investment, $				
Surface condenser (water-cooled)		10,000	20,000	31,000
Cooling tower		12,000	29,000	43,000
Contact condenser	2,900			
Incinerator	9,500	13,000	18,000	20,000
Total capital investment	12,000	35,000	67,000	94,000
Operation, h/yr	2,140	3,200	3,200	3,200
Operating costs, $/yr				
Fuel	1,000	4,600	9,200	15,000
Water	720	350	1,000	1,600
Sewage	2,280	570	1,700	2,500
Electricity	40	450	1,300	2,200
Total direct cost,[b] $	4,000	6,000	13,000	21,000
Depreciation (10-yr straight line), $/yr	1,200	3,500	6,700	9,400
Interest, taxes, insurance (10%), $/yr	1,200	3,500		
Maintenance (3%), $/yr	650	1,400	2,600	3,400
Total annual cost,[b] $/yr	7,100	14,000	29,000	43,000

[a] Derived from Osag and Crane.[35 (pp. 6-18 and 6-19)]

[b] Total has been rounded to two significant figures.

the following average emission rates for rendering plants controlled by condenser-incinerator systems.[49]

Contact condenser and incinerator:
- Three plants with cookers and blood-dryers: 63 o.u./scf
- Six plants with cookers: 27 o.u./scf

Surface condenser and incinerator:
- Four plants with cookers and blood-dryers: 99 o.u./scf
- Three plants with cookers using boiler incineration: 27 o.u./scf

One plant with cooker, press, and processing tank: 85 o.u./scf

Scrubbing

A control technique that has received increasing attention in recent years is chemical scrubbing. Chemical scrubbing is essentially a gas-absorption technique whereby one or more constituents of a gas stream are removed by being dissolved in a selective liquid solvent. In addition to being dissolved, the absorbed gases may react chemically with the scrubbing liquid. Scrubbing offers economic advantages over incineration methods in the treatment of large volumes of air containing malodorous contaminants at relatively low concentrations and saturated airstreams.

One limitation on the use of chemical scrubbers has been the inlet concentration of malodorous gases. Odor concentrations greater than about 10,000 or 20,000 o.u./ft^3 have complicated the problem of providing adequate gas-liquid contact time in the scrubber.[35] This restriction would preclude the use of scrubbers on a basis similar to incinerators, which are generally designed to treat low volumes of highly concentrated odors. The usual solution to the odor-inlet limitation of scrubbers has been to reduce the odor concentration of highly odorous streams to 10,000–20,000 o.u./ft^3 by mixing them with percolator-pan ventilation air, expeller exhaust, and general plant-ventilation air. Scrubbing systems are therefore designed to treat large volumes of air that include most of the rendering-plant airstreams. One three-stage system, however, is currently being installed to treat high-intensity odors.

Scrubbers are designed to provide thorough contact between the gas and liquid streams to allow interphase diffusion of the gases being absorbed. The required degree of contact can be provided by several types of equipment: bubble-plate columns, jet scrubbers, packed towers, spray chambers, and venturi scrubbers have all been used for gas-absorption work.

Acid, alkaline, and strong oxidizing solutions have been used to control rendering-plant malodors with various degrees of success. It is conceivable that alkaline or acid scrubbers could be effective control devices if all odorous compounds reacted in the same manner, but the mixture of malodorous gases encountered during the rendering process is not homogeneous, from an acid-base standpoint. Some success has been reported for a system using both acid and alkaline scrubbing solutions in a two-stage, spray-chamber unit.

Strong oxidizing solutions—such as chlorine dioxide, sodium hypochlorite, and potassium permanganate—are reported to be effective means of eliminating odors. Given sufficiently vigorous reaction conditions, potassium permanganate can oxidize most organic compounds. Under the relatively mild conditions in most scrubbing systems, however, not all organic compounds will readily react. Thus, odorous compounds that are susceptible to oxidative degradation under relatively mild conditions are aldehydes, reduced sulfur compounds, unsaturated ketones and hydrocarbons, phenols, amines, hydrogen sulfide, and sulfur dioxide.[2] Among the compounds that resist oxidation under these conditions are saturated organic acids and hydrocarbons, ketones, and some nitrogen ring compounds. Potassium permanganate scrubbing solutions have the disadvantage of being more expensive and requiring more extensive sewage treatment than sodium hypochlorite solutions. Sodium hypochlorite has a greater reaction rate than potassium permanganate and is effective against most of the same odorants.[12]

A scrubbing system for high-intensity odors was evaluated by Prokop.[37] The system consisted of a venturi scrubber and two packed towers treating air at 7,500 cfm that entered the system at 150°F (65°C). The venturi used trisodium phosphate at a pH of 8; the first packed tower, phosphoric acid at a pH of 3; and the second tower, sodium hypochlorite and sodium hydroxide at a pH of 10. This system achieved odor reduction equivalent to or better than incineration. Prokop made the point that, as natural gas becomes increasingly expensive and less available, scrubbing devices will become more popular. The installed cost for the scrubber system evaluated by Prokop[37] was $70,000, and the operating cost was $6.80/h.

The decision of whether to install a full two-stage system in a specific rendering plant would depend on the nature of the gases being scrubbed. A two-stage system would be required to treat gas streams with a high particle loading or a high odor concentration, i.e., the exhaust from blood-dryers or cooker off-gases. In the treatment of general ventilation air, percolator-pan ventilation air, or expeller exhausts, the venturi scrubber could be eliminated.[35]

Available information indicates that scrubbing with a solution of either potassium permanganate or sodium hypochlorite may reduce odor concentrations from 25,000 o.u./ft^3 to 50–200 o.u./ft^3.[35] The 11 plants that were tested had flow rates ranging from 6,000 to 55,000 ft^3/min. An EPA observer who participated in tests at one of these 11 plants believed that the above performance was demonstrated in the case that he observed.

The capital investment, direct operating cost, and total annual cost of a chemical scrubbing system are presented in Table 7-18. Condenser costs must be added to these figures to obtain the total costs of control.

SUMMARY

There are over 700 rendering plants in the United States. By the nature of the rendering process, odors are produced that require treatment if complaints are to be avoided. Odors are produced by the cooker exhaust

TABLE 7-18 Capital Investment, Direct Operating Cost, and Total Annual Cost of a Chemical Scrubbing System[a]

	Model B	Model C	Model D
Raw material process rate, lb/h	5,200	15,000	22,500
Capital investment, $	30,000	42,000	42,000
Operation, h/yr	3,200	3,200	3,200
Operating costs, $/yr			
Water	420	660	660
Sewage	1,300	2,000	2,000
Chemicals (sodium hypochlorite)	540	710	710
Electricity	1,400	2,200	2,200
Labor	2000	2,000	2,000
Total direct costs,[b] $	5,700	7,600	7,600
Depreciation (10-yr straight line), $/yr	3,000	4,200	4,200
Interest, taxes, insurance (10%), $/yr	3,000	4,200	4,200
Maintenance (3%), $/yr	900	1,300	1,300
Total annual cost,[b] $/yr	13,000	17,000	17,000

[a] Reprinted from Osag and Crane.[35 (p. 6-26)]

[b] Total has been rounded to two significant figures.

and from the screw press, because these operations involve heated material. Other sources of less intense odors require strategic placement of suction pickup vents. The age of the raw material is important in determining the intensity of the odors produced.

An extensive technology exists for the control of rendering-plant odors; when properly installed and operated, the equipment provides satisfactory odor control. Treatment systems for odorous emission generally include one or more of the following: condensers (direct-contact or surface), incinerators, and scrubbers. Each of these devices has advantages for particular plant sizes and designs.

Several states and metropolitan areas have odor-control regulations governing rendering plants. Most of these regulations are currently based on a specified time and temperature of natural-gas incineration or an equivalent degree of discharge.

FISHERIES AND FISH-PROCESSING

Choice parts of fresh fish are consumed directly or are frozen or canned for human consumption after filleting and packing. The remainder of the edible fish and inedible species are converted to byproducts not for human consumption that include canned catfood, fish oil, and a dry, defatted meal consisting of ground flesh, skin, and bone that is used as a high-protein animal feed supplement. Byproduct reduction plants that cluster around filleting, freezing, and canning plants to process scrap material called gurry and inedible fish brought to land by commercial fishermen, including menhaden, are responsible for most of the malodorous air contamination attributed to food-fish preparation. It is clear that, at the start, gurry has the same quality of freshness and odorlessness as the food portions from which it was separated.

Fish offal is a waste product of the packers that they would find it difficult and costly to dispose of in a sanitary manner if the byproduct processors did not take it off their hands. Therefore, they receive very little recompense for their waste and are disinclined to ice or otherwise refrigerate it while it is in storage at their plants or in transit to a processing plant. The undesirable consequences of this are twofold: the offal becomes a major source of malodors at the fresh-fish-processing plant if allowed to accumulate in warm weather, and the waste is already putrefying when it arrives at the byproduct-processing plant, where its malodorous emission increases substantially when it is again stored without refrigeration. Similar economic considerations press on all the fishermen whose principal effort is directed toward catching menhaden and trash fish for the byproduct-processing plants. The result is that, during warm weather, these catches arrive at the byproduct-processing plant in an advanced state of putrefaction, and this greatly complicates the odor-control program of the processing plant. The solution to this odor problem is obviously refrigeration, but economic considerations make it unacceptable to producers.[18] Therefore, fish byproduct-processing plants start off with a built-in obstacle to odor control, although it does not seem to affect adversely their processing regimen or the quality of their final product.

Freezing plants generally start by filleting the fresh fish, trimming the pieces to size, and packaging for freezing in the raw state or after cooking. Although the odor of fresh fish frying in fresh oil is generally considered pleasant, commercial frying is frequently a source of odor complaints, because of the intensity of the odor and its unremitting nature. Odor reduction requires the application of off-gas incineration or chemical scrubbing. Fish-canning plants use one of two basic processes: in the "wet-fish" method, raw trimmed fish are cooked directly in unsealed cans; and in the "precooked" process, whole, eviscerated fish are cooked in steamers, and the cooked parts are hand-packed into cans that are then capped and sterilized with steam. In the wet-fish canning process, after cooking, the cans are drained of condensed steam, fish juices, and oils and refilled with tomato sauce, oil, or other liquid before being sealed and sterilized. In both processes, the drained fish-cooking liquids, called stickwater, are centrifuged to separate oil from the soluble protein, and both streams are combined with the much larger byproduct-processing operations that are carried out for raw fish offal and rejected, cooked fish parts.[7] The canning processes for fresh or frozen fish that is intended for human consumption do not themselves produce inherently unpleasant odors, beyond the likelihood of high intensity, and, at least during the height of the canning season, the exposure may be unremitting. Most of the volatile fish-cooking odors produced during canning are accompanied by copious amounts of steam, so process off-gases can be satisfactorily deodorized by cold-water scrubbing to induce condensation and absorption. Addition of chlorine, chlorine dioxide, or ozone to the scrubbing water is highly desirable to prevent microbiologic degradation of the soluble fish protein and thereby avoid the production of malodors within the scrubber itself.

The highly putrescible nature of the raw and cooked products and byproducts associated with filleting, freezing, and canning of fish make scrupulous attention to plant sanitation a requirement of highest priority, if malodorous emission from these facilities is to be avoided. Inasmuch as human-food processing is under strict sanitary surveillance by health and other government authorities, this part of the process is seldom a serious source of foul odors, but the handling of waste byproducts may be separated from food-production areas and maintained under far less supervision and less rigid processing controls. This is especially true for plants devoted exclusively to byproduct-processing, and these operations have "been associated with air pollution wherever a plant is located near a population center."[18]

Large quantities of the inedible portions of fresh fish and whole fish that are not up to standards for human consumption are converted to canned catfood. Sanitation standards in catfood plants are not up to those used in the production of human food, and odor complaints occur, although not as frequently as for fishmeal plants that use byproducts that even the catfood canners reject. Because the canned fish product must satisfy the pet owner, as well as the cat, the fish products that are used are in a reasonable state of preservation when received; but poor plant sanitation and poor handling of wastes are the major sources of malodors from petfood processors. Installation of fish-cooker off-gas scrubbers that use chlorine-containing water and strict attention to plant sanitation, including prompt removal of wastes, are usually effective for avoiding neighborhood odor complaints about catfood canners.

Although fish-byproduct-processing factories are generally referred to as "fishmeal plants," the byproducts produced usually include fish oil and fish solubles ("the molasses-like concentrate containing soluble proteins and vitamins that have been extracted from fish flesh by cooking processes"[2]), in addition to fishmeal ("a solid product consisting largely of protein obtained by removing most of the water and some oil from fish or fish waste"[51]). If the dried fishmeal is to be used as a human food supplement (not permitted in the United States when offal and unrefrigerated trashfish and gurry are used as the starting materials), it will be completely defatted by organic-solvent extraction to remove rancid oil flavors.

As already noted, the fish products may be in an advanced state of decomposition when delivered to the fishmeal plant and usually become further degraded by prolonged holding in unrefrigerated raw-fish storage silos. Therefore, means must be provided to exhaust-ventilate raw-fish delivery hoppers, conveyor transfer points, and storage silos and to pass the exhaust air through a scrubber to which a powerful oxidizing substance (such as chlorine, chlorine dioxide, or ozone) is added, to destroy absorbed odorous gases and particles and to prevent microbiologic fermentation that produces additional malodors. Many malodorous compounds have been found in the gases coming from rotting fish. Among the most prominent are ammonia, trimethylamine, triethylamine, acrolein, butyric acid, indole, methyl mercaptan, dimethyl sulfide, and hydrogen sulfide—all notoriously foul-smelling and, except for ammonia, all detectable in concentrations measurable in parts per billion.

The processing off-gases are warm, are saturated with moisture, and contain condensed steam droplets. Because they are low in noncondensable gases, their volume can be further reduced by passage through a shell-and-tube condenser with seawater as the coolant. Large amounts of water condense here and trap a major portion of the solids and water-soluble gases that were volatilized by the steam treatment. The condensed liquid is returned to the stickwater steam. The noncondensable gases have now been sufficiently reduced in volume that they can be used as combustion air for the boilers that generate steam for the fish cookers and multiple-effect evaporators, thereby destroying residual odors by incineration.

The moist dewatered and defatted solids from the press are sent to a meal grinder and then to the fishmeal dryer, from which the most serious odor emission originates. The worst odors are produced by direct-flame rotary dryers in which the wet fish solids flow concurrently with hot combustion gases generated by a gas or oil burner at one end of the rotary dryer. "Hot products of combustion are mixed with air to provide a temperature of 400° to 1,000°F [about 200 to 540°C] at the point where the wet meal is initially contacted."[7] In addition to volatilization of typical rotten-fish odors as the fish dries, the temperature of the combustion gases is difficult to regulate in a direct-flame dryer, and some of the flesh inevitably becomes scorched; this gives rise to a whole new spectrum of intense burnt-fish malodors. In addition to water vapor and malodorous gases, substantial amounts of solid meal are entrained in the flue gases, and these must be separated in a cyclone or other particle collector before the gases can be sent to a scrubber for deodorization for later release to the atmosphere. Direct-flame dryers are favored by meal manufacturers, because they are rapid and fuel-efficient; but experience demonstrates that it is exceedingly difficult to prevent emission of foul odors and finely divided solid meal from this device. Therefore, the indirect meal-dryer is preferred by air-pollution control officials. In this device, heat is supplied through steam pipes arranged longitudinally inside the rotary dryer. Drying air is blown through the dryer countercurrently to the direction of the meal. Even though less meal becomes entrained in the indirect-dryer exit gases and scorching of the meal from overheating is reduced, such dryers also are a major source of odors from fishmeal plants.

Typical fishmeal dryers exhaust moisture-laden air containing a very high concentration of malodorous substances and finely divided solids at 10,000–20,000 cfm. For example, under average conditions, a direct-fired rotary dryer operating at high temperature will produce 10 tons (9 t) of dried meal per hour and exhaust air at 15,000–20,000 cfm that requires dilution with 40,000 times that volume to reach the odor threshold. Control of this emission has occupied air-cleaning engineers all over the world for the last 50 yr. Current wisdom decrees that effective odor control of off-gases from meal-dryers shall consist of multistage chemical scrubbers that use oxidizing and neutralizing scrubbing liquors in series. Other devices that have proved effective in service have been described earlier. Although gas incineration has been used for fishmeal off-gas odor control, the current cost of fuel has shifted attention from combustion to scrubbing as the most cost-effective deodorization process for this industry. Gaseous effluent from other steps in the process—e.g., evaporators, grinders, and centrifuges—is conveniently deodorized by increasing the capacity of the dryer off-gas cleaner.

A typical off-gas scrubber for a typical fish-byproduct plant has been described[7] as follows: The process depends largely on the reaction of chlorine gas with odorous compounds at dryer exit temperatures. Gases from the dryer are first directed through a cyclone separator to remove fine particles. Chlorine is then added at a rate calculated to provide a concentration of 20 ppm by volume in the gas stream. The reaction is allowed to proceed at about 200°F (95°C), the dryer exit temperature, in the ductwork for approximately 0.6 s before the stream is chilled and scrubbed with seawater in a packed tower. Gases pass up through the packing countercurrently to the seawater.

In Figure 7-2, odor concentrations from the scrubber exit are plotted against the chlorine addition rate at constant gas and seawater throughput. As can be seen from the curve, odors reach a minimum when the chlorinate is at about 20 ppm. When more than 20 ppm is added, chlorine odors become readily detectable in treated gases, and odor concentrations tend to increase. All the odor measurements used to draw this curve were made on dryer-gas samples taken between 170 and 205°F (77 and 96°C) when there was essentially no overheating of meal in the dryer. This method provides an overall odor reduction of 95–99% when fresh fish scrap is being processed in the dryer. Chlorination itself provides a 50–80% reduction in odor concentration. Scrubbing reduces the remaining odor concentration by another 50–80%. Condensation provides a 12–22% reduction in volume, depending on the original moisture content of the gases.

The exact mechanism of the chlorination reaction is uncertain, but it is assumed that chlorine reacts with odorous compounds, probably amines, to form additional products that are less odorous than the original compounds. Chlorine is not considered to be a sufficiently strong oxidizing agent to oxidize fully the odorous organic materials present in the dryer gases.

FIELD-BURNING

Field-burning, sometimes called "open burning," was widely practiced, before the current concern for clean air, by farmers, road-builders, those engaged in land-clearing for building construction, and municipalities for trash disposal. Widespread burning of domestic trash by individual households was, at one time, thought to be a primary cause of smog in Los Angeles County and was outlawed in the 1950's. Although this regulation failed to cure the smog problem (it was later found to result from photochemical reactions among automobile emissions), prohibition of trash-burning was beneficial in eliminating the foul odors and smoke that resulted from the incomplete combustion of household wastes. Gradually, open field-burning of municipal waste collections was outlawed in most organized communities. This prohibition was assisted by the identification of vast burning dumps as the cause of the widespread detrimental effects of "Yokohama asthma" and "New Orleans asthma." Lesser burning operations, including fall leaf-burning, were slowly coming to a halt when the passage of the Clean Air Act of 1970 imposed a federal ban on all open field-burning. Because open field-burning had few friends, this ban was

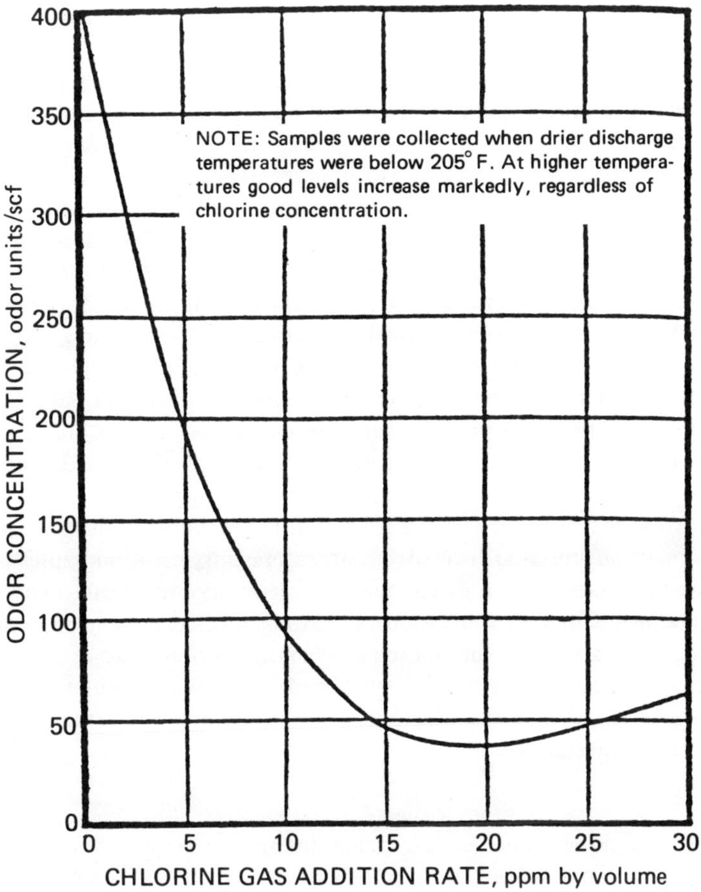

FIGURE 7-2 Exit odor concentrations from a chlorinator-scrubber as a function of the chlorine gas addition rate. Temperatures of gas discharged from dryer are less than 205°F (96°C). Reprinted from Danielson.[7]

widely hailed as a prompt benefit of the new clean-air legislation.

Sanitary landfilling replaced burning dumps, although truly sanitary landfills remain few to this day. Burial of tree stumps, large brush, and wood chips replaced field-burning in the preparation of road-building and construction sites; but it was soon found that land-clearing waste made poor landfill, because it could not be properly compacted and the fill soon settled and caved, creating dangerous conditions.

The burning of agricultural waste was stoutly defended as an important means of destroying insect pests and preventing them from carrying over from one growing season to the next in large numbers. Therefore, it gradually became evident to air-pollution control officials that an absolute ban on open field-burning would have to be rescinded, and attention shifted to working out methods by which essential field-burning, following a successful petition for variance to the regulation, might be conducted in a manner that would maximize official control and minimize air-pollution effects. Generally, these methods involve stacking the waste to promote vigorous open-flame burning and minimize smoldering with the production of amines, reduced sulfur compounds, and other foul-smelling products that are generated during incomplete combustion of mixed wastes; orienting the piles or windrows of waste so as to obtain maximal oxygenation by the wind during burning; and permitting field-burning only on days and during the hours when meteorologic conditions are favorable for rapid dispersion and dilution of combustion products and when the wind direction favors the minimizing of human exposure to the volatile combustion products.

Malodorous emission from field-burning results from incomplete combustion products formed during the burning of organic matter. The kinds of malodorous combustion products that are formed during open dump-burning and burning of land-clearing debris are very similar, differing only in the quantities of each that are produced. It has been estimated that the burning of large quantities of land-clearing debris emits the following quantities of pollutants:[17]

	lb/ton burned
Total organic gases, excluding methane	166
Total olefins	30
Total oxygenates	59
Total aromatics	11

Although specific compounds have not been identified, it is clear that many of the most malodorous substances fit into these categories, and it is well known that open field-burning usually evokes odor complaints.

Hydrocarbon emission from typical agricultural burning is as follows:[3,8]

	lb/ton burned
Fruit prunings	13.9
Barley straw	18.2
Grasslands	10.6

An improved method of open field-burning intended to reduce air pollution and the nuisances normally associated with open burning has been developed and put into effect in Connecticut. The method depends on strict separation of demolition and clearing material into combustible and noncombustible portions. Metals, asphalt shingles, tile, and siding are considered to be noncombustible and are removed. Piles of material are allowed to dry before being burned.[44] Burning is facilitated when the material is allowed to dry to about 50% moisture content. Unfortunately, dry fuel and brisk winds "are decidedly not favorable from a fire control standpoint."[13] Piles to be burned are placed on flat ground with access from all sides. Smokeless combustion-boosters are used to ignite and maintain fires. Material is stockpiled no more than 10 ft (3 m) high in cigar-shaped windrows, with the length of the pile parallel with the wind direction. Ignition takes place high in the pile and at the downwind end, to permit smoke to be burned in the flames by a secondary combustion effect. Portable fans are used to feed air to the fire and control its progress through the pile, making more complete burnout possible. Stoking the fires aids complete combustion and speeds the rate at which material can be burned, but increases fly-ash emission. A crawler tractor with a bucket loader has been found best for this purpose. If the practice is allowed, there must be strict operational, as well as meteorologic, control.[44]

Meteorologic control of open field-burning faces a number of difficulties in addition to the proper preparation and ignition of the piles.[13] It is difficult or impossible to predict with certainty when weather conditions will be satisfactory, so that burning and supervisory crews can be notified in adequate time to assemble on the site. Contractors' work-scheduling problems and penalties associated with failure to complete a project on time make prolonged delays, while workers wait for satisfactory weather conditions, extremely stressful. ("Experience shows that as the specified conditions become more appropriate from an air pollution control standpoint, they become more meteorologically unique."[13]) Prolonged delays during which large amounts of combustible debris are piled up ready for burning are likely to be extremely stressful for the fire marshall. And weather conditions that are optimal for protection of close-in areas are likely to be less than optimal for those farther downwind, and vice versa. Nevertheless, these difficulties must be endured if open field-burning is to produce the least possible damage to the atmospheric environment, including odor complaints.

A form of open burning has been conducted at sea in locations more or less remote from inhabited land on the premise that nighttime off-shore breezes will blow the combustion products out to sea and, if this does not occur, the added distance will tend to make them innocuous by dilution. This policy has resulted in the use of open barges and LST's for burning of demolition and construction wastes at sea off Boston, New York City, and other coastal cities. Open barge-burning in lower New York Bay produced downwind particle concentrations of 250 $\mu g/m^3$ for the first 5 h of burning. Therefore, lower-bay burning was replaced by burning aboard LST's 12 miles (about 20 km) from shore. Open barge-burning off Boston, within 2–3 miles (about 3–5 km) of shore, proved unsatisfactory from the standpoint of on-shore air pollution and was abandoned. For all these operations, sea dumping of residues is prohibited, although wind action often produces considerable surface litter in the vicinity of open vessels.

Open burning far enough off shore in the open ocean appears to be an effective and innocuous method for disposing of bulky combustible wastes, but serious safety problems are associated with this type of operation, because it generally requires about 72 h for a fire to burn out and for the temperature of the vessel to fall to a safe point. During this period, unfavorable weather can produce serious dangers for an unmanned vessel. This kind of operation has application only for coastal communities. If it is to continue, improvements in material-handling, burning technology, and administrative controls are urgently required.[44] A more satisfactory alternative is to develop economically and aesthetically acceptable methods for salvaging and recycling useful components from agricultural residues, brush and trees from land-clearing, demolition waste, and bulky solid-waste collections to replace open field-burning. Such treatment steps as grinding, shredding, chipping, and composting would have to be followed by mechanical automatic separations to produce clean, concentrated stocks for reuse.

REFERENCES

1. American Society of Agricultural Engineers. 1978 Agricultural Engineers Yearbook, p. 465. St. Joseph, Mo.: American Society of Agricultural Engineers, 1978.

2. Anderson, C. E. Chemical control of odors. Poll. Eng. 4(5):20–21, 1972.

3. Boubel, R. W., E. F. Darley, and E. A. Schuck. Emissions from burning grass stubble and straw, J. Air Poll. Cont. Assoc. 19:497–500, 1969.

4. Burgwald, T. A. Identification of Chemical Constituents in Rendering Industry Odor Emissions, p. 41. Project No. C8172. Chicago, Ill.: IIT Research Institute, 1971.

5. Burnett, W. E., and N. C. Dondero. The Control of Air Pollution (Odors) from Animal Wastes—Evaluation of Commercial Odor Control Products by an Organoleptic Test. ASAE Paper No. 68-909. Presented at the American Society of Agricultural Engineers Winter Meeting, Chicago, Ill., Dec. 10–13, 1968. 25 pp.

6. Danielson, J. A., Ed. Air Pollution Engineering Manual, County of Los Angeles Air Pollution Control District. Publ. No. 999-AP-40. Cincinnati: U.S. Department of Health, Education, and Welfare, Public Health Service, National Center for Air Pollution Control, 1967. 919 pp.

7. Danielson, J. A., Ed. Air Pollution Engineering Manual, County of Los Angeles Air Pollution Control District. 2nd ed. Publ. No. 999-AP-40. Cincinnati: U.S. Department of

Health, Education, and Welfare, Public Health Service, National Center for Air Pollution Control, 1973.

8. Darley, E. F., F. A. Burleson, E. H. Mateer, J. T. Middleton, and V. P. Osterli. Contribution of burning of agricultural wastes to photochemical air pollution. J. Air Poll. Cont. Assoc. 16:685–690, 1966.

9. Day, D. L. Liquid hog manure can be deodorized by treatment with chlorine or lime. Ill. Res. 8(3):16, 1966.

10. Denmead, O. T., J. R. Simpson, and J. R. Freney. Ammonia flux into the atmosphere from a grazed pasture. Science 185:609–610, 1974.

11. Development Planning and Research Associates. Economic Impact of Water Pollution Controls on Selected Food Industries. Vol. III, Costs, Capabilities, and Economic Impact of Water Pollution Controls on Feedlots Industry. Contracts No. WQ5AC011 and No. WQ5AC063. Washington, D.C.: National Commission on Water Quality, November 1975. 390 pp.

12. Doty, D. M., R. H. Snow, and H. G. Reilich. Investigation of Odor Control in the Rendering Industry. Contract No. 68-02-0260. Washington, D.C.: U.S. Environmental Protection Agency, Office of Research Monitoring, 1972. 142 pp.

13. Duckworth, S. The meteorologically scheduled open burn. J. Air Poll. Cont. Assoc. 15:274–277, 1965.

14. Eliott, L. F., G. E. Schuman, and F. G. Viets, Jr. Volatilization of nitrogen-containing compounds from beef cattle areas. Proc. Soil Sci. Soc. Amer. 35:752–755, 1971.

15. Faith, W. L. Odor control in cattle feed yards. J. Air Poll. Cont. Assoc. 14:459–460, 1964.

16. Faith, W. L. Odor Control in the Inedible Rendering Industry, p. 7. Report to the Fats and Protein Research Foundation. Des Plaines, Ill.: Fats and Proteins Research Foundation, 1967.

17. Feldstein, M., S. Duckworth, H. C. Wohlers, and B. Linsky. The contribution of the open burning of land clearing debris to air pollution. J. Air Poll. Cont. Assoc. 13:542–545, 1963.

18. First, M. W. Control of systems, processes, and operations, pp. 4–39. In A. C. Stern, Ed. Air Pollution, 3rd ed., Vol. IV. New York: Academic Press, 1977.

19. Ford, J. P., and W. L. Ulich. Odor control for confined beef cattle feedlots, pp. 189–204. In Proceedings of the First Annual Symposium on Air Pollution Control in the Southwest. College Station, Tex.: Texas A & M University, 1973.

20. Hill, D. T., and C. L. Barth. Quantitative prediction of odor intensity. Trans. Amer. Soc. Agr. Eng. 19:939–944, 1976.

21. Hill, D. T., and C. L. Barth. Removal of gaseous ammonia and methylamine using ozone. Trans. Amer. Soc. Agr. Eng. 19:935–938, 944, 1976.

22. Hutchinson, G. L., and F. G. Viets, Jr. Nitrogen enrichment of surface water by absorption of ammonia volatilized from cattle feedlots. Science 166:514–515, 1969.

23. Ingram, S. H., R. C. Albin, C. D. Jones, A. M. Lennon, L. F. Tribble, L. B. Porter, and C. T. Gaskins. Swine fecal odor as affected by feed additives. J. Anim. Sci. 36:207, 1973. (abstract)

24. Koelliker, J. K., and J. R. Miner. Desorption of ammonia from anaerobic lagoons. Trans. Amer. Soc. Agr. Eng. 16:148–151, 1973.

25. Kromer, G. W. U.S. tallow and grease production and marketing trends. Fats and Oils Situation. FOS-260:17–25, November 1971.

26. Lindvall, T., O. Norén, and L. Thyselius. Odor reduction for liquid manure systems. Trans. Amer. Soc. Agr. Eng. 17:508–512, 1974.

27. Luebs, R. E., A. E. Laag, and K. R. Davis. Ammonia and related gases emanating from a large dairy area. Cal. Agr. 27(2):10–12, 1973.

28. Miner, J. R. Evaluation of alternate approaches to control of odors from animal feedlots. Final report. National Science Foundation Grant No. ESR74-23211. Moscow, Idaho: Idaho Research Foundation, 1975. 83 pp.

29. Miner, J. R. Management of odors associated with livestock production, pp. 378–380. In Managing Livestock Wastes, Proceedings of the Third International Symposium on Livestock Wastes. ASAE Publ. PROC-275. St. Joseph, Mo.: American Society of Agricultural Engineers, 1975.

30. Miner, J. R. Odors from confined livestock production. Report No. EPA-660/2-74-023. Washington, D.C.: U.S. Environmental Protection Agency, 1974. 130 pp.

31. Miner, J. R., and T. E. Hazen. Ammonia and amines: Components of swine building odor. Trans. Amer. Soc. Agr. Eng. 12:772–774, 1969.

32. Miner, J. R., M. D. Kelly, and A. W. Anderson. Identification and measurement of volatile compounds within a swine building and measurement of ammonia evolution rates from manure-covered surfaces, pp. 351–353. In Managing Livestock Wastes, Proceedings of the Third International Symposium on Livestock Wastes. ASAE Publ. PROC-275. St. Joseph, Mo.: American Society of Agricultural Engineers, 1975.

33. Miner, J. R., and R. C. Stroh. Controlling feedlot surface odor emission rates by application of commercial products. Trans. Amer. Soc. Agr. Eng. 19:533–538, 1976.

34. Moncrieff, R. W. An instrument for measuring and classifying odors. J. Appl. Physiol. 16:742–749, 1961.

35. Osag, T. R., and G. B. Crane. Control of Odors from Inedibles-Rendering Plants. Report No. EPA-450/1-74-006. Research Triangle Park, N.C.: U.S. Environmental Protection Agency, Office of Air Quality Planning and Standards, July 1974. 60 pp.

36. Peters, J. A., and T. R. Blackwood. Source Assessment: Beef Cattle Feedlots. Report No. EPA-600/2-77-107. Washington, D.C.: U.S. Environmental Protection Agency, June 1977. 102 pp.

37. Prokop, W. H. Wet Scrubbing of High Intensity Odors from Rendering Plants, pp. 153–164. In State-of-the-Art of Odor Control Technology II—Specialty Conference. Pittsburgh: Air Pollution Control Association, 1977.

38. Rowe, N. R. Odor control with activated charcoal. J. Air Poll. Cont. Assoc. 13:150–153, 1963.

39. Sagebrush for odor control: In the feed or in the manure? Feedlot Manag. 14(5):74, 1972.

40. Seltzer, W., S. G. Moum, and T. M. Goldhaft. A method for the treatment of animal wastes to control ammonia and other odors. Poultry Sci. 48:1912–1918, 1969.

41. Sweeten, J. M. Environmental protection requirements for swine operations. Fact sheet. No. L-1302. College Station, Tex.: Texas Agricultural Extension Service, 1974. 4 pp.

42. Sweeten, J. M., D. L. Reddell, and H. B. H. Cooper. Odor measurement for livestock feeding operations, pp. 55–80. In Proceedings of the Specialty Conference on Control Technology for Agricultural Air Pollutants. Memphis, Tenn.: Air Pollution Control Association, 1974.

43. Sweeten, J. M., D. L. Reddell, L. M. Schake, and B. Garner. Odor intensities at cattle feedlots, pp. 205–239. In Proceedings of the First Annual Symposium on Air Pollution Control in the Southwest. College Station, Tex.: Texas A & M University, 1973.

44. TS-2.4 Refuse Committee. Disposal of bulky solid wastes. J. Air Poll. Cont. Assoc. 22:858–864, 1972.

45. Turner, D. B. Workbook of atmospheric Dispersion Estimates (Revised). Publ. No.

AP-26. Cincinnati: U.S. Department of Health, Education, and Welfare, Public Health Service, Division of Air Pollution, 1970.

46. U.S. Department of Agriculture. Agricultural Statistics—1966. Washington, D.C.: U.S. Government Printing Office, 1967.

47. U.S. Department of Agriculture. 1977 Handbook of Agricultural Charts. Agricultural Handbook No. 524. Dorothy Mayes, Ed. Washington, D.C.: USDA, 1977. 124 pp.

48. U.S. Environmental Protection Agency. Air Pollution Aspects of Odors. 2 vols. February 1971. (unpublished)

49. U.S. Environmental Protection Agency. Background Information for Proposed New Source Performance Standards for Rendering Plant Odors. Technical Report Number 10. U.S. EPA, Research Triangle Park, N.C., 1973. (unpublished)

50. Willrich, T. L., and J. R. Miner. Litigation experiences of five livestock and poultry producers, pp. 99–101. In Livestock Waste Management and Pollution Abatement, Proceedings of the International Symposium on Livestock Wastes. ASAE Publ. PROC-271. St. Joseph, Mo.: American Society of Agricultural Engineers, 1971.

51. Working Party on the Suppression of Odours from Offensive and Selected Other Trades. Odours: Report of the Working Party on the Suppression of Odours from Offensive and Selected Other Trades, Part 1: Assessment of the Problem in Great Britain. Stevenage, England: Warren Springs Laboratory, 1974. 116 pp.

8 Control of Industrial Odors

This chapter discusses the application of odor-control technology to various industrial enterprises. These examples were chosen because they are representative of the industrial applications of this technology, and their inclusion does not imply any ranking as sources of odor pollution.

SEWAGE

Wastewater treatment and disposal systems, by the nature of the materials handled, are notorious generators of foul odors. Although fresh raw domestic sewage in the highly diluted condition characteristic of U.S. wastewater effluents is not markedly offensive in odor character or intensity, sewers become a source of especially foul odors when they cease to operate under aerobic conditions.[6] This occurs whenever liquid flowrate or grade is inadequate to prevent settling of solid putrescible materials to the bottom of the flow channels. The gradient must be such that the flow velocity is above 0.6 m/s to prevent settling, but below 4 m/s to avoid rapid erosion.[23] Whenever ground levels make it impossible to attain minimal gradients, systems are designed so that sewage gravitates to collecting chambers from which it is pumped to treatment plants. Sewage may be retained in these collecting chambers (lift stations) for many hours between pumpings, become septic, and evolve foul odors. Continuous pumping and frequent removal of bottom sludge and slime growths on the chamber walls are required to control odor emission. These conditions are especially likely to occur in combined sewers that are designed to carry off storm waters, as well as sewage. During dry periods, the normal sewage flow alone is inadequate to convey all the solids to the treatment station, and anaerobic conditions develop within the settled sludge that produce foul gaseous emission high in hydrogen sulfide. The addition of large amounts of putrescible domestic wastes from garbage-grinders and industrial wastes from dairies, cheese factories, breweries, etc., increases the danger of producing septic conditions and the generation of foul odors in the wastewater transport system. Malodors from the sewer system can be avoided by designing them for a fast flow, avoiding low places during construction where solids can accumulate, and providing adequate ventilation to maintain aerobic conditions throughout the entire system of flow channels.[23] If these good design standards are not adhered to, frequent mechanical cleaning of the sewage conduits and sterilization with chlorine must be resorted to, to cope with foul odors emanating from this part of the wastewater disposal system.

At the sewage-treatment plant, primary treatment includes removal of large solids by screening, grinding to small size, and return of this waste to the main sewage inflow. Screening is followed by settling of coarse solids (grit) in low-flow-rate holding basins. If the sewage is already septic (putrefactive) when it arrives at the treatment plant, it continues to give off malodorous anaerobic-decomposition gases that include a full range of low-molecular-weight mercaptans, organic disulfides, organic acids, and methylamines, plus indoles, skatoles, and much larger amounts of hydrogen sulfide and ammonia. The primary screenings and settled grit represent an enormous reservoir of putrescible materials high in sulfur and nitrogen, and hence a continuing source of foul gases, if they are not removed from the treatment basins promptly and continuously for burial, incineration, or biologic treatment. This may be done conveniently by mechanical sludge-ejectors that scrape the bottom of the basins with broad scoops attached to an endless chain and lift the drained sludge out of the basin at one end for deposit in a sludge bin or onto an automatic conveyor that conducts the sludge to the next stage.

Control of septic conditions in screening and especially in settling basins, other than by sludge removal, is not practical, for two reasons: the turbulence that would be produced by aeration of the raw sewage to maintain aerobic conditions is incompatible with a need to maintain the quiescent conditions required for efficient gravitational sedimentation; and the addition of such agents as chlorine and ozone to the raw sewage to destroy the microorganisms that produce the foul gases as a byproduct of their metabolic process would eventually affect adversely the organisms that are essential for the next step in the sewage-treatment process. Therefore, whenever a sewage-treatment facility is close to incompatible land uses—e.g., residences—good practice requires enclosure of the settling basins with a structure, ventilation of the space, and treatment of the exhausted air by chemical scrubbing to remove or reduce malodorous components.

Addition of ozone to the exhausted air, without other treatment, has been widely practiced for gas-phase deodorization, but its advocates have failed to demonstrate effectiveness other than by the well-known ability of ozone to destroy odor perception. Addition of chlorine or ozone to the exhausted air followed by caustic scrubbing is recommended as an effective and least-cost method for deodorizing the gases evolved from raw-sewage screening and settling basins.

Additional solids may be removed from raw sewage by continuous vacuum filtration through a filter cloth that retains particles down to microscopic size and prepares them for destruction by incineration by dewatering them to 65–85% moisture content before discharge from the filter. Incineration of solids usually takes place in a multiple-hearth furnace adapted from ore-processing technology. The furnace consists of "a number of annular-shaped hearths mounted one above the other. There are rabble arms on each hearth that are driven from a common center shaft. The feed is charged at the center of the upper hearth. The arms move the charge outward to the periphery where it falls to the next hearth. Here, it is moved again to the center from which it falls to the next hearth. This continues down the furnace."[13] Combustion gases rise in the furnace countercurrently to the downward flow of sludge, giving maximal fuel economy and, with proper operation, total destruction of organic matter. However, the raw sludge discharged to the uppermost hearth is exposed to flue gases that have been cooled sufficiently by countercurrent flow of downward-moving sludge to volatilize malodorous gases trapped within the raw sludge without sufficient heat intensity to destroy them. In addition, waxes, fats, and oils contained in the raw sludge tend to be volatilized in the cool upper stages of the furnace and to condense in the air, producing an opaque white plume after leaving the stack. For both these reasons, it is customary to scrub the flue gases with a high-energy scrubber before they are released to the atmosphere. Fluidized-bed sludge combustion has been suggested as a thermal destruction method that can avoid the release of malodorous compounds without the need for a flue-gas cleaning device.[1]

An alternative treatment method for the sludge removed by screening and settling is anaerobic digestion at slightly increased temperature (35°C) to favor the production of methane, which can be burned as a fuel to produce power for lift pumps and for similar uses at the sewage plant. When the anaerobic-digester off-gases are burned, malodorous, reduced sulfur and nitrogen gases that accompany methane are destroyed in the oxidizing flame, and this serves as an odor-control method as well as an energy-conservation step. After digestion, the liquid can be returned to the inlet of the treatment plant and the digested sludge spread on the ground, buried, or incinerated. "Heavy applications of digested sewage sludge to land may . . . lead to a localized slight odour problem, especially in warm weather after a period of rain."[23]

Secondary treatment, in almost all cases, is biologic and depends on the ability of microorganisms of the aerobic type, largely aerobic bacteria, to extract from the settled sewage most of the residual dissolved and suspended organic matter to be used as a microbiologic feedstock. Two processes are in widespread use: the activated-sludge method, in which "starter organisms" from a previous batch are added to the settled sewage and the combination vigorously sparged with air to provide ample oxygen for optimal biologic oxidations and to prevent the sewage from turning septic (which would generate foul-smelling gases and endanger the survival of the biologic mass of aerobic organisms); and the trickling-filter method, in which a deep bed (2 m) of crushed stone of carefully graded sizes supports on all its surfaces a thick biologic slime layer that contains some of the same types of aerobic microorganisms that are used for the activated-sludge process and through which the settled sewage trickles in thin films that remain in intimate contact with the slime when the operation is properly conducted. Sewage is added to the top of the trickling filter by spray so as to entrain large amounts of air with the falling films and keep the bed under strong oxidizing conditions throughout its depth. "Provided that the sewage entering the plant is not septic, neither type of biological process is likely to cause odour nuisance unless overloaded to the extent that the supply of dissolved oxygen and nitrate become insufficient and anaerobic conditions occur."[23] This makes it clear that imposition of operational controls is the most effective way to prevent foul odors from secondary sewage-treatment processes.

Tertiary sewage-treatment processes that include removal of nitrate or phosphate ions from the clarified liquid product of the secondary treatment stage and chlorination, to destroy residual pathogens and avoid a reversion to septic conditions after discharge, are not an important source of malodorous gaseous releases when properly conducted. Here, also, operational controls are certain to be the most effective and economical of odor-control methods.

The increase in the mass of microbiologic sludge during the aerobic-treatment stage represents an important potential odor source, because it can turn septic rapidly after separation from the clarified liquid. After a portion is recycled to the aeration tanks to seed the newly arrived sewage, the remainder must be disposed of rapidly, if malodors are to be avoided, by application to the ground without further treatment, application to the ground in dried and screened form, burial, incineration, or anaerobic digestion. Land disposal of untreated sludges is the method most likely to cause neighborhood odor nuisances from the decomposition of undigested sewage solids. Whereas no method of sludge disposal is without serious problems, including potential malodors, burial in a sanitary landfill appears to be the most satisfactory from the standpoint of avoiding an odor nuisance.

Odor emission from specific industrial wastewater-treatment facilities is discussed in the sections of this report devoted to specific industries. It has been observed that, as municipal wastewater systems become larger and as the treatment methods used become more comprehensive and more effective, economies of scale, as well as the need to engage the services of highly skilled technical specialists, have resulted in the acceptance of increased diversion of trade wastes into municipal systems. These include "wastes from abattoirs, meat cooking, food processing, brewing, tanning, and wool scouring."[8] At least one combined advanced wastewater-treatment system for handling paper-making waste and domestic sewage in a single process has been constructed and operated successfully for a number of years.[3] A characteristic of the processes used in advanced treatment plants that accept sizable amounts of individual wastewater is the addition of an activated-carbon treatment stage for the removal of odor-producing chemicals that are characteristic of specific industrial wastes.

RUBBER-PROCESSING

The rubber industry in the United States is dominated by tire manufacturers in so far as rubber use and manpower are concerned (about two-thirds of total rubber consumption), but tire-working plants tend to be large, and non-tire-making rubber-processors are more numerous and their products cover a wide range of familiar articles—baby cribsheets, elastic bands, bowling balls, etc.

Natural rubber, obtained principally from the sap of the hevea tree, competes with a number of synthetic rubbers on the basis of price and, for some special uses, on the basis of physical and chemical properties. For example, the synthetic rubber neoprene has unusually good resistance to degradation by organic oils and solvents, and the synthetic butyl rubber has excellent high-temperature resistance.[10] Because natural and synthetic rubbers, as well as mixtures of various proportions of each, are processed in the same manner and in the same machinery, odor emission is little affected by the specific composition of the rubber or rubber-like materials.

Rubber-producing trees are not grown in the United States, so the production of dried crude rubber and of rubber latex from this source need not be discussed here. When the dried rubber product arrives from overseas, it is nearly odorless, and it remains in that condition during storage and cold handling. Ammonia is added to natural latex to prevent coagulation during shipping and storage. Therefore, whenever sealed containers are opened and when latexes are processed, ammonia will be released. Odor control is by dilution to below the odor threshold or by scrubbing with an acidified solution.

Synthetic rubbers are manufactured by polymerization of a variety of monomers. Large quantities of styrene and butadiene are copolymerized in emulsion form to produce GR-S, a synthetic rubber that has wear-resistance qualities superior to those of natural rubber for tire treads. Neoprene, a flame-resistant synthetic rubber, is formed by homopolymerization of chloroprene. Others are manufactured by similar processes with the aid of polymerization catalysts.

The raw materials are transported, transferred, and stored in closed systems, because most have high vapor pressures at ambient temperature, and the polymerization steps have to be carried out in closed vessels for the same reason. Unreacted monomers are recovered from the spent emulsions and recycled after purification. Under ideal operating conditions, there should be no loss of monomers to the environment, but fugitive emission of volatile organic vapors from leaking pipes, valves, pumps, flanges, and doors occurs frequently. Many of the chemicals have an unpleasant odor and a low odor threshold—e.g., styrene; and even minor leaks can evoke odor complaints. The remedy is preventive maintenance guided by the liberal use of a portable flame-ionization or other detector to "sniff out" incipient leaks. Storage tanks "breathe"—an increase in internal pressure occurs as the contents warm, vapors are discharged to avoid overpressurization of the tank structure, and vapors are displaced when tanks are filled. This can be associated with a severe fugitive-emission loss and provoke odor complaints, unless the storage tanks are equipped with a floating roof or a solvent-recovery system. Recovery systems consist of refrigerated vapor condensers and adsorbent activated-charcoal canisters in the vapor vent lines that retain up to 99% of the vapors that would otherwise escape. These devices make it possible to recover the solvent vapors for reuse. Another source of fugitive odor emission is the opening of reactor vessels for manual cleaning after the discharge of the reacted chemicals. Until this cleaning function is performed automatically in closed reactors, it is necessary to exhaust-ventilate the interior of the vessels (drawing in air from the workroom) whenever a hatch is removed, and then to remove the vapors from the exhaust air with activated charcoal or to destroy them by incineration before discharging the residual effluent gases to the atmosphere. Generally, the solid reaction product will be discharged from this process in the form of washed rubbery crumbs.[18] They do not constitute a malodor problem in this form, but small amounts of residual solvents are released during washing, dewatering, and drying of the crumbs, and these add to the fugitive solvent emission (and odor) unless these operations are carried out in exhaust-ventilated enclosures and the evolved solvent vapors are treated in the same manner as those released during reactor cleaning and maintenance.

Although neither natural nor synthetic dry rubber by itself is a source of unpleasant odors in the cold state, the addition of compounding chemicals (including volatile solvents and oils), followed by heating, releases the foul-smelling gases and vapors that are identified as "rubber odors" and "burnt-rubber smell." In their pure state, natural and synthetic rubbers have unwanted physical and chemical properties that make them useless for all the purposes associated with rubber use, except such uses as adhesives and sealants. Therefore, many minerals, oils, solvents, and organic chemicals are added to crude rubber to change it into desirable forms. Uniform dispersion of these additives into the rubber base requires heating to make the rubber more plastic. All the heating necessary is generally provided by friction generated in mechanical mixers and homogenizers known as Banbury mixers and rolling mills. Plasticity may be improved and hastened by the addition of softening oils, such as pine oil. When the mixture gets hot, rubber volatiles and degradation products, volatile oils, and products of reaction with the elemental sulfur that is always added as a curing agent (vulcanization) are evolved and released to the atmosphere. Even more unpleasant-smelling products are evolved during vulcanization when the compounded and shaped rubber is heated to 250–300°F (120–150°C) for up to 45 min to produce the cross-linkages of the rubber molecule that are important in giving the final product such desirable properties as abrasion resistance, freedom from tackiness, and resistance to solvents. Vulcanization is sometimes carried out in steam autoclaves, sometimes in dry ovens, and sometimes in steam-heated molds—as in curing tires and impressing the tread pattern into them. In either process, air is excluded, and reduced volatile sulfur compounds are formed and released with copious quantities of steam when the autoclaves of steam-heated molds are opened to remove the cured rubber pieces. This mixture of reduced sulfur and other organic gases plus water vapor is intensely odorous and foul-smelling. It is the most important source of odorous releases from plants that manufacture rubber products.

Significant malodors of lower intensity and quantity are associated with rubber-cement solvents used to fabricate complex structures, such as rubber footware and automobile tires. The odors are associated with the evaporating solvent, which is likely to be petroleum naphtha or toluene. Similar odors are associated with the coating of fabrics with a layer of rubber that is applied as a thin film of compounded rubber dissolved in an organic solvent. After application, the solvent is evaporated, and the rubber-coated fabric is vulcanized by heating. The evaporated solvent has a potential for evoking odor complaints if it is released to the atmosphere in large quantities; during vulcanization, additional releases of unpleasant volatile reduced sulfur compounds tend to make things worse. Other foul-smelling substances released from compounded rubber when it is heated are derived from aldehyde amines used as vulcanization accelerators and secondary aromatic amines and substituted phenols used as antioxidants to improve aging properties.

Another widely used process in rubber-manufacturing is the dipping of articles in solvent solutions of rubber or latex emulsions to coat them. An example is the rubber coating of cotton, rayon, or nylon fibers to form tire cord. Evaporation of volatile vapors from the surface of the dipping bath and from the cord-drying ovens are important potential odor-emission sources at tire-making plants.

Odorous organic-solvent emission from the manufacturing operations that have been described may be controlled with solvent-recovery systems that use activated-carbon adsorption to trap the solvents, steam desorption, condensation of the mixed vapors, and recovery of solvent by decantation from the condensed steam. These systems are capable of removing 98–99% of solvent vapors and are seldom designed to give less than 90% removal efficiency. Although thermal incineration requires much simpler equipment, it is equally effective for solvent reduction. If the solvent vapor concentration is high enough, incineration may become self-sustaining without the addition of other fuel, especially if catalytic combustion is used. Incineration with or without added fuel becomes more attractive than solvent recovery for odor control whenever the heat generated can be used productively for process steam or for space-heating. Incinerator fuel requirements can be kept within acceptable bounds by reducing the odorous off-gas volume to a minimum with fully enclosed processes and by substituting nonvolatile or nonodorous chemicals to the extent possible. An important trend is the substitution of hydrosols for organic solvent plastisols in the making of coated fabrics.

Control of malodorous emission from vulcanization is more difficult, because the volatile materials have no important reclamation value and are in insufficient quantities to reduce the fuel costs of incineration significantly. Furthermore, the use of catalytic combustion is contraindicated, because sulfur in any form is a potent catalyst poison. When vulcanization takes place in a steam autoclave, there is no emission during the curing cycle, but at the conclusion the autoclave must be returned to atmospheric pressure by the release of steam before it can be opened, and it must be purged of steam and cooled before workers can enter. In both operations, high concentrations of malodorous gases and vapors that are generated from the rubber by heating and steam elution during vulcanization are released with the steam. If discharged to the atmosphere untreated, they can impart foul odors to large downwind areas, in spite of the fact that the total volume of the discharges will not be large. The odor intensity will be high, and the characteristic "burnt-rubber" odor will be offensive to most people. Deodorization of the gases and vapors released during autoclave blowdown and purge begins with cooling of the off-gases to reduce total volume and odor by condensation of steam and high-boiling-point odorous organic compounds. The noncondensable malodorous gases and vapors that remain can be destroyed by direct-flame incineration. When conducted carefully, the process is effective and not excessively costly in fuel or cooling water. However, process control is essential to concentrate odorous substances in the smallest possible volume of air, because "the cost of control equipment is based principally on the volume of gas that must be handled and not on the amount or concentration of the substances that must be removed. Also, most air and gas removal equipment is more efficient when handling higher concentrations of contaminants, all else being equal."[7]

When vulcanization is conducted in dry ovens at atmospheric pressure (without the use of steam)—as is the practice for such objects as rubberized cribsheets, paint brushes, and automobile brakeshoes—there is a continuous evolution of burnt-rubber odors over the entire curing period, which, at the low temperatures required to prevent degradation of the more heat-sensitive components, may last for several hours. For heat-conservation purposes, it is customary to recirculate the oven atmosphere through open-flame gas burners, introducing only sufficient makeup air to provide the oxygen needed for proper burning of the fuel. Although this reduces the volume of oven gases discharged to the atmosphere over what would be the case with a "single-pass" system, cycling of the oven gases, with their substantial burden of volatile organic products derived from the curing rubber, through open-flame burners forms increased quantities of foul-smelling degradation products. Direct-flame incineration is the only practical method available for deodorizing these oven off-gases other than atmospheric dilution. Substitution of products is one route to odor control for these operations. An example is "substitution of cold-setting synthetic resin for rubber in the manufacture of paint brushes. Before this change, it was necessary to vulcanize the rubber bond at the base of the bristles for a period of many hours, causing severe odor nuisances in the vicinity of the factory. The cold-setting resins selected as rubber substitutes produce no odors and completely eliminate air polluting emissions from this operation."[7] Where material substitutions are not yet possible, and gas incineration is the control measure of last resort, fuel economies must be sought by redesigning the curing ovens so that all the heat produced by off-gas incineration is used in place of the oven burners to keep the ovens at the desired high temperature. By this stratagem, each unit of fuel can be made to do double duty, i.e., off-gas deodorization and process heating.

The most offensive rubber process, from the standpoint of odorous emission, is rubber-reclaiming. The raw material is almost exclusively discarded tires, with some additions of rubber mill scrap. With tires, the process begins with the removal of the imbedded wire bands that form the firm attachment and seal between the tire and the wheel rim; this is followed by shredding and grinding to produce granules of approximately millimeter size. Presumably, steel cords will also have to be removed from steel-belted radial tires before shredding and grinding. Burnt-rubber odors are emitted from the grinders whenever the stock is overheated because of excessive material throughput or poor machinery operation and maintenance, such as failure to maintain sharp cutting surfaces. Odor control for this phase of the process can be maintained satisfactorily solely by good process and maintenance procedures.

The next step is devulcanization of the ground tire carcasses to remove cord scraps and to restore plasticity to the recovered rubber. This is done by adding large amounts of softening oils, such as mineral and pine oils, to the ground rubber and subjecting it to heat and pressure. This may take place in a screw device called a "reclaimator" or in an autoclave at steam pressures of 500–1,000 psi.[18] In both processes, the lighter fractions of the softening oils are driven off with large quantities of organic products heavy in reduced sulfur compounds, amines, and acrolein that are evolved from the rubber undergoing processing. Inasmuch as the objective of devulcanization is to degrade the stable cross-linked rubber product and restore it, insofar as possible, to its original unpolymerized state, it is clear that a great deal of molecular rearrangement occurs in this medium, which is high in the sulfur and nitrogen compounds that were added during original manufacture to impart the specific properties to cured rubber that are needed for satisfactory service. Therefore, it is not surprising that large amounts of extraordinarily evil-smelling compounds of a poorly defined chemical nature are evolved during the devulcanization process; when uncontrolled or poorly controlled, they produce a severe odor nuisance wherever there are reclaiming plants.

Control of malodorous emission from devulcanizers begins with condensation to remove steam and the malodorous water-soluble gases and vapors associated with it and to remove the condensable softening oils and the malodorous oil-soluble gases and vapors associated with them. The condensed oils are salvageable and, after decanting, may be reused in the devulcanization process. The water layer can be treated with a strong oxidizing agent, such as hypochlorite, to destroy malodorous components and then sent to waste. When the devulcanizer steam blowdown is treated in this manner with a multistage series of fin-coil coolers and spray washers containing caustic liquor, the volume of insoluble, noncondensable gases that pass through it is negligible, and the condenser-scrubber behaves like an infinite-retention-time absorber with a collection efficiency close to 100%. During the autoclave purge phase, increasing quantities of air mix with the steam atmosphere, and the retention efficiency of the condenser-scrubber gradually declines. But by then, the fraction of volatile devulcanization products in the purge gases is also declining, so the quantity of malodorous products escaping to the atmosphere with the inert fixed gases remains small. Gas incineration is effective as an odor-control step for the devulcanizer off-gases, but the cost of fuel and the loss of salvageable softening oils make this method less attractive. The use of simple washer-condensers, such as allowing the blowdown and purge gases to bubble through a tank of water, are futile odor-control measures for this manufacturing step.

From the devulcanizer, the hot rubber granules are conveyed to washer screens to cool them and free them of cord fragments, inert filler materials, and excess oils and then placed on trays and passed through a drying oven. After drying, the granules are passed through a strainer to remove residual solids and then to rolling mills, where the reclaimed rubber is homogenized and sheeted out with added retarder chemicals to prevent revulcanization during storage and shipping. All the operations that heat the rubber or that handle hot rubber (e.g., conveyors) release malodors similar to those generated in the devulcanizer, but in smaller amounts. If these operations are carried out in the open, severe fugitive emission results. If they are carried out in exhaust-ventilated equipment, the volume of odorant gases that must be treated is too large for economical incineration and the gases are too dilute for efficient scrubbing or condensation. Therefore, odor control calls for extensive use of totally enclosed, automatic processing machines that make it possible to incinerate the small volume of inert gases that must ultimately be exhausted to the atmosphere. The hot drying-oven exhaust gases can be incinerated and recycled to the drying oven in the manner that was described for deodorization of dry off-gases from oven vulcanization.

Rubber-reclaiming activity rises and falls in response to the value of reclaimed rubber in the marketplace. Enormous numbers of discarded tires (some 10^8 per year) glut waste-disposal facilities throughout the nation, because the value of the reclaimed rubber is not now sufficient to cover the cost of reclamation plus the cost of satisfactory air-pollution control facilities. This resource-recovery process is very difficult to bring into full compliance with current air-pollution control regulations.

Tire retreading and recapping shops are generally small enterprises and are widely dispersed throughout populated areas. They are frequently a source of unpleasant odors. The first step in the process is to remove the old worn tread from the tire casing with a motor-driven rasp that grinds it off rapidly. Temperatures at the rasp head increase with the rate of removal of old rubber and result in the generation of burnt-rubber odors. Although this type of equipment "can be operated at a sufficiently slow speed not to cause smoking, such slow-speed operation is considered uneconomical by most firms."[4] Odor-control methods include the spraying of cooling water on the tire during rasping to prevent overheating and the use of cutters, instead of rasps, to remove the old tread. The remaining operations are like those of tire-making—applying rubber adhesive to the buffed tread surfaces, cementing on the new tread strip, and vulcanization in a steam-heated tire mold. The same odor-control methods used for tire-manufacturing plants are applicable here.

Just as the rubber industry is very diverse with respect to its production operations and its products, so too is its odorous emission. Generally speaking, odor problems arise whenever rubber or rubber-based products are heated to temperatures in excess of 200°F (about 95°C); and, the higher the temperatures, the more disagreeable are the volatile odorous products that are evolved. Although the "burnt-rubber odor" is difficult to define chemically, it is easily recognized and long remembered by members of the public, who universally dislike it heartily. Low-temperature condensation of semivolatile malodorous compounds and chemical scrubbing of noncondensable gases and vapors are effective odor-control methods for malodorous off-gases from some rubber-processing operations and, where effective, they should be applied vigorously. But for many important operations, only gas incineration can be effective; for these, operational controls, material substitutions, and process enclosure are essential to reduce to the absolute minimum the volume of gases that must be incinerated. Technologically, adequate means are available to prevent emission of rubber-processing odors, but the methods are costly, and the entire industry needs improved malodor-control techniques that are less energy-intensive.

STEEL INDUSTRY

Odorous emission associated with steel-making evolves primarily from the coking process. Coke is produced mainly by the high-temperature (900–1,000°C) distillation of bituminous coal heated in the absence of air. This high-temperature carbonization (pyrolysis) also produces a variety of chemical compounds, collectively called "coke-oven" byproducts or "coal chemicals." Among these products are coal gas (coke-oven gas), tar, light oil, and ammonia or ammonium sulfate. Coal gas and tar volatiles include hydrogen sulfide, mercaptans, organic sulfides, and naphthalenes. Similar fugitive emission occurs from blast furnaces and slag piles.

PROCESS DESCRIPTION

The coke oven is a narrow, tapered (for ease of discharge) chamber with a capacity of 12–16 tons (10.8–14.4 t) of coal. Each oven is operated intermittently; but inasmuch as a battery contains 30–90 ovens that are charged and discharged at different times, the overall operation is continuous. The coal charge in the ovens is heated through both sidewalls, so heat travels toward the center of the charge. The heat is supplied by burning gases in the flues between the ovens, and no burning takes place in the ovens. Producer or coke-oven gas is burned to furnish the necessary heat. When coke-oven gas is used, about 35% of that produced is required to heat the ovens.

The conventional coking process has the following potential sources of air pollution:[16]

- Handling, crushing, and blending of coal; handling and screening of coke.
- Charging of coal into the slot ovens.
- Underfiring of the slot ovens with coke-oven gas.
- Sealing of the slot ovens during early stages of coking.
- Discharging of the newly produced coke (pushing).
- Quenching of the hot coke.

The second, fourth, and fifth of those operations are discussed below in order of diminishing importance for potential odorous emission. Other operations are not discussed, because they are not usually associated with odorous emission, although quenching produces voluminous gaseous emission. A technique known as "dry cooling" is rapidly replacing quenching throughout the world.[2]

Charging

In the charging operation, a blend of the desired coal types is transported in a lorry car to the coke oven. This method of operation is described sequentially in Table 8-1. As soon as the coal becomes heated above 400°F (about 200°C), pyrolysis of the high-molecular-weight hydrocarbons forms smoke, tar vapors, and gases. Oven-charging emission can be controlled by charging on the main, pipeline charging, or direct collection with incineration and wet scrubbing.

Charging on the main is the common practice of drawing off and venting the accumulated charging smoke to the collector main of the byproduct system. A steam ejector on top of an ascension pipe leading from the top of the oven is activated throughout the charging operation to create a suction draft through the oven. It should be noted that charging on the main is a control technique to minimize emission from existing batteries.

Equipment must be designed so that gases can pass across the oven to the ascension pipe without obstruction. An adequate amount of steam must be supplied to the ascension pipe to create the necessary suction. Four major schemes that have been investigated display various degrees of success, problems, and expense:

- *Double Collecting Mains*: An additional collecting main on the battery can increase the draft suction capacity. Aside from the added expense of an additional main, the arguments against using this scheme include the doubling of the maintenance required on the mains and the increased possibility of leaks from the standpipe. The size of the mains does, however, block ventilation from the top of the oven, trapping the heat.
- *Breeches Pipes*: This scheme involves ventilation pipes to connect one end of the oven with the end of another oven served by a separate ascension pipe and steam ejector. The overall effect is that of a double collecting main. The small increase in improving the gas collection may not justify the need for extra ports, handling, and maintenance.
- *Conical Relief*: Another suggested control possibility is the redesigning of the charging port to distribute the coal within the oven. Gases will be obstructed from passing to the ascension pipe if piles of coal are allowed to build up in the oven. The idea is to feed the coal through ports shaped like inverted funnels.
- *Sequential Charging*: Lorry-car hoppers can be used so that a sequence of charging and leveling can occur automatically. Coal piles that block the passage of gases to the ascension pipe are leveled, and the number of open ports and doors is minimized. This complex system, with

TABLE 8-1 Sequence of Charging Operations[a]

Elapsed Time,[b] min	Operations	Emissions
0.0	Lorry car is filled at blended-coal bunker and weighed. Oven doors are replaced. If lids of charging ports are removed by hand, this is done now.	Dust at bunker discharge
2.0	Lorry car moves to position over the charging ports. (Automatic lid lifters, if provided, lift lids.) Drop sleeves are lowered into ports.	Generally none
2.7	Steam ejector in standpipe is turned on to draft oven. Coal begins to flow from hoppers on lorry car and first rush of gases ignites within the oven.	Smoke emission begins slowly at ports
2.8	Heavy rush of smoke, dust, and gas emerges around charging ports as steam formation increases oven pressure rapidly. Hoppers continue to discharge.	Brown to black smoke, with some flashes of flame
4.0	Hoppers discharge more slowly as coal in oven rises to bases of the charging sleeves. Pusher operator goes to open the chuck door for leveling.	Continued smoke
5.0	Coal discharge completed, drop sleeves raised. On signal from lorryman, leveler bar is pushed all the way across oven, then cycled back and forth 3 to 5 feet to level the coal.	Smoke continues from ports and from the chuck door
5.7	After 5 to 7 strokes of the leveler bar, it is withdrawn. Lorryman operates lid-replacement system, or if none, he moves the lorry car and crew begins to replace the lids by hand. Chuck door is closed and ejector is turned off.	Smoke continues until chuck door and port lids are well seated
6.5	Charging completed	Smoke via leakage only

[a] Reprinted from U.S. Environmental Protection Agency.[16]

[b] Attainable with mainly hand operations if coordination between lorryman, top crew, and pusherman is good. Charging of tall ovens or with sequenced discharge of hoppers may take longer, and some time is generally wasted between steps.

intricate timing and added expense, may require too much maintenance to be justified. Most recent innovations include the programing of the entire charging process.

Ascension-pipe design affects the efficiency of the draft suction obtained. The availability of steam, the quantity of suction desired, and the degree of cleaning and maintenance required have promoted experimentation with different steam-jet arrangements and ascension-pipe geometry. In practice, steam availability and cost of equipment changes dictate which design is most appropriate. In the past, steam ejectors were in the ascension pipe, where material corrosion would be minimized, and no thought was given to increasing the efficiency of gas collection.

Preheating and pipeline charging involve the feeding of coal to the coke oven through an enclosed pipeline. A prototype battery of 24 15-ton (13.5-t) ovens was installed in 1970 by the Semet-Solvay Division of Allied Chemical Corporation in its Ironton, Ohio, plant. The method is now being widely used in Britain, France, Japan, and Sweden. Jones and Laughlin and Inland Steel also use the process.[2]

Coal is fed from a hopper to a preheating section. A stream of oxygen-free hot gas heats the wet coal and carries it to cyclone separators. The preheated coal is collected in a large receiving bin. The receiving bin feeds a measuring bin, which accurately meters the charge to the ovens. Coal is transported to the ovens through pipes by means of steam jets spaced along the bottom.

The gaseous phase from the cyclone separators is divided into two streams: one is recycled to the preheating chamber, and the other is cleaned in a wet scrubber and discharged to the atmosphere.

The greatest advantage of this method is the virtual elimination of oven-charging emission, which is estimated to be about 60–70% of the total coke-oven emission. Emission normally encountered in the pushing operation is also minimized, as a result of preheating through improved control of coking-process variables. The scheme is attractive from an operational standpoint. Coking time is reduced by 40–50% by preheating the coal to about 500°F (260°C); therefore, production can be increased so long as the capacity does not overextend the limits of the gas collection, the battery machinery, or the coal- and coke-handling facilities.

With this method, production can be increased at a capital cost lower than that of adding equivalent capacity by conventional means. Pipeline charging and preheating are adaptable for installation on sound existing batteries where additional production is desired, as well as in construction of new or replacement batteries for older units, where conversion would not be economically feasible.

Direct collection with incineration and wet scrubbing is different from the other two practices. An external draft is used to draw smoke into special shrouds around the charging ports. The shrouds serve as combustion chambers for the volatile gases collected, and an adjacent scrubbing system treats the unburned dust and condensates. The "cleaned" gases are vented through stacks on top of the lorry car.

Many operational and maintenance problems have been encountered. Several manufacturers have altered the two basic designs slightly to create collection and scrubbing systems with fewer defects and greater efficiency. In particular, Bethlehem Steel Corporation is making technologic improvements in the Shalker-Eisenhutte lorry-car gas-cleaning system on its No. 1 Coke Battery at the Burns Harbor Plant.

Sealing

Emission from poor sealing is much less than that from the charging process. Poor sealing can be a result of design, but is more a function of mediocre operation and maintenance. The seals around the charging ports and around the big doors used for leveling and pushing are the most critical. Special care must be taken to clean the doors and ports after each charging operation. All lids must be closed immediately at the appropriate times. Failure to clean doors and ports results in carbon buildup that prevents the complete sealing of the apertures. Gases then leak out during the entire charging and coking period. Another problem is damage to seals caused by denting during the replacement of the heavy doors after pushing.

The best control technique for poor sealing is automation. Emission is decreased by eliminating the possibility of human error. Automatic lid-lifters, door-cleaners, and jamb-cleaners greatly improve efficiency and reduce emission. Redesigning of automatic equipment is constantly in progress, because no existing examples are completely satisfactory.

Pushing

The coke-discharging operation occurs after a heating time of 15–40 h. The nature of this operation is such that effective emission control by hooding or shrouding is difficult. Incomplete carbonization results in "green" coke, which can be formed by uneven heating, by a breakdown in the heating system, by the presence of coal adjacent to unheated doors, or simply by premature pushing of the charge. Gas is emitted to the open air when this insufficiently coked coal is pushed out of the oven. Controls for pushing are currently not as satisfactory as those for charging. The most highly developed and promising control device is a large hood extending over the hot-coke car to catch the rising gas emitted during the pushing operation and to direct it to wet scrubbers for treatment; however, three prototypes of the hood system have never functioned on a continuing basis.

Another coke-pushing control combines the pushing and quenching operations, so that emission control can be improved by completely enclosing the system. In this way, the volatile gases can be collected and treated before being vented to the atmosphere.

Maintenance and control of the heating system and a uniform pushing schedule are probably the best operational control techniques for pushing. The emission problem from pushing actually depends on the degree to which the charged coal has been coked.

Continuous Coking

New coking processes are at least 20 yr away from entirely displacing conventional coke ovens; however, some development work has been completed, so some of the new potential sources of emission from these processes can be predicted. The intent of the new processes is to lower coking time by introducing a continuous process that will produce high-strength blast-furnace coke from low-grade coals. The blast-furnace coke product will be sized and shaped into briquettes or pellets called "formcoke." Formcoke is made in successive steps by blending together powdered coal and some type of pitch binder; shaping the blend by pelletizing, briquetting, or pressing; and devolatilizing the formed pieces by heating. Odors may be emitted from the following unit operations that occur in some of the formcoke production processes:

- Fluidized-bed operations for drying coal and for oxidizing and devolatilizing coal.
- Forming balls in a hot retort.
- Hot briquetting.
- Curing of briquettes.
- Final coking in gas-fired vertical-shaft furnaces.
- Intermediate and final coking in units that recirculate hot sand.

In addition to these operations, which are individual phases of overall continuous processes for making blast-furnace coke, the following complete coking scheme proposals are suspected sources of emission and odors.

TABLE 8-2 Estimated Costs of Emission Control Systems for Coke Oven Cases Studied[a]

Controls Installed	Capital Costs, $		Operating, Maintenance and Repair Costs, $/year	
	2 Batteries	4 Batteries	2 Batteries	4 Batteries
1. AISI staged charging	1,500,000	3,000,000	200,000	400,000
2. Coke side shed				
(a) with scrubber	3,000,000	6,000,000	450,000	900,000
(b) with precipitator	6,000,000	12,000,000	200,000	400,000
3. Charging air with scrubber	1,500,000	3,000,000	225,000	450,000
4. Enclosed coke pushing and quench car with scrubber	8,000,000	16,000,000	800,000	1,600,000
5. Pipeline charging (retrofit)	20,000,000	40,000,000	2,000,000	4,000,000
6. Dry quenching	10,000,000	20,000,000	650,000	1,300,000

[a] Using updated literature data and equipment vendor quotations. Reprinted with permission from Kenson *et al.*[9]

- Traveling-grate cokers.
- Rotary-hearth cokers.
- Batch-type sole-flue ovens.

Process descriptions are not provided here, because these new coking processes are far from reality.

COSTS OF EMISSION CONTROL

A study reported in 1976[3] presented estimated costs of coke-oven emission controls. A summary of the costs presented in the study is given in Table 8-2. Figure 8-1, also reproduced from the study, shows the cost-effectiveness of coke-oven controls in terms of particles. No similar data are available for odors, but the effectiveness of coke-oven emission controls for odors should be proportional to that for particulate emission.

Total capital costs for air and fugitive-emission control (which directly affects odorous emission) in the steel industry are estimated at $4.81 billion for 1977–1985.[11] Annual operating and maintenance costs for this equipment are estimated at $1.85 billion. Capital recovery charges add $715 million per year.

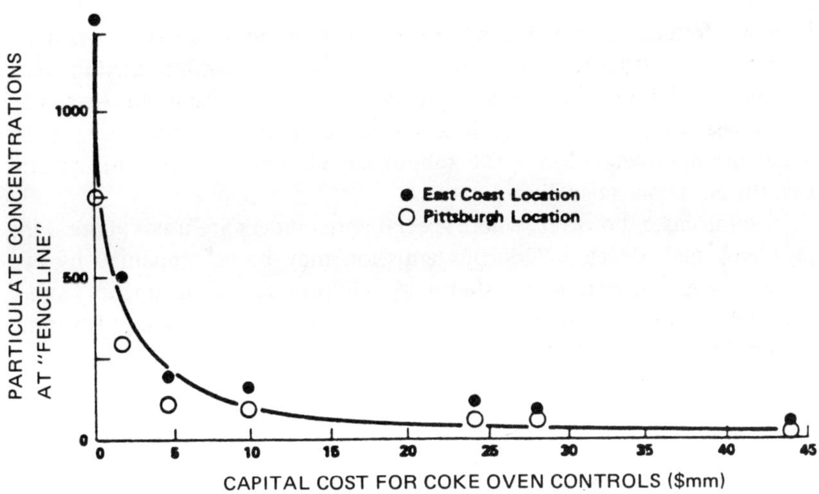

FIGURE 8-1 Cost effectiveness of coke-oven emission controls. Reprinted with permission from Kenson *et al.*[9]

RECOMMENDATIONS

Expenditures of these huge sums will reduce odors in the vicinity of steel mills, but will not eliminate them. Studies are needed to establish community annoyance caused by various concentrations of "steel-mill odorants" and establish the area associated with unacceptable odors downwind of steel mills as functions of production capacity, generic controls, and meteorologic conditions, so that it can be determined whether there is a national steel-mill odor problem and, if so, what corrective actions are possible.

FOUNDRIES

Founding, or casting, of metals is a widespread industrial activity. In 1960, there were 4,200 iron foundries and 2,000 nonferrous-metal foundries in the United States.

The gray-iron foundry industry is engaged in the production of cast iron from scrap and pig iron, usually in a coke-fired cupola furnace, sometimes in electric furnaces. The principal air-pollution problem of the industry is the emission of smoke and particulate matter. The major odor problem of the industry arises from the manufacture of cores used in the sand molds in which molten iron is cast.

EMISSION FROM CORE OVENS[16]

In many foundries, core ovens are used in the sand molds. Cores usually contain binders that require baking to develop the strength needed to resist erosion and deformation by metal during the filling of the mold. As a rule, the emission from core ovens does not cause major odor problems. If the ovens are operated below 400°F (about 200°C) and are fired with natural gas, the emission is usually tolerable.

In some cases, however, where special core binders are used, major odor problems may develop. Odorous emission may be accompanied by eye irritants, e.g., aldehydes and other oxidation products. Odor impact from a "typical" foundry is usually limited to within approximately 1/4 mile (about 0.4 km) from the foundry.

CORE BINDERS

In practice, core binders may be classified into three types: type 1 binders harden on heating, type 2 binders harden on cooling after being heated, and type 3 binders adhere on heating.

Type 1 binders harden by chemical action, partial oxidation, and polymerization. Binders of this type are drying oils (e.g., linseed oil) and resins (e.g., phenolformaldehyde). Commonly used binders are mixtures of both types combined with corn flour and kerosene. The baking process must be controlled carefully. If heating is excessive, burning occurs and degradation products are evolved. Vapors evolved are kerosene hydrocarbons and the light ends of the core oil used. Oxidation of the organics in the core produces aldehydes, acids, and other degradation products, which, as mentioned previously, are odorous and often irritating. The amounts of various contaminants evolved from tests of three different core ovens are shown in Table 8-3. Note that only a slight odor was emitted even from the uncontrolled oven (test 1), which was operated at 380°F (about 195°C). Operations below 400°F (about 200°C) seldom, if ever, need controls. The further one goes above this temperature, the more likely it is that undesirable odors will be emitted.

Pitch, resin, and other materials of type 2 binders are solids at room temperature; on heating, they melt and flow around the sand grains. When the sand-binder mixture cools, the binder solidifies and holds the grains together. The binders often are dispersed in a solvent; when baked, the solvent is driven off as an air contaminant. Generally, no odor problem develops.

Binders of type 3 are mixed with sand in the dry state. Water is added to gelatinize the binder and impart green strength to it. When the mixture is baked, the binder hardens and holds the grains of sand together. No air contaminants are released.

CONTROL OF ODOR EMISSION

Odors from core ovens may be reduced to tolerable amounts by modifying the composition of the core binder and by lowering baking temperatures. When neither of these changes is feasible, some type of afterburner is necessary. The afterburner may be of either the direct-flame or the catalytic-combustion type. The latter type is particularly adaptable to the larger core ovens. Inlet temperatures of 600–650°F (315–345°C) are usually required.

SHELL MOLDING

Shell molding is a technique for making castings in a mold having a wall thickness of only 1/8–3/8 in. (about 0.3–0.9 cm). The shell-molding process consists of dumping (investing) a mixture of fine sand and a thermosetting plastic binder (resin) on a metal pattern. This causes the

TABLE 8-3 Air Contaminant Emission from Core Ovens[a]

Test No.	1	2		3	
Oven date					
Size	6 ft 2 in. W × 7 ft 11 in. H × 19 ft L	3 ft 10 in. W × 5 ft 3 in. H × 18 ft L		4 ft 2 in. W × 6 ft 8 in. H × 5 ft 9 in. L	
Type	Direct gas-fired	Direct gas-fired		Indirect electric	
Operating temp., °F	380	400		400	
Core binders	1 to ½% phenolic resin	3% linseed oil		1% linseed oil	
Weight of cores baked, lb	700	1,600		600	
Baking time, h	11	2½ to 3		6	
Afterburner data					
Size		10 in. dia. × 7 ft 6 in. H		3 ft dia. × 4 ft H	
Type	None	Direct flame		Direct flame	
Burner capacity, Btu/h		200,000		600,000	
Air contaminants from:	*Oven*	*Oven*	*Afterburner*	*Oven*	*Afterburner*
Effluent gas volume, scfm	100	140	260	250	440
Effluent gas temp., °F	380	400	1,400	400	1,780
Particulate matter, lb/h	0.13	0.2	0.013	0.27	0.02
Organic acids, lb/h	0.068	0.008	0.000	0.44	0.87
Aldehydes, ppm	52	10	10	377	4
Hydrocarbons, ppm	124	—	<10	158	<19
Opacity, %	0	—	0	—	0
Odor	Slight	—	Slight	—	None

[a] Reprinted from U.S. Environmental Protection Agency.[16]

binder to set, thus holding the sand grains in place. The excess sand and resin are removed from the pattern, which with adhering material is put into an oven at 600°F (315°C) for about 1 min. The short, high-temperature treatment fully cures (hardens) the sand-binder mixture. The shell mold is then stripped from the pattern and is ready to be assembled.

The primary odor problem arises from emission of phenol. Hexamethylene tetramine and free formaldehyde vapors emitted are said to constitute secondary odor problems.

The odorous emission may be controlled by hooding the molding operation to collect the odor-bearing air and treating the collected air by flame incineration, catalytic combustion, or wet scrubbing (usually with an oxidizing reagent).

PULP AND PAPER INDUSTRY

PROCESSES AND EMISSION

There are two distinct phases involved in the manufacturing of paper: pulping of the wood, in which the pulp wood is separated into its two principal components, cellulose and lignin; and pressing of the cellulose fibers into various types of sheet or "cake" suitable for use in paper or paperboard products. The pulping process is by far the major source of odorous emission. Odors from the second step—paper manufacturing—are usually associated with solvents used to apply various coatings.

There are four primary chemical pulping techniques: kraft or sulfate, sulfide, semichemical, and soda. The kraft or sulfate process is the most widely used, accounting for over 80% of the chemical pulp produced annually in the United States.[17] New mills almost invariably use the kraft process.

In the kraft process, wood chips are cooked (digested) at high temperature and pressure in "white liquor," an aqueous solution of sodium sulfide and sodium hydroxide. The white liquor chemically dissolves lignin from the wood. The remaining cellulose (pulp) is filtered from the spent cooking liquor and washed with water. Usually, the pulp proceeds through intermittent stages of washing and bleaching, after which it is pressed and dried into the finished product. The balance of the process is designed to recover the cooking chemicals and heat[19] from the lignin-containing, or "black," liquor.

The characteristic kraft-mill odor is caused principally by a variable mixture of hydrogen sulfide, methyl mercaptan, dimethyl sulfide, and dimethyl disulfide. These sulfur-bearing compounds are referred to as total

TABLE 8-4 TRS Emission from an Uncontrolled 907-t/day (1,000-ton/day) Kraft Pulp Mill[a]

Source	Typical Exhaust-Gas Flow Rate, m^3/s(acfm)	TRS Emission		Average[b] TRS Emission Rate		
		ppm	g/kg ADP (lb/ton ADP)	ppm	g/s(lb/h)	g/kg ADP (lb/ton ADP)
Recovery furnace	212(450,000)	18-1,303	0.75-31(1.5-6.2)	550	79(625)	7.5(15.0)
Digester system	3(6,200)	1,525-30,000	0.24-5.3(0.47-10.5)	9,500	8(63)	0.75(1.5)
Multiple-effect evaporator system	1(2,200)	92-44,000	0.015-3.2(0.03-6.3)	6,700	5(42)	0.5(1.0)
Lime kiln	37(79,200)	3-613	0.01-2.1(0.02-4.2)	170	4(33)	0.4(0.8)
Brown-stock washer system	71(150,000)	—	0.005-0.5(0.01-0.9)	30	2(13)	0.15(0.3)
Black-liquor oxidation system	14(30,000)	3-335	0.005-0.37(0.01-0.73)	35	0.5(4)	0.05(0.1)
Smelt-dissolving tank	27(58,100)	5-811	0.007-1.9(0.013-3.70)	60	1(8)	0.1(0.2)
Condensate-stripper system	2(4,000)	—	—	5,000	10(83)	1.0(2.0)

[a] Uncontrolled-emission data for condensate strippers from U.S. Environmental Protection Agency. Data for all other sources from Duffee *et al.*[5]

[b] Average values listed are calculated from data in Duffee *et al.*[5] Insufficient information was available in Duffee *et al.* to evaluate the operation of the units for which data were reported.

reduced sulfur (TRS), and emission from kraft mills is usually expressed and regulated in terms of TRS concentrations.[19]

Typical TRS emission from an uncontrolled kraft mill producing 1,000 tons (907 t) of air-dried pulp (ADP) per day is shown in Table 8-4.[19 (p. 5-5)]

As to odor emission, relatively few studies have measured directly the odor, in terms of dilution to threshold, of the various emission sources. One study, however, conducted at a mill in Maine selected by the National Council of the Paper Industry for Air and Stream Improvement as representing a well-controlled plant of that time (1973–1974), produced the relationship between odor and TRS values shown in Figure 8-2. These data are in conflict with laboratory results with hydrogen sulfide and

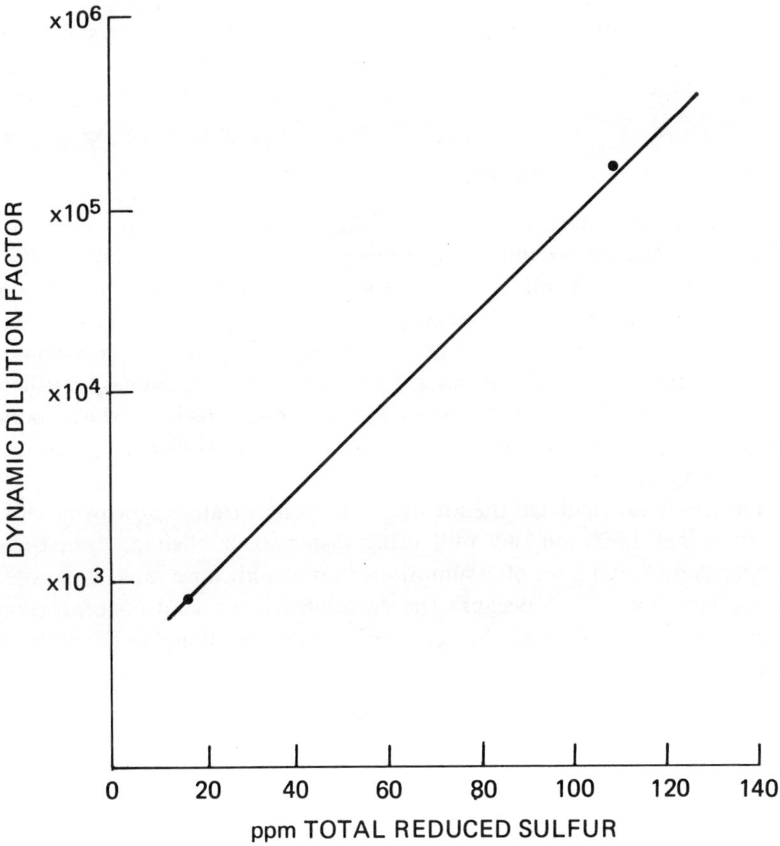

FIGURE 8-2 Odor dilution ratio and total reduced sulfur, recovery boiler. Reprinted with permission from Duffee *et al.*[5 (p. 67)]

methyl mercaptan mixtures shown in Figure 8-3. An equation was developed in this study that related the odor levels with the measured values for the concentrations of total sulfur in the recovery boiler:

$$\text{odor-dilution factor} = \text{total sulfur (ppm)} \times 870.$$

The correlation coefficient for this equation was 0.98.

Typical odor emission from kraft-mill sources is summarized in Table 8-5, compiled from data presented by Duffee *et al.*[5]

The reported odor-detection thresholds for hydrogen sulfide, methyl mercaptan, dimethyl sulfide, and dimethyl disulfide are shown in Tables 8-6 and 8-7.[19] In the study at the Maine mill, the odor threshold for methyl mercaptan was determined to be 0.00085 ppm—approximately an order of magnitude lower than the value in Table 8-7. In a comparison of the detection-threshold values with either the TRS emission rates from an uncontrolled plant or the average emission values shown in Table 8-4, the *potential* impact of kraft-mill emission on community odors is readily seen.

INDUSTRY CHARACTERISTICS

The average emission rates shown in Table 8-4 total 10.45 lb of TRS per ton of ADP. Total daily production of kraft pulp is 105,567 tons (about 96×10^3 t) as shown in Table 8-8.[19] Consequently, total national TRS emission from kraft production is approximately 1.1 million pounds of TRS per day.

As shown in Table 8-8, a large portion of this emission, approximately 47%, occurs in five southern states: Florida, Georgia, Alabama, Louisiana, and Mississippi. However, because there are kraft mills in 28 states, odors associated with pulp and paper production cannot be considered a regional problem of the south.

The EPA has calculated the maximal TRS concentration downwind of an uncontrolled 1,000-ton/day mill using dispersion modeling, "worst-case meteorology," and a set of assumptions that would cause rapid downwash of the emission.[19 (p. 3-5)] Because the calculated downwind concentrations from a Gaussian dispersion model are directly proportional to the emission strength, it is possible to derive the equivalent downwind concentration for an average 1,000-ton/day kraft plant. These values are shown in Table 8-9.

Assuming an average odor-detection threshold of 0.01 ppm, in the "worst case," the odor in the immediate vicinity (0.3 km) of a typical kraft mill can be seen to be up to 1,000 times the threshold.

As indicated earlier, however, "worst-case" meteorology is a variable set of conditions and depends on the site for which calculations are made. Kraft-mill odors have been followed 50 miles (80 km) from the mill under

CH_3SH	H_2S	CH_3SH/H_2S	OSSD Odor U.	Odor Quality	Odorant Conc. at Threshold
6.0 ppm	0 ppm	100/0	6500	Cabbage	0.00085
0.28 ppm	6.6 ppm	4/94	2200	Cabbage	0.0001
0.28 ppm	8.0 ppm	3.4/96.6	3500	Rotten Egg	0.0025
0 ppm	5.6 ppm	0/100	2400	Rotten Egg	0.0023

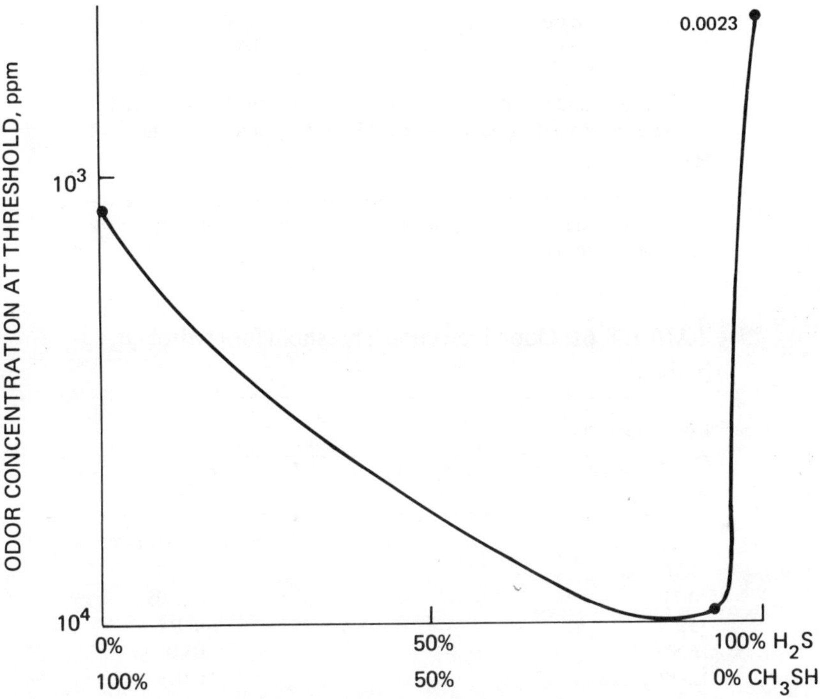

FIGURE 8-3 Threshold concentration of hydrogen sulfide (H_2S), methyl mercaptan (CH_3SH), and mixtures. Reprinted with permission from Duffee *et al.*[5 (p. 40)]

meteorologic conditions that would be ideal in minimizing ground-level TRS concentrations in the immediate vicinity of the mill.[24]

CONTROL METHODS FOR MALODOROUS SULFUR COMPOUNDS

The various methods for controlling TRS emission from kraft mills, efficiencies, and costs have been very well presented and analyzed by the EPA.[19 (pp. 6-1–6-28, 8-1–8-39)] The following discussion is a summary of that presentation.

TABLE 8-5 Kraft Pulp Mill Odor Emission

Source	Measured TRS Value	Measured Odor, Dynamic Dilution Factor
Recovery boiler exhaust	18[a]	800
	110-120[b]	15,000
Evaporator vent gas	—	8,500
Lime kiln scrubber ex.	18	480

[a] Odors were undetectable *beyond* the plant and only barely detectable on the plant grounds. Recovery-stack TRS values were unvarying at 18 ppm.

[b] Plant personnel deliberately set recovery boiler up to produce the reported TRS values. Typical kraft odors readily noticed 3 miles from plant with this condition.

TABLE 8-6 Odor-Detection Threshold for Hydrogen Sulfide[a]

Odor Threshold	
$\mu g/m^3$	ppmv
9-45	0.007-0.032
7.1[b]	0.005
0.71[c]	0.0005
15	0.011
6.8[d]	0.005
12-30	0.009-0.022

[a] From U.S. Environmental Protection Agency.[19 (p. 2-7)]

[b] Hydrogen sulfide from sodium sulfide.

[c] Hydrogen sulfide gas.

[d] Mean value ratio of highest to lowest odor-threshold concentration detected by all observers in successive tests is 3.18.

The TRS-emission sources subject to control in a kraft mill are the recovery furnace, digester system, multiple-effect evaporator system, lime kiln, brown-stock washer system, black-liquor oxidation system, smelt-dissolving tank, and condensate-stripping system. Table 8-10 summarizes the control techniques and corresponding TRS concentrations achievable for each source of TRS.[19]

TABLE 8-7 Odor Thresholds of Reduced Sulfur Compounds Other Than Hydrogen Sulfide[a]

Compound	Odor Threshold $\mu g/m^3$	Odor Threshold ppm
Methyl mercaptan, CH_3SH	4.5	0.0021
Dimethyl sulfide, $(CH_3)_2S$	2.9	0.0010
Dimethyl disulfide, $(CH_3)_2S_2$	23.7	0.0056

[a] From U.S. Environmental Protection Agency.[19 (p. 2-7)]

Recovery Furnace

For the recovery furnace, the process controls indicated in Table 8-10 refer to the design of the furnace. Other furnaces may not incorporate manufacturers' improvements, such as new means of introducing air, distributing air in the furnace, and variable adjustment of air velocity at injection ports. Since 1964, installing a membrane between the wall tubes in front of the furnace-wall insulation has been standard.

Black-liquor oxidation and use of a noncontact evaporator are the two techniques used to reduce TRS emission from the direct-contact evaporator. In black-liquor oxidation, the sodium sulfide, which is the source of the hydrogen sulfide normally generated, is oxidized to sodium thiosulfite, $Na_2S_2O_3$, in the black liquor before it enters the direct-contact evaporator. With a noncontact evaporator, contact between furnace gases and black liquor is eliminated, thereby preventing formation of hydrogen sulfide.

Digesting and Multiple-Effect Evaporator Systems

Noncondensable gases from these two sources are normally combined for treatment. Incineration of the gases in the lime kiln is the most common method. Special gas-fired incinerators are occasionally used either as a backup for the kiln or as a full-time control system. Scrubbers are also used at some mills with white liquor as the scrubbing medium. This removes hydrogen sulfide and methyl mercaptan, but not dimethyl sulfide or dimethyl disulfide. Alkaline-scrubber efficiencies are, therefore, roughly 50%.

TABLE 8-8 Summary Industry Statistics: States, Numbers of Mills, and Capacity[a]

State	No. Mills	% of U.S. Total	State Mill Capacity Megagrams Per Day	State Mill Capacity Tons Per Day	% of U.S. Total
Alabama	13	11	9,325	10,280	10
Arizona	1	1	545	600	1
Arkansas	6	5	4,925	5,430	5
California	4	3	1,732	1,910	2
Florida	8	7	8,400	9,260	9
Georgia	11	9	12,250	13,505	13
Idaho	1	1	860	950	1
Kentucky	2	2	835	920	1
Louisiana	11	9	10,570	11,655	11
Maine	6	5	3,583	3,950	4
Maryland	1	1	603	665	1
Michigan	2	2	750	825	1
Minnesota	2	2	785	865	1
Mississippi	4	3	4,270	4,707	4
Montana	1	1	1,090	1,200	1
New Hampshire	1	1	635	700	1
New York	1	1	535	590	1
North Carolina	5	4	5,125	5,650	5
Ohio	1	1	490	540	1
Oklahoma	1	1	1,450	1,600	2
Oregon	7	6	5,357	5,906	6
Pennsylvania	3	3	780	860	1
South Carolina	4	3	4,983	5,494	5
Tennessee	2	2	1,156	1,275	1
Texas	6	5	4,145	4,570	4
Virginia	4	3	4,127	4,550	4
Washington	7	6	5,310	5,854	6
Wisconsin	4	3	1,140	1,256	1
TOTALS	119		95,750	105,567	

[a] Reprinted from U.S. Environmental Protection Agency.[19 (p. 3-5)]

Lime Kiln

TRS emission from the lime kiln, principally hydrogen sulfide, is controlled by maintaining proper process conditions. Exhaust discharge temperature, oxygen content of the lime mud feed, and the pH and sulfide content of the water used in a particulate scrubber on the exhaust gases are the primary

TABLE 8-9 Impact of Various TRS Emission Sources on Ambient TRS Concentrations from an Uncontrolled and Average 1,000-Ton/Day Kraft Pulp Mill

	Emission Factor, lb/TADP		Maximal Ambient Concentration at 0.3 km, ppm	
Source	Uncontrolled	Typical	Uncontrolled	Typical
Recovery furnace	15.0	2.5	6.4	1.07
Digester	1.5	0.64	16.0	6.83
Multiple-effect evaporator	1.0	0.43	3.0	1.28
Lime kiln	0.8	0.62	0.64	0.49
Brown-stock washer system	0.3	0.3	0.30	0.30
Black-liquor oxidation system	0.1	0.1	0.25	0.25
Smelt-dissolving tank	0.2	0.2	0.45	0.45
Condensate-stripping system	2.0	0.22	11.2	1.12

factors to be controlled. A caustic scrubber on the exhaust gases may be used to reduce hydrogen sulfide emission from the kiln.

Brown-Stock Washer System

Vent gases from this system are usually exhausted uncontrolled. In a few mills with newer furnace systems, these gases are fed to the recovery furnace as combustion air; the furnaces were designed for this purpose. Incineration in recovery furnaces not designed for this purpose could be ineffective and even dangerous.

A chlorination-caustic scrubbing system reportedly has been used effectively.[19] In this system, the chlorine absorbs and oxidizes the dimethyl sulfide and dimethyl disulfide, which are the principal constituents of brown-stock washer gases. Chlorine-gas injection reportedly has also been used effectively.[19]

TABLE 8-10 Techniques for Controlling Kraft Pulp Mill TRS Emission[a]

Source	Control Technique	Achievable TRS Level
Recovery furnace	Process controls + black-liquor oxidation	20 ppm (old-design[b] furnaces)
		5 ppm (new-design[c] furnaces)
		25 ppm (cross-recovery furnaces)
	Process controls + conversion to noncontact evaporator	20 ppm (old-design furnaces)
		5 ppm (new-design furnaces)
		25 ppm (cross-recovery furnaces)
Digester system	Caustic scrubbing	7,000 ppm[d]
	Incineration	5 ppm
Multiple-effect evaporator system	Caustic scrubbing	350 ppm[d]
	Incineration	5 ppm
Lime kiln	Process controls	40 ppm
	Process controls + good mud washing	20 ppm
	Process controls, good mud washing + caustic scrubbing	8 ppm
Brown-stock washer system	Incineration	5 ppm
Black-liquor oxidation system	Molecular oxygen	0 ppm
	Incineration	5 ppm
Smelt-dissolving tank	Freshwater use	0.0084 g/kg BLS
Condensate-stripping system	Caustic scrubbing	—
	Incineration	5 ppm

[a] From U.S. Environmental Protection Agency.[19 (p. 6-2)]

[b] Old-design furnaces are defined as furnaces without welded-wall or membrane-wall construction or emission-control-designed air systems.

[c] New-design furnaces are defined as furnaces with *both* welded-wall or membrane-wall construction and emission-control-designed air systems.

[d] Calculated on basis of scrubber's removing only hydrogen sulfide and methyl mercaptan and using Duffee *et al.*[5] to determine percent of hydrogen sulfide and methyl mercaptan present in vent stream.

Black-Liquor Oxidation (BLO) System

Vent gases from BLO systems are usually emitted directly to the atmosphere. Incineration in the recovery furnace or power boiler might be feasible, provided that sufficient moisture could be removed by condensers before burning. The condensate would be odorous, and incineration of the moist gases would increase corrosion problems.

A second control technique is the use of molecular oxygen instead of air. This is a closed system with no vent gases. Two mills in the United States use this method.

Smelt-Dissolving Tank

TRS emission from this source is governed by the presence of reduced sulfur compounds either in the smelt or in the water. The principal control option is the choice of the water used in the dissolving tank or particulate control scrubber.

Condensate-Stripping System

In a few mills, dissolved sulfides and other volatile compounds are stripped from the digester and evaporator condensate before discharge to treatment ponds. Steam or air is used as the stripping medium. Stripper gases may be incinerated in the lime kiln, the recovery furnace, or a separate incinerator. One mill uses steam stripping followed by white-liquor scrubbing, which reportedly is less effective than incineration.[19]

COSTS OF CONTROL SYSTEMS

The EPA has prepared emission guidelines for TRS-emission control that may be achieved by application of best adequately demonstrated technology to *existing* facilities. These are listed in Table 8-11.[19]

The EPA has also evaluated six alternative systems to meet these guidelines. The basic characteristics are summarized in Table 8-12.[19] The incremental costs for equipping existing mills with these systems, in the 28 states that have TRS-emission regulations, are summarized in Tables 8-13[19] and 8-14.[19]

TABLE 8-11 TRS Emission Guidelines for Existing Kraft Pulp Mills[a]

Affected Facility	Emission Guidelines,[b] ppm
Recovery furnace[c]	
Old-design furnaces[d]	20
New-design furnaces[e]	5
Cross-recovery furnaces	25
Digester system	5
Multiple-effect evaporator system	5
Lime kiln	20[f]
Brown-stock washer system	No control
Black-liquor oxidation system	No control
Condensate-stripping system	5
Smelt-dissolving tank	0.0084 g/kg BLS

[a] From U.S. Environmental Protection Agency.[19 (p. 1-7)]

[b] Four-hour averages.

[c] Three percent of all 4-h TRS averages above the specified level are not considered to be excess emission.

[d] Furnaces without welded-wall or membrane-wall construction or emission-control-designed air systems.

[e] Furnaces with both welded-wall or membrane-wall construction and emission-control-designed air systems.

[f] Two percent of all 4-h TRS averages above 20 ppm are not considered to be excess emission.

IMPACT OF GUIDELINES ON GROUND-LEVEL TRS CONCENTRATIONS

The calculated maximal ambient TRS concentrations in the vicinity of a 1,000-ton/day plant were calculated by EPA for each of the control schemes.[19 (p. 9-11)]

The 10-s value represents the value of concern for odor. These values show that, for best available control technology, maximal ground-level TRS concentration would be 97 $\mu g/m^3$. Assuming the TRS to be methyl mercaptan (odor threshold, 0.002 ppm), the odor could be estimated at 97 $\mu g/m^3 \times 4.5 \times 10^{-10} m^3/\mu g = 0.04$ ppm, or 20 times the odor-detection threshold. For the systems deemed most cost-effective, the equivalent odors would be 0.19 ppm, or 100 times the detection threshold.

RECOMMENDATIONS

These estimates represent maximal values at distances of 0.3 km from the plant or less. At a distance of 2 km, the concentration would be 16–20% of the maximum, depending on the elevation of the source, or 20 times the odor-detection threshold. This would be only slightly odorous and perhaps sufficiently low to prevent complaints of malodors from the public. Studies of odors, in terms of dilution to threshold values, at various distances from a model plant are needed to determine the effectiveness of the proposed guidelines on odors and to indicate whether any additional regulations are desirable. As part of this study, the relationship between odor detectability dilution ratio and odor acceptability (complaints) needs to be investigated. Some sulfur-based odors, especially that of hydrogen sulfide, rapidly desensitize the olfactory mechanism of those continuously exposed to them. Therefore, complaints, if any, are likely to come from transients, not residents. Community odors less than 10 times the detection threshold are usually adequate to preclude complaints. Hydrogen sulfide odor, however, is an exception.

PETROLEUM REFINERIES

Petroleum-refining is one of the most common and most complex major industrial odor sources in the United States. As indicated in Table 8-15, there are refineries (of various sizes and characters) in 38 of the 50 states. The raw crude starting material is odorous and most of the refined products are odorous, but much of the process odorous emission is derived from impurities in the crude, especially sulfur.

SOURCES OF REFINERY ODORS

Quantification of odor-emission sources in a refinery is difficult, because almost all materials are odorous. The quantity and type of emission depend, in part, on the refinery size, type of crude processed (whether sweet or sour), types of processes used, presence of petrochemical processes, air-pollution control measures in use, maintenance standards for such equipment as pump and compressor seals, and general housekeeping standards.

There are many available texts, references, and articles that describe the general processes used in a modern refinery. Among the best sources of information are *Hydrocarbon Processing*, which biannually publishes a list of essentially all refining processes, and *Oil and Gas Journal*, which discusses new process developments once or twice a year.

TABLE 8-12 Summary of Retrofit Control Techniques for Alternative Control Systems on Existing Kraft Mills[a]

Source	Control Systems 1	2	3	4	5	6
Recovery furnace	Replace furnaces >10 yr of age. Add 2nd stage black-liquor oxidation (including furnaces <10 yr of age). Also improve furnace air distribution.	Same as 1	Replace furnaces >20 yr of age. Add 2nd stage black-liquor oxidation for all other furnaces. Also improve furnace air distribution.	Same as 3	Same as 3	Same as 1
Digester system	Incineration	Incineration	Incineration	Incineration	Incineration	Incineration
Multiple-effect evaporators	Incineration	Incineration	Incineration	Incineration	Incineration	Incineration
Lime kiln	Increase lime mud washing capacity. Increase fan capacity and monitor oxygen and temperature (kiln). Add caustic to kiln scrubber	Same as 1	Same as 1	Increase lime mud washing capacity. Increase fan capacity and monitor oxygen and temperature (kiln)	Increase fan capacity and monitor oxygen and temperature (kiln)	Same as 5
Brown-stock washer system	Incineration	No control	No control	No control	No control	No control
Black-liquor oxidation system vents	Molecular oxygen	No control	No control	No control	No control	No control
Smelt-dissolving tank	Substitute fresh water for condensate	Same as 1	Same as 1	Same as 1	Same as 1	Same as 1
Condensate-stripping system	Incineration	Incineration	Incineration	Incineration	Incineration	Incineration

[a] Reprinted from U.S. Environmental Protection Agency.[19] (p. 8-14)

The most important refinery processes are crude-oil distillation (single- or two-stage) with the addition of light ends recovery, catalytic cracking, catalytic reforming, polymerization, alkylation, hydrocracking, isomerization, blending and sweeting, and asphalt production (air blowing, etc.). Besides these relatively standard processes, many refineries contain sulfur-recovery plants, sulfuric acid plants, and various chemical and petrochemical plants, e.g., for ammonia and synthetic-detergent manufacture. All refineries contain air- and water-pollution control facilities, such as oxidation ponds, sulfide oxidizers, oil-water separators, and phenol-recovery units.

All the processing and waste-treatment units discharge some hydrocarbons to the atmosphere. Although most are odorous, the paraffins and other saturated hydrocarbon molecules, such as alkanes, which make up the bulk of refinery emission to the atmosphere, are less odorous, with respect to perceived intensity at equal concentrations, than the sulfidic emission from refineries.

Table 8-16 lists approximate emission factors for a petroleum refinery. Table 8-17 lists common types of odorous emission and the most likely potential sources from an oil refinery. The most unpleasant odor-formers are the mercaptans and other sulfur-hydrocarbon compounds, and the most common sources of these compounds are the processing units, drains, tanks, vents, wastewater separators, oxidation ponds, and barometric condenser pumps.

Other important odorous chemicals released in a refinery are ammonia, hydrogen sulfide, and sulfur dioxide. In a modern refinery, the largest source of ammonia and hydrogen sulfide is often the hydrocracking unit. Other sources are catalytic-cracking and catalytic-reforming units. Sulfur dioxide is emitted in the stack gases of furnaces burning relatively high-sulfur fuel and from Claus sulfur-manufacturing plants.

There are no readily available published data on the quantities of odorant emitted from individual sources in refineries, nor on the resulting community odor intensities downwind of the facilities. Certainly, odorous emission from some refineries produces detectable and recognizable "refinery odors" miles from the refineries.

As shown in Table 8-16, the major *process* sources of hydrocarbon emission are the miscellaneous process units, which include storage tanks, transfer operations, and process-vessel off-gases. These are followed closely by the fluid catalytic units and process drains as primary odor-emission sources. However, preliminary results from an EPA-funded study indicate that, in a well-controlled refinery, fugitive sources—such as emission from flanges, pump seals, transfer operations, and transportation vehicles—are the major sources of volatile organic compounds.

TABLE 8-13 Incremental Retrofit Control Costs for a 1,000-TPD Modern Mill (Built After 1965)[a] —Location: State with Typical Regulations[b]

	Capital Costs, $1,000	Annualized Costs, $1,000/yr	Unit Annualized Costs, $/t	Capital Costs, $1,000	Annualized Costs, $1,000/yr	Unit Annualized Costs, $/t	Capital Costs, $1,000	Annualized Costs, $1,000/yr	Unit Annualized Costs, $/t
	No. 1			No. 2			No. 3		
Recovery furnace									
Direct contact	1,000	242	0.74	1,000	242	0.74	0	0	0
Indirect contact	0	0	0	0	0	0	0	0	0
Batch digesters and multiple-effect evaporator	0	0	0	0	0	0	0	0	0
Brown-stock washers[c]	2,500	1,670	5.08	0	0	0	0	0	0
Black-liquor oxidation system vents[c] (direct contact only)	560	420	1.28	0	0	0	0	0	0
Lime kiln[d]	830	252	0.77	830	252	0.77	830	252	0.77
Condensate stripper	0	0	0	0	0	0	0	0	0
TOTAL COSTS									
Direct contact	4,890	2,584	7.87	1,830	494	1.50	830	252	0.77
Indirect contact	3,330	1,922	5.85	830	252	0.77	830	252	0.77

	No. 4			No. 5			No. 6		
Recovery furnace									
Direct contact	0	0	0	0	0	0	1,000	242	0.74
Indirect contact	0	0	0	0	0	0	0	0	0
Batch digesters and multiple-effect evaporator	0	0	0	0	0	0	0	0	0
Brown-stock washers[c]	0	0	0	0	0	0	0	0	0
Black-liquor oxidation system vents[c] (direct contact only)	0	0	0	0	0	0	0	0	0
Lime kiln[d]	830	246	0.75	100	107	0.33	100	107	0.33
Condensate stripper	0	0	0	0	0	0	0	0	0
TOTAL COSTS									
Direct contact	830	246	0.75	100	107	0.33	1,100	349	1.07
Indirect contact	830	246	0.75	100	107	0.33	100	107	0.33

[a] Reprinted from U.S. Environmental Protection Agency.[19 (p. 8-24)]

[b] A typical state is assumed to require 20 ppm for the recovery furnace and incineration (5 ppm) of TRS emissions from digesters, multiple-effect evaporators, and condensate strippers.

[c] Destruction in separate incinerator.

[d] Low retrofit penalty.

TABLE 8-14 Incremental Retrofit Control Costs for a 1,000-TPD Old Mill (Built Before 1965) — Low Retrofit Penalty[a] — Location: State with Typical Regulations[b]

	Capital Costs, $1,000	Annualized Costs, $1,000/yr	Unit Annualized Costs, $/t	Capital Costs, $1,000	Annualized Costs, $1,000/yr	Unit Annualized Costs, $/t	Capital Costs, $1,000	Annualized Costs, $1,000/yr	Unit Annualized Costs, $/t
	No. 1			No. 2			No. 3		
Recovery furnace	23,300	4,000	12.18	23,300	4,000	12.18	0	0	0
Batch digesters and multiple-effect evaporators	0	0	0	0	0	0	0	0	0
Brown-stock washers[c]	2,500	1,670	5.08	0	0	0	0	0	0
Black-liquor oxidation system vents[c]	560	420	1.28	0	0	0	0	0	0
Lime kiln	830	252	0.77	830	252	0.77	830	252	0.77
Condensate stripper	0	0	0	0	0	0	0	0	0
TOTAL COSTS	27,190	6,342	19.31	24,130	4,252	12.95	830	252	0.77
	No. 4			No. 5			No. 6		
Recovery furnace	0	0	0	0	0	0	23,300	4,000	12.18
Batch digesters and multiple-effect evaporators	0	0	0	0	0	0	0	0	0
Brown-stock washers[c]	0	0	0	0	0	0	0	0	0
Black-liquor oxidation system vents[c]	0	0	0	0	0	0	0	0	0
Lime kiln	830	246	0.75	100	107	0.33	100	107	0.33
Condensate stripper	0	0	0	0	0	0	0	0	0
TOTAL COSTS	830	246	0.75	100	107	0.33	23,400	4,107	12.51

[a] Reprinted from U.S. Environmental Protection Agency.[19 (p. 8-27)]

[b] A typical state is assumed to require 20 ppm for the recovery furnace and incineration (5 ppm) of TRS emissions from digesters, multiple-effect evaporators, and condensate strippers.

[c] Destruction in separate incinerator.

TABLE 8-15 Distribution of Petroleum Refineries in the United States[a]

State	No. Refineries	Crude Capacity, bbl/calendar day
Alabama	3	49,375
Alaska	3	73,000
Arizona	1	4,000
Arkansas	4	60,400
California	36	2,297,385
Colorado	3	64,200
Delaware	1	140,000
Florida	1	5,700
Georgia	2	19,750
Hawaii	2	99,000
Illinois	11	1,181,550
Indiana	7	561,650
Kansas	11	453,918
Kentucky	4	164,470
Louisiana	20	2,036,950
Maryland	2	28,500
Michigan	6	150,050
Minnesota	3	216,800
Mississippi	5	328,541
Missouri	1	107,000
Montana	7	156,181
Nebraska	1	5,000
New Jersey	5	645,000
New Mexico	8	119,020
New York	2	107,000
North Dakota	3	58,658
Ohio	7	589,950
Oklahoma	12	546,825
Oregon	1	14,000
Pennsylvania	11	804,920
Tennessee	1	43,900
Texas	49	4,193,072
Utah	9	158,425
Virginia	1	53,000
Washington	7	366,900
West Virginia	3	19,450
Wisconsin	1	45,400
Wyoming	12	188,630

[a] Data from U.S. Environmental Protection Agency.[16]

TABLE 8-16 Odorant Emission Factors for Petroleum Refinery[a]

Process	Dimensions of Emission Factor	Emission Factor
Boilers and process heaters	lb hydrocarbon/1,000 bbl oil burned	140
	lb hydrocarbon/1,000 ft^3 gas burned	0.026
	lb NO_2/1,000 bbl oil burned	2,900
	lb NO_2/1,000 ft^3 gas burned	0.23
	lb HCHO/1,000 bbl oil burned	25
	lb HCHO/1,000 ft^3 gas burned	0.0031
Fluid catalytic units	lb hydrocarbon/1,000 bbl of fresh feed	220
	lb NO_2/1,000 bbl of fresh feed	63
	lb HCHO/1,000 bbl of fresh feed	19
	lb NH_3/1,000 bbl of fresh feed	54
Moving-bed catalytic-cracking units	lb hydrocarbon/1,000 bbl of fresh feed	87
	lb NO_2/1,000 bbl of fresh feed	5
	lb HCHO/1,000 bbl of fresh feed	12
	lb NH_3/1,000 bbl of fresh feed	5
Compressor internal-combustion engines	lb hydrocarbon/1,000 ft^3 fuel gas burned	1.2
	lb NO_2/1,000 ft^3 of gas burned	0.86
	lb HCHO/1,000 ft^3 of fuel gas burned	0.11
	lb NH_3/1,000 ft^3 of fuel gas burned	0.2
Miscellaneous process	lb hydrocarbon/1,000 bbl refinery capacity	
With control		5
Without control		300
Process drains	lb hydrocarbon/1,000 bbl wastewater	
With control		8
Without control		210
Vacuum jets	lb hydrocarbon/1,000 bbl vacuum distillation capacity	
With control		neg.
Without control		130
Cooling towers	lb hydrocarbon/1,000 gal cooling-water capacity	6
Pipeline valves and flanges	lb hydrocarbon/1,000 bbl refinery capacity	28
Vessel relief valves	lb hydrocarbon/1,000 bbl refinery capacity	11
Pump seals	lb hydrocarbon/1,000 bbl refinery capacity	17
Compressor seals	lb hydrocarbon/1,000 bbl refinery capacity	5
Others (air blowing, blend changing, and sampling)	lb hydrocarbon/1,000 bbl refinery capacity	10

[a] Data from U.S. Environmental Protection Agency.[16]

TABLE 8-17 Potential Sources of Specific Emissions from Oil Refineries[a]

Emission	Potential Sources
Oxides of sulfur	Boilers, process heaters, catalytic-cracking unit regenerators, treating units, H_2S flares, decoking operations
Hydrocarbons	Loading facilities, turnarounds, sampling, storage tanks, wastewater separators, blowdown systems, catalyst regenerators, pumps, valves, blind changing, cooling towers, vacuum jets, barometric condensers, air-blowing, high-pressure equipment handling volatile hydrocarbons, process heaters, boilers, compressor engines
Oxides of nitrogen	Process heaters, boilers, compressor engines, catalyst regenerators, flares
Aldehydes	Catalyst regenerators
Ammonia	Catalyst regenerators
Odors	Processing units (air-blowing, steam-blowing), drains, tank vents, barometric condenser pumps, wastewater separators

[a] Derived from U.S. Public Health Service.[22]

CONTROL OF REFINERY ODORS

Many methods are available for controlling hydrocarbons and organic emission, and there are many reviews of them. One of the best is *Control Techniques for Hydrocarbons and Organic Solvent Emissions from Stationary Sources.*[21] Some of the methods for specific chemicals used in refineries are described here.

Control of Ammonia and Hydrogen Sulfide Odors

Ammonia and hydrogen sulfide often occur together, particularly in hydrocracking and reforming operations. In these two processes, the most common means of removal from process vapor streams is water injection, which results in an ammonium sulfide water stream.

This particular waste stream is usually treated by sour-water distillation (stripping in the presence of acid or base) followed by incineration of the ammonia or hydrogen sulfide stripped from the water.

The ammonia and hydrogen sulfide produced in the catalytic-cracking units are also partially removed by water injection. This technique of water injection for control is applicable only when ammonia and hydrogen sulfide are present together. If the gas stream contains only hydrogen sulfide, the water solubility is so low as to make removal by water injection (scrubbing) impractical. In these instances, some other process is used, most commonly scrubbing with monoethanolamine (MEA) or diethanolamine (DEA). However, a host of other processes for hydrogen sulfide removal are used in refineries and in gas-treating plants. These include the proprietary Rectisol process, the Sulfinol process, and many others.[16]

Fluidized-Bed Incineration

This is essentially a hot bed of fluidized sand to which the waste stream is fed for combustion. It has been suggested for use in refineries to burn acid sludges from various drains, sewers, and treatment steps.

Control of Odors in Waste-Collection Pits

The DEA-Scholven refinery at Karlsruhe has reported an ingenious method of reducing substantially the odors emitted from the "collection pit" for drainage water. It spreads a blanket of Allplas balls on the surface of the water in the collecting pit. These balls are hollow moldings of polypropylene that is resistant to almost all chemicals. The main use of these balls is to prevent heat loss and evaporation (and hence reduce odor). Tests showed that a single layer of Allplas reduced evaporation by up to 88% and that additional layers provided further marginal reduction. At DEA-Scholven, the balls were used in a triple layer. The operators reported that there had been no complaints of smell from the collecting pit since the balls were brought into use and that, even when they leaned over the pit, the characteristic smell of oil was faint. The identity and location of the odor observers were not reported.

REGULATIONS

A number of performance standards have been promulgated by the EPA for various refinery processes. Two of these, although not specifically addressed to odorous emission, could have a pronounced effect on such emission.

The first, a standard promulgated in March 1974 (39 FR 9308), limits emission of sulfur dioxide from process heaters, boilers, and waste-gas disposal systems that burn process gas generated in the refinery. It limits hydrogen sulfide concentrations of fuel gas to 230 ppm, unless the resulting combustion gases are treated in a manner equally effective in preventing the release of sulfur dioxide to the atmosphere. To comply with this standard or to satisfy state and local emission codes, refineries "sweeten" the fuel gas, or remove hydrogen sulfide from it, before burning it in process heaters and boilers.

Removal of hydrogen sulfide is by scrubbing with solutions that absorb it (e.g., MEA and DEA). Regeneration of the solutions involves stripping a side stream to produce concentrated hydrogen sulfide with smaller amounts of carbon dioxide, water vapor, and hydrocarbons. These gases are processed in a recovery facility, such as a Claus plant, to produce elemental sulfur.

Claus sulfur capacity in U.S. refineries totaled 8,000 long tons/day in April 1973.[16] Over 1,000 LT/day, or 13% of existing refinery sulfur-plant capacity, was scheduled for completion in 1974. The trend is toward larger facilities, averaging 107 LT/day. Accordingly, in September 1976, the EPA issued proposed standards of performance for petroleum-refinery sulfur-recovery plants.

Depending on the type of emission-control system installed to comply with these standards, residual emission released to the atmosphere will consist of sulfur dioxide or reduced sulfur compounds—i.e., hydrogen sulfide, carbonyl sulfide, and carbon disulfide. The standards, therefore, limit either the concentration of sulfur dioxide or the concentration of hydrogen sulfide, carbonyl sulfide, and carbon disulfide in the gases discharged into the atmosphere from new or modified refinery sulfur-recovery plants. Specifically, emission is limited to either sulfur dioxide at 0.025% by volume on a dry basis and oxygen at 0%, or hydrogen sulfide at 0.0010% by volume, reduced sulfur compounds at 0.030% by volume on a dry basis, and oxygen at 0%.

The standards also require continuous monitoring of the concentration of sulfur dioxide or hydrogen sulfide and reduced sulfur compounds in the gases discharged into the atmosphere. This is to ensure proper operation and maintenance of the emission-control systems.

This may be compared with existing state regulations summarized in Table 8-18.

Two alternative emission-control systems are proposed as a basis for performance standards. These systems are the low-temperature Claus reactor system and the tail-gas-scrubbing system. As part of the proposed standard, the EPA has predicted the impact of these alternatives with a standard Gaussian dispersion model. The effects of different averaging times have been estimated with simple peak-to-mean ratios. The results are summarized in Table 8-19.

For these calculations, it was assumed that emission is from a single stack, that fugitive emission does not exist, and that downwash conditions do not occur.

The results indicate that only alternative II, with reduction scrubbing, would result in transient concentrations in excess of the odor threshold for hydrogen sulfide, carbonyl sulfide, and carbon disulfide.

TABLE 8-18 State Regulations for Sulfur-Recovery Plants[a]

State	Regulation: Existing Plants	Regulation: New Plants
Alabama	0.16 lb S/lb S input	0.08 lb S/lb S input
Arkansas	ground level concentrations only	
Colorado	ground level concentrations only	
Connecticut	0.01 lb S/lb S input	0.01 lb S/lb S input
Delaware	2,000 ppmv SO_2	2,000 ppmv SO_2
Florida	0.004 lb SO_2/lb S input	0.004 lb SO_2/lb S input
Louisiana	0.01 lb S/lb S input	0.01 lb S/lb S input
New Hampshire	0.01 lb S/lb S input	0.01 lb S/lb S input
New Jersey	15,000 ppmv SO_2	15,000 ppmv SO_2
Ohio	0.01 lb S/lb S input	0.01 lb S/lb S input
Oklahoma	—	0.01 lb S/lb S input
Pennsylvania	[b]	[b]
Texas	[c]	2,200 ppmv SO_2[d]
Utah	—	[e]
Virginia	0.05 lb S/lb S input	0.05 lb S/lb S input
West Virginia	0.06 lb S/lb S input	0.06 lb S/lb S input

[a] Reprinted from U.S. Environmental Protection Agency.[20]

[b] According to $A = 0.32\ E^{-0.5}$ where E = plant rating in LT/D and A = allowable emission in lb SO_2/lb S input.

[c] 214 lb SO_2/h/1,000 scfm effluent flow rate.

[d] At incinerator outlet, calculated for 50% excess air.

[e] 80% control for new sources with uncontrolled sulfur emission $\geqslant$250 ton/yr.

The costs for each alternative, normalized for a 100-LT/day sulfur plant and adjusted to June 1975, are summarized in Table 8-20.

Table 8-18 summarizes current state emission-control regulations for both existing and new sulfur-recovery plants. The most stringent regulations for sulfur-plant emission are found in Florida, Los Angeles County, and Philadelphia. Florida regulations specify maximal allowable sulfur emission as 0.004 lb of sulfur dioxide per pound of sulfur input or a 99.8% sulfur recovery (equivalent to 500-ppmv total sulfur on an undiluted basis). Los Angeles County restricts emission to 500-ppmv sulfur calculated as sulfur dioxide equivalent, and Philadelphia restricts refinery sulfur plants to 500-ppmv sulfur dioxide.

State codes applicable to sulfur-recovery plants are generally expressed in pounds of sulfur input. A majority of states with sulfur-emission regulations for sulfur-recovery plants have the same regulation: 0.01 lb of sulfur per pound of sulfur input, or 99% sulfur recovery.

The format of standards in Table 8-18 shows some states regulating sulfur dioxide only and several with standards for sulfur emission. The intent of sulfur-emission codes is not specified, but is assumed to be sulfur dioxide regulation only, in that emission of hydrogen sulfide or carbon disulfide at the allowed rates would incur severe odor problems.

RECOMMENDATIONS

Because refinery odors are widespread, an inventory of odorous emission from each source type in a refinery should be developed. This should include odor intensity, quality or character (compound identity if feasible), and relative concentration. The relation between odor emission rate and community annoyance needs to be established. The results could be used in zoning and siting studies and in specifying control alternatives.

CHEMICAL INDUSTRY

The chemical industry can be broadly described as including industrial operations that are in the Standard Industrial Classification (SIC) Code 28 category. The broad range of industrial processes covered by this SIC category makes it a formidable task to identify all odorous-emission sources that can be encountered. This survey, therefore, used two criteria for arriving at a reasonable list of chemical processes for evaluation:

- The sector of the industry involved (petrochemicals, etc.), so that a broad range of process types could be represented.
- The potential for low-odor-threshold emission from the process. This was based on both the threshold and the potential total emission (large number of plants or emission sources).

PETROCHEMICALS

Phthalic Anhydride

In 1975, approximately 351,000 tons (about 318,000 t) of phthalic anhydride (PA) was produced. This PA traditionally goes into three markets—plasticizers, alkyd resins, and glass-reinforced polyester. PA is usually produced by vapor-phase air oxidation of either naphthalene or *o*-xylene. This process uses a catalyst based on vanadium pentoxide with some type of promoter, such as potassium oxide. The current status of PA production is described elsewhere (Spitz[14] and Betz, cited in U.S. EPA[16]).

PA plants are notorious for their pollution problems, and in particular

TABLE 8-19 Estimated Maximal Ambient Air Pollutant Concentrations[a]

Pollutant	Stack Concentration, ppm	Ambient Air Concentration, $\mu g/m^3$							
		2-10 s	5 min	15 min	30 min	1 h	8 h	24 h	1 yr
Base case	9,250	—	—	—	—	860	345	175	15
Alternative I, SO_2	2,000	—	—	—	—	180	80	25	5
Alternative II									
Oxidation, SO_2	300	—	—	—	—	25	11	4	<1
Reduction									
H_2S	10	25	2	1	1	1	—	—	—
COS	100	380	30	15	7	7	—	—	—
CS_2	95	540	35	15	10	10	—	—	—

[a] Reprinted from U.S. Environmental Protection Agency.[20]

[b] 100-LT/day refinery sulfur plant. Overall sulfur recovery of 95%.

TABLE 8-20 Comparison of Alternative Emission Control System Costs for a 100-LTD Sulfur Plant (Cost Adjusted to June 1975)

Control System	Total Costs			Differential over Preceding Case			
	Investment, \$	Annual Operating Cost, \$/yr	Emission Rate Total Sulfur as SO_2, lb/h	Investment, \$	Annual Operating Cost, \$/yr	Emission Rate Total Sulfur as SO_2, lb/h	Unit Cost, \$/ton
Base case	2,783,000	(265,900)[a]	932	—	—	—	(5)
Alternative I	3,828,000[b]	(15,800)[b]	187	1,045,000	250,100	745	80[c]
Alternative II							
Oxidation	4,521,000[b]	314,600[b]	19	693,000	330,400	168	468[c]
Reduction	5,561,000[b]	462,800[b]	19	1,733,000	478,600	168	678[c]

[a] Denotes gain.

[b] Includes costs of base case Claus sulfur recovery plant.

[c] Incremental costs per incremental ton of SO_2 recovered.

for their odor-control problems. In the vapor-phase oxidation of either naphthalene or *o*-xylene, not all the feedstock is converted to PA; many side reactions occur. To avoid explosions, the hydrocarbon content of the airstream entering the reactor is kept at about 1 mol%. Conversion of the feedstock is complete, with some of the feed burned completely to carbon oxides (including carbon monoxide) and water and some byproduct formation. In the case of naphthalene, the byproducts include quinones (mostly naphthoquinone) and maleic and benzoic acids. *o*-Xylene gives somewhat more maleic and benzoic acids, maleic anhydride, cyclohexane, and phthalide. The PA-containing airstream (and, of course, the undesired and odor-causing byproducts) leaving the reactor is cooled and enters a bank of condensers, where the product is sublimed on the cooled surface of specially designed finned tubes. The crude molten phthalide is batch heat-treated at atmospheric pressure; this converts some of the impurities to compounds that are readily removed during a vacuum-distillation step.

The off-gas stream from this final recovery step has a very strong odor; in particular, the quinones have a strong characteristic odor. Three main methods are used to treat these gases: wet scrubbing, with an estimated efficiency of 50–90%; catalytic oxidation, 99%; and thermal oxidation, 99%. The use of scrubbers is not necessarily completely effective, because the compounds being removed have very strong odors, even in trace amounts. In addition, the use of a scrubber tends primarily to convert the air-pollution problem to a water-pollution problem. Catalytic fume incineration (catalytic oxidation) is the preferred method. Thermal oxidation (incineration) is higher in operating cost, but can be very effective in controlling these odors. One article is available[16] on odor control by catalytic combustion (see also the material on incineration in Chapter 6).

Acrylates

In 1964, the U.S. consumption of acrylates was 73,500 tons (about 66,700 t). Acetylene is the preferred raw material, although processes based on propylene have been used. The Reppe process for acrylate ester production uses a nickel (actually, nickel carbonyl) catalyst to convert acetylene to acrylic acid. The acid is esterified with the desired alcohol to make an acrylate ester. Most of these acrylate esters are used to produce thermoplastic resins. Acrylate resins find use as protective coatings on textile, paper, and leather, as well as for plastics.

The usual practice for odor control is tight control of process operations to prevent release of acrylates into the atmosphere. Acrylates are strong odorants and can generate unpleasant odors. Vent gas from the first-stage reactors is flared to prevent release of these potentially flammable (as well as odorous) gases. Any process vent can be the source of acrylate odors. The control of acrylates can potentially be accomplished by wet scrubbing, with an estimated efficiency of 99%; catalytic oxidation, 99%; thermal oxidation, 99%; or carbon adsorption, 95%.

The use of scrubbers may be effective, because the acrylates have reasonable solubility in water, but care must be taken not to convert an air-pollution problem into a water-pollution problem. Thermal or catalytic oxidation is potentially very effective, but safety reasons may prevent their use in explosion-hazard areas, which are sometimes encountered in acrylate plants. Carbon adsorption could be the technology of choice for odor control, if 95% control of acrylates can be shown to bring emission below the odor threshold for the surrounding community.

Phenol

About 775 tons (about 700 t) of phenol was produced in 1975, almost all from oxidation of cumene. In a typical process to prepare phenol and acetone coproduct, cumene is oxidized with air in an alkaline medium to produce cumene hydroperoxide. The resulting peroxide is decomposed by sulfuric acid catalyst to phenol and acetone at 140°F (60°C) in an agitated vessel. Peroxide concentrations are kept to 1% by use of a recycle phenol and acetone stream. Separation of acetone, cumene, and α-methylstyrene leaves the final undistilled phenol to be obtained overhead in a vacuum tower and recrystallized to produce phenol.

Phenol is used primarily in the production of phenolic resins, although some phenol is used as an intermediate in the production of more complex cyclic organic compounds, such as nonyl phenol and *o*-nitrophenol.

One major concept for odorous cumene and phenol emission control is the use of cryogenic condenser systems on process vents. These are not designed as much for odor control as for control of material losses.

Any process vent can be the source of small quantities of odorous cumene or phenol emission, which can lead to community odor problems. Phenol has a very characteristic and heavy odor.

The wastewater discharges from a plant producing phenol and acetone from cumene could be a source of odor problems from phenol degassing.

Potential odor-control concepts for phenol and cumene odors include catalytic oxidation, with an estimated efficiency of 99%; thermal oxidation, 99%; and wet scrubbing, 50–95%. Thermal or catalytic oxidation could be very effective, but safety considerations may prevent their use in a petrochemical complex with potential explosion hazards. Scrubbing has the disadvantage of producing a phenol water-pollution problem, if the blowdown is not treated with chemicals.

Styrene

About 2,335 tons (about 2,120 t) of styrene was produced in 1975. Styrene is produced primarily from the alkylation of ethylene and benzene to form ethylbenzene, with later catalytic dehydrogenation to form styrene. The alkylation of benzene requires use of an aluminum chloride catalyst, and ferric oxide is used as a catalyst for the dehydrogenation of ethylbenzene.

Styrene is used primarily in the production of styrene polymers, including polystyrene, ABS (acrylonitrile, butadiene, styrene), and SAN (styrene-acrylonitrile).

Distillation of the styrene monomer and transfer-system leaks, as well as process vents, can be sources of styrene odors. The control of all styrene-odor sources in a plant can be difficult, because of its high volatility and low odor threshold.

Potential odor-control concepts for styrene include carbon adsorption, with 95% efficiency, and thermal oxidation, with 99% efficiency. Tall-stack dispersion (stack height, over 500 ft, or 150 m) has been used in Michigan in relatively flat terrain. Detectable odors are still recognized on occasion downwind of the plant.

AGRICULTURAL CHEMICALS

Ammonia

In 1975, approximately 15,781,000 tons (about 14.3 million tonnes) of synthetic ammonia was produced. Synthesis of ammonia requires catalytic reaction of nitrogen and hydrogen over a nickel catalyst promoted with iron and vanadium.

Ammonia finds its biggest use in the fertilizer industry as a plant nutrient, in the form of anhydrous ammonia, urea, or ammonium nitrate. Ammonia plants can have significant odor problems from process leaks, venting of tanks and product-transport systems, and spills of product during transfer. Ammonia plants are usually in industrial or rural sites, and this tends to mitigate odor complaints. However, such mitigation may be only temporary, as housing and commercial enterprises approach plant boundaries in many growth areas in the United States.

Leaks and odors are usually controlled by preventive maintenance and equipment design. Tank vents can be ducted to tall stacks to ensure adequate dispersion of releases. Leak patrols and preventive-maintenance programs can help to avoid fugitive emission. For well-defined sources, sulfuric acid scrubbers are the only effective control measure. Their efficiency can approach 95% if acid is used, but 50% if water alone is used as a scrubbing liquid.

Urea

Urea production in 1975 was some 3,800,000 tons (3,450,000 t). In a typical urea process, ammonia and carbon dioxide react to form ammonium carbonate, which is thermally rearranged to produce urea. The urea is sent through a prilling tower to produce a prilled solid product, or it is diluted with water to form a solution for use as a fertilizer. Urea has two significant end uses; the main one is as a plant nutrient, and the other is as a starting material for melamine and urea-formaldehyde thermoset resins.

Emission sources include ammonia storage and transfer, as well as the prill tower or urea-dissolver system vents. The low odor threshold of ammonia could result in significant odor problems from urea plants near urban areas. Siting does, however, tend to be in industrial and rural areas.

Preventive maintenance and good equipment design are the best control practices for ammonia emission from process and storage operations. Prill towers and urea dissolvers require particulate-emission controls (baghouses and venturi scrubbers) followed by sulfuric acid scrubbers to control the odors from ammonia and urea particles. Control efficiencies can approach 95% for odorous emission from these sources.

Pesticides

In 1975, approximately 600 tons (540 t) of pesticides was produced in the United States. These include insecticides, fungicides, and rodenticides; and the total number of chemically distinct pesticides produced is very large. Processing operations, however, are very similar. To take an example, in a process for production of Treflan (trifluralin), a weed- and crabgrass-killer, *p*-chlorobenzotrifluoride is nitrated in a batch reactor, then dinitrated in a second batch reactor. The dinitro intermediate is dissolved in chloroform and aminated with dipropylamine. The raw trifluralin is purified by vacuum distillation, then treated with aromatic naphtha and emulsifiers to formulate the product.

Odor sources in pesticide manufacture include the transfer of solvents, intermediates, and reactants in and out of batch reactors; depressurizing and venting of reactors; and vacuum-pump and steam-ejector discharges. In final product formulation, spray dryers, granulators, and bagging operations can be significant odor sources. Where volatile products are involved, product transfer and storage can also be significant odor sources.

The odors from pesticide manufacture can be partially controlled through proper maintenance and equipment design. In many cases, enclosure or hooding of operations is required to capture fugitive odors, such as those from reactor vents. Control of point sources of emission can be achieved by wet scrubbing, with an estimated efficiency of 50–90%; thermal oxidation, 90–99%; or baghouses (for particles only), 50–90%. Scrubbers must be capable of neutralizing acidic or basic odorous compounds in many instances. In some cases, the use of baghouses must be followed by wet scrubbing or thermal oxidation, to remove gaseous odorous chemicals. Because of the refractory nature of some pesticide intermediates, thermal oxidation may not be effective unless long residence time (> 3 s) and high temperature (> 1,800 °F, or 980°C) are maintained in the incinerator system.

SPECIALTY CHEMICALS

Flavors and Fragrances

No production figures are broken out for this chemical manufacturing sector, but thousands of compounds are produced yearly in quantities of pounds to tons for commercial use. In manufacture, batch reactors are the general rule, and product separation-purification is an important part of the process. Because the compounds are to be used for their flavor or fragrance, there is high potential for odor problems.

Odor sources in the processing of flavor and fragrance chemicals include charging, venting, and discharging of reaction vessels; separation and purification of the final product; and transfer and storage of products and reactants. Spray dryers for powdered-product manufacture can also be odor sources. The intermittent and ill-defined nature of most odor sources makes their control very difficult.

Improved housekeeping and maintenance constitute one method of reducing odor emission from flavor and fragrance chemical manufacture. Controls that can be considered include wet scrubbing, incineration (catalytic and thermal), and carbon adsorption. Organic precipitators can also be applied where oils are being processed. The estimated control efficiencies are as follows: wet scrubbing, 50–95%; catalytic oxidation, 90–99%; thermal oxidation, 95–99%; carbon adsorption, 80–99%; and organic precipitation (oils), 50–90%. Because odor-source characteristics vary widely, the choice of odor-control concept may be vastly different for different processes in the same manufacturing facility.

Dyes

Approximately 103,000 tons (84,000 t) of dyes is produced in the United States each year. Several thousand different dyes are manufactured by companies in the United States, in the form of pastes, powders, lumps, and solutions. Concentrations vary from 6% to 100% active material. The process technology usually involves multistage synthesis starting with common chemicals. Although intermediates are sometimes produced by continuous reaction processes, most dye-manufacturing processes are batch processes. In some plants, the operation is integrated, starting with petrochemical bases, progressing to intermediates, and finally producing dye products. In many facilities, dye production starts with the intermediates.

The process for indigo dye manufacture starts with benzene, which is nitrated to form nitrobenzene and hydrogenated to form aniline, an important intermediate for many dyestuffs. Aniline then reacts to form phenylglycine, which reacts to produce indigo paste, which is 20% indigo dye.

Odor sources in dye-processing include charging, venting, and discharging of reaction vessels; purification and separation of products; and storage and transfer of solvents and reactants used in the dye-making process. Such odor sources are intermittent and sometimes ill-defined. Because the same processing equipment can be used for many different dyes, the odor-control problems can be very challenging. The odor-control concepts that can be applied include wet scrubbing, with an estimated efficiency of 50–95%; catalytic oxidation, 90–99%; thermal oxidation, 95–99%; and carbon adsorption, 80–99%. As with flavor and fragrance chemicals, the odor-source characteristics vary widely, so the choice of odor-control concept may be vastly different for different processes in the same manufacturing facility.

Soaps and Detergents

There are about 90 manufacturers producing approximately 2 million tons (1.8 million tonnes) of detergent and 0.5 million tons (0.45 million tonnes) of soap each year. Dentifrices, shampoos, and other liquid cleaners are also produced in the same plants with detergents as raw material, but no odor problem results from their production.

In the soap-making process, four types of fats or oils are processed: coconut oil, tallow, soya soap stock, and cottonseed soap stock. The coconut oil and tallow are triglycerides, which are split by a Twitchell reaction. This process takes place in large steam-heated vats. The fatty acid is then ready for introduction into the Sharples soap-making process, in which it is converted to the sodium salt, which is soap. The soya soap stock and the cottonseed soap stock do not require that treatment, because they already contain a significant percentage of free fatty acids. The soya or cottonseed soap stock is fed directly into the Sharples process to produce soap. High-speed centrifuges (15,000 rpm) are used to separate the soap and aqueous phases. Neat soap from this process is pumped to reactors and then to the top of a tower, where it is sprayed into the tower. Air at 450°F (230°C) is blown from the bottom. Drying occurs, and the soap granules drop to a hopper at the bottom of the tower, where they are removed. Moisture-laden air, with some soap fines, is exhausted at 175°F (80°C) from the top of the tower.

An obnoxious, fish-like odor was found to be prevalent when the above process was used. The three sources responsible for the odors were the storage and Twitchell tanks, the spray-tower dryer, and the centrifuging operation. The control measures in these sources generally consisted of careful sealing and leading the vapors to steam generators, where they were burned at approximately 2,400°F (1,315°C). Other types of control measures found successful were adsorption on activated charcoal, oxidation by chlorine, and scrubbing of the odor-laden air with sulfuric acid mist.

Although the predominant form of soap is bars, small quantities of flakes and granules are still made. The production of soap normally involves the hydrolysis of fats to fatty acids, followed by distillation and purification. The unloading and storage of flash-dried fat stock and the distillation and purification of higher-molecular-weight fatty acids produce odor bodies. The odors that are discharged from vacuum steam ejectors may be controlled by spray condensers or surface condensers. Dust emission from the manufacture of soap products is primarily an in-plant environmental-dust problem.

Liquid detergents are generally produced by a small batch sulfonation process. Any acid fumes vented from the sulfation tanks can be handled effectively by small acid scrubbers. The intermittent emission of acid fumes would occur during a short period.

From an air-pollution and odor standpoint, the major interest is in the spray-drying of synthetic detergents. The drying occurs normally in large towers—e.g., 20 ft (6 m) in diameter and 100 ft (30 m) high—rather than in a multitude of small units. The large volume of water evaporated is discharged from the dryer exit stack and condenses in the atmosphere to form a dense white plume. Malfunctioning of collection equipment can also lead to white plumes, which may generate erroneous impressions of odor and dust, especially when a plume dips to the ground.

The odor of the tower exhaust may be 4 or 5 times the detection threshold level. The smell of the tower-exhaust odor is similar to that of detergent used in the household. Even with large manufacturing units, the odor is diluted and diffused; thus, even if noted, it is generally unobjectionable. In many communities, it would be highly desirable to discharge the exhaust from a tall stack at a high velocity. The top of the stack should be well above nearby buildings, to prevent turbulent downwash on their leeward side, which can frequently bring the plume to ground level. Although the plume will be more visible from farther away when carried aloft, it usually will generate fewer complaints.

Material unloading and storage of raw materials are necessary adjuncts to the manufacture of soap and detergents. Raw materials handled include liquids—such as linear alkylate, fatty alcohol, caustic, and sulfuric acid or oleum—and dry materials, such as phosphates and other builders. The hydrocarbons and caustics have very low vapor pressures, and unloading and storage of these materials present no particular air-pollution control problem. The dry material may be unloaded in any of several ways. Vacuum conveyors and fluidizers are sometimes used, and these systems could present a potential air-pollution and odor problem. The technique of separating the conveyed product from the conveying air with a cloth filter prevents a dust-emission problem, but does not prevent the odor problem.

The unloading of the acid may be done with self-priming pumps or, in some cases, air blows. Compressed air or nitrogen is used to provide the motive power to lift the acid out of the tank cars and trucks to the storage tanks. The emission problem occurs as the car is emptied. The resistance to flow decreases, and a large volume of compressed air blows through the car, sweeps it free of vapors, and passes into the tank and out the vent in a noticeable cloud of mist. This is particularly true in the unloading of oleum. This cloud can drift beyond the plant boundaries and create a noticeable odor and a visible plume. A minor problem of acid-mist emission can also occur from the filling of storage tanks. The displaced air can result in a carryover of a noticeable acid mist, although this will usually be a problem only within plant boundaries. One way of handling the problem is venting through a high stack, where sufficient diffusion and dilution can occur. However, this method is not recommended unless the occurrence is infrequent and the quantity to be vented is very small.

Where small vents are used, it is important to heat-trace and insulate them, to prevent condensation of the acid mist on the vent walls in winter. Such condensation would polymerize and build up on the surfaces of the vent and rapidly plug it. A more positive method of controlling displaced vapors from the filling of oleum storage tanks is the scrubbing of the vent gases in a small packed tower with 98% sulfuric acid.

INORGANIC CHEMICALS

Chlorine Chemicals

In 1975, U.S. production of chlorine reached 9,265,000 tons (8,400,000 t). Hydrochloric acid and calcium chloride, derivatives of chlorine, reached 1,973,000 tons (1,790,000 t) and 1,062,000 tons (963,000 t), respectively. In 1975, chloroorganic chemical production included ethylene dichloride and vinyl chloride—2,970,000 tons (2,690,000 t) and 2,098,000 tons (1,903,000 t), respectively.

Chlorine is produced by electrolysis of salt brines in a cell with an asbestos-covered cathode and graphite anodes. A diaphragm permits flow of brine from the anode to the cathode, but reduces diffusion of products from the electrodes. Chlorine is produced at the anode and piped to a drying condensing system and finally to storage tanks or cylinders. Chlorine is handled in a closed system and transported by barge, tank car, or tank truck to its point of use. In spite of all attempts to control chlorine emission during production and use, there are small leaks of chlorine, and they can produce stinging fumes that are readily detectable. Spills during storage and transfer can also be appreciable odor sources, or even imminent health hazards.

Control systems for chlorine leaks include vigilant inspection and maintenance programs, as well as safety systems to divert or stop chlorine flow at critical parts of the chlorine processing and distribution system. Where small amounts of chlorine are emitted, caustic scrubbers are useful. They can remove 50–90% of the chlorine when chlorine concentrations are kept low.

Chlorine spills are best controlled by hosing down with water and then neutralizing with caustic. Holding tanks can be used to divert chlorine flows to prevent spills in many cases. Chlorine is extremely corrosive, so proper choice of construction materials can strongly affect the potential of chlorine-using systems for leakage.

Carbides and Carbon

In 1975, carbon-black production reached 1,378,000 tons (1,250,000 t). No information was available on silicon carbide production.

Carbon production is a furnace process. Silicon carbide processing is similar to that for carbon, and a furnace process is the basis for its manufacture. Carbon black finds its prime uses in automobile tires and as an electrode material for aluminum, caustic-chlorine, and steel production by electrolytic processing. Storage batteries are another major use. Silicon carbide is an abrasive with wide use in industry.

Odor problems from carbon manufacture are confined to two sources—the baking furnace and the graphitizing furnace. Only at high temperatures is there enough volatile material to cause odor problems. Some of the odors are carried on particulate emission from these furnaces, which leaves the process buildings through roof monitors. In the case of silicon carbide, the furnace is again the only significant odor source. The odors also tend to be associated with the particulate emission, which usually leaves through the building roof monitors.

The only technically feasible odor-control concept for either process would be a baghouse to remove particles. Odor-control efficiency could be speculated to be 10–50%. The size and cost of a baghouse for these processes would constitute a substantial capital investment. The improved odor in the community from such controls would not be substantial, considering the high cost of the control system. No other odor-control systems are considered feasible for these applications.

POLYMERS/FILMS

Viscose Rayon

Rayon is produced by the contact and steeping of sulfite (or sulfate) wood-pulp sheets in a caustic solution that dissolves the alpha cellulose. The alkali cellulose sheets are then reduced to crumbs in a shredder and aged for 1–2 days. The crumbs are put in contact with carbon disulfide to produce the xanthate. The cellulose xanthate balls are then dissolved in dilute sodium hydroxide, and the final product, to which pigments or titanium dioxide may be added, is a golden brown viscous liquid. The solution is deaerated and then spun by discharge into a spinning bath that contains sulfuric acid and generally salts, such as zinc sulfate and sodium sulfate. This permits coagulation of the xanthate to the final fiber.

The major emission problem is carbon disulfide, which comes off a number of process steps and especially from the aging tanks. Carbon disulfide is most frequently recovered or handled with activated carbon. Indeed, the fluidized-bed activated-carbon absorber was developed primarily for this purpose. Odor-control problems are caused primarily by carbonyl sulfide and hydrogen sulfide emission. Control of these odors by caustic scrubbing may be practical. Carbonyl sulfide and hydrogen sulfide can be recovered and reused if the concentrations are high enough. In all cases, carbon disulfide, carbonyl sulfide, and hydrogen sulfide can be controlled by incineration, but at high cost, because corrosion will require frequent incinerator maintenance and high capital costs. Carbon adsorption (for carbon disulfide) has an estimated efficiency of 75–95%; caustic scrubbing (for carbonyl sulfide and hydrogen sulfide), 50–90%; and thermal oxidation, 90–99%.

Tall-stack releases are the primary type of odor control used in the industry. They are singularly unsuccessful, however, because rayon-cellophane plant odors are detectable many miles downwind of the plant. In Virginia, reliable reports of rayon-plant effluent odors come from as far as 20 miles (32 km) from the source. This is due partially to the atmospheric oxidation of carbon disulfide to carbonyl sulfide, which has an odor similar to that of hydrogen sulfide at high concentrations and a much lower odor threshold than carbon disulfide.[15]

Fiberglass

Glass fibers are derived from two fundamental spinning processes: blowing, for the production of glass wool; and drawing, for the production of glass textiles. The major problem arises from the use of phenolics as binders in the production of glass wool. Phenolic odors have caused serious problems, because the phenolics are passed through an oven. Both phenolic and burnt-phenolic odors are produced. It has been found that the burnt-phenolic odors may be controlled and eliminated by flame incineration. More difficulty has been encountered in the control of phenolic odors. A number of techniques have been used, including process changes, alkaline scrubbing, catalytic combustion, and ozonation. Finally, it was necessary to make a process change: the substitution of air-forming for steam-forming of the fiber mat products. This eliminated a dense water fog that previously had scavenged and concentrated the odorous materials. The substitution has greatly reduced the problem, but has not eliminated it.

The primary odor problem arises from emission of phenol. Emitted hexamethylene tetramine and free formaldehyde vapors are said to constitute secondary odor problems.

Odorous emission may be controlled by hooding the molding operation, to collect the odor-bearing air, and subjecting the collected air to flame incineration with an estimated efficiency of 95–99%, catalytic combustion (90–99%), or wet scrubbing (50–95%).

Resins

Resins are classified, according to their broad general application, into thermosetting, thermoplastic, oil-soluble, and protein products. On the basis of derivation, the resins are grouped as natural resins, cellulose derivatives, protein products, and synthetic resins. The chemical conversion in resin manufacture is polymerization, either by condensation or by addition. Polymerization takes place in a number of ways—bulk, solution, emulsion, and suspension.

Resins are used as binders in the manufacture of plastics. Most resins, particularly vinyl polymers and copolymers, require plasticizers to improve workability or to modify natural properties. In addition, pigments are added, and the resin is eventually fed into a calender, which blends the mixture on heated rollers into film of uniform thickness. In the preparation of many products, a vinyl film or sheet is laminated onto a cloth backing, sometimes with a vinyl-foam layer between, passed through a curing oven, and held at about 230°C, to bond the plies and expand the foam.

During high-temperature heating, substantial amounts of the plasticizer are volatilized, to mix with hot air that is circulated through ovens as a heat-exchange medium. Under these conditions, the air can become saturated with volatilized material. Processing ovens must be purged to prevent the occurrence of explosive mixtures and to prevent condensation and dripping of high-boiling-point fractions onto the product. When the vented gases are allowed to cool in ducts leading from curing ovens or are discharged to the atmosphere, the plasticizer vapors tend to condense in the gas as a fog of submicrometer-sized droplets. Dioctyl phthalate (DOP) and diisooctyl phthalate (DIOP), in particular, can produce a 0.3-μm aerosol smog with powerful light-scattering properties. Particles of this size seem too small to permit the use of inertial effects for collection and too large to permit the use of diffusional forces. Plasticizers produce unpleasant odors, also. Thus, exhaust air must be treated to remove the oil fog and the associated odors.

Potential control devices include electrostatic precipitators, scrubbers, filters, and incinerators. Two-stage low-voltage electrostatic precipitators are used widely for particulate control in this application. Efficiency for particles is 95%; for odor control, it is much lower. Fog and venturi scrubbers, filters, and direct and catalytic incinerators are also used to control plasticizer fog. These have a greater efficiency in reducing odors.

The control techniques are described for a plant that prepares vinyl-covered flooring products. The process involves preparing a continuous blank backing sheet with a coating of vinyl polymer, then printing the surface with a pattern, and finally coating the sheet with a thin, wear-resistant layer of vinyl polymer. The coating is applied to the continuous sheet with a cylinder that rotates in a bath of fluid containing vinyl monomers dissolved in mineral-oil solvent. The wet sheeting is then transported through an oven, in which the solvent is driven off, and monomers are polymerized to yield the clear wear-resistant coating. The exhaust gases from these ovens are the source of air pollution and odor.

In the system used for control of this emission, the initial step involves the recovery of the solvent for reuse. Exhaust gases from the oven are sent first to a scrubber-condenser, in which the less volatile components condense on the surface of the water-cooled coils. Some of the condensate is recycled to the spray heads. The spray is used to prevent condensate from coating the coils. The remainder of the condensate is pumped to tanks, where the oil and water fractions are separated. The gases leaving the scrubber contain fine oil droplets or mist, less than a micrometer in diameter. This is controlled by the use of Brink mist-eliminator elements. These fiber filter elements are widely used to trap particles less than 3 μm in diameter, at high collection efficiencies.

The liquid drainings from the mist eliminators are mixed with condensate and sent to settling tanks, where a heavier water fraction separates from the lighter oil fraction. Water is drained periodically, and the oil is reused.

Estimated efficiencies of control devices are as follows: electrostatic precipitation, 75–95%; wet scrubbers, 50–95%; filtration, 75–95%; thermal oxidation, 95–99%; and catalytic oxidation, 90–99%. Recovery of 90–95% of the volatilized organic materials has been obtained. The Brink mist-eliminators are claimed to remove at least 99% of the mists from the gases.

Paints and Varnishes

Paint and varnish manufacturing is widely distributed throughout the United States. There are approximately 2,000 plants, with small ones predominating. The products of the industry are used almost entirely as decorative and protective coatings, although in some instances they may be used for their electric insulating or chemical-resistance properties. The coatings are made by closely controlled chemical reactions and accurately proportioned formulations, which may include natural and synthetic drying oils, pigments, volatile solvents, resins, driers, thinners, plasticizers, and antioxidants.

The principal operations of the industry are mixing and cooking. Pollutant evolution from these operations arises from vaporization of the various organic compounds being mixed or from volatile products of chemical reactions. Many of the vaporized compounds are odorous, some intensely so. There may also be small losses of volatile organics during handling and storage, but generally odors from this source are minor.

Paints Paint may be defined as a mixture of pigment with a suitable liquid that will form a solid adherent covering when spread thinly on a surface. Enamels are paints that form an especially smooth and glossy film.

Paint manufacture involves the following sequence of operations: mixing the pigment with a sufficient portion of the vehicle (organic compounds that form the binder) to make a paste, grinding the paste on a mill to reduce the size of pigment particles, diluting the ground paste with remaining liquid materials, tinting to a desired color, testing, and straining the paint and filling containers. Most of these operations are done in batches. Many of the products are almost custom-made, with each requiring raw materials of differing natures and amounts.

The two sources of odor in paint manufacture are grinding (during which the batch may heat up considerably, which leads to vaporization of organic compounds) and thinning (during which thinners may vaporize). Losses of odorous organic materials to the atmosphere from paint manufacture are estimated to be 0.5% of the weight of the paint. During the mixing operation, 1–2% of the solvent may be lost by evaporation.

The volatile compounds used in paint manufacture include aliphatic and aromatic hydrocarbons, alcohols, ketones, and esters. Although most of these materials are odorous, the odors are seldom obnoxious and are therefore commonly tolerated. In any case, during paint application and drying of the paint film, similar materials evaporate into the air.

Methods of controlling odorous emission to the atmosphere during paint manufacture are condensation and adsorption by scrubbing (either with water or with acidified or alkaline solutions), scrubbing and adsorption with activated charcoal or other adsorbents, combustion, dispersion from high stacks, and production of water-base coatings. One control method is satisfactory for all applications.

Varnishes Varnish is commonly defined as a transparent, homogeneous, heat-processed blend of drying oil, resin, drier, and solvent. When the varnish is applied as a thick film, the solvent evaporates, and the remaining materials oxidize and polymerize to form the varnish coating. In the paint and varnish industry, the term "volatile" is used to describe any base or vehicle for pigmented coatings.

Historically, "varnish" meant one type of product: a resin-in-oil, or oleoresinous, coating material that dried to a hard finish. Now, there are many types of varnish, such as spirit varnish (a resin with little or no oil in a volatile solvent), alkyd resin varnish (a solution of alkyd resin in a volatile solvent with added drier), asphalt varnish (asphalt in a volatile solvent), and lithograph varnish (partially oxidized and polymerized linseed oil blended with drier and resin).

Most of the odor problems of the varnish industry are related to the cooking process. Cooking performs a great many functions, including polymerization, depolymerization, melting, esterification, isomerization, distillation, and evaporation.

Varnish cooking is carried out in two types of kettles: the old type of portable, open kettle and the more modern, totally enclosed, stationary kettle.

The portable open kettle, which is rapidly disappearing, is a cylindric flat-bottom vessel (capacity, 150–370 gal) mounted on a three- or four-wheel platform. The kettle is charged in the loading room; wheeled to a fire pot, where it is heated by an open flame until cooking is completed; and then moved elsewhere for cooling and addition of drier and thinner.

The totally enclosed modern varnish kettle is similar to an autoclave, usually built of stainless steel, and automatically controlled. Heat may be furnished by the burning of natural gas or oil, by electric coils, or by circulation of Dowtherm or hot oil in the jacket. Liquid raw materials are pumped directly into the kettle; solids are added through a manhole in the top, which may be closed tightly. During the cooking operation, the kettle is provided with an inert atmosphere of nitrogen or carbon dioxide. Cooling coils in the kettle allow the product to be cooled quickly and may also be used to prevent runaway exothermic reactions during the cooking cycle.

Cooker emission varies greatly in composition and amount, depending on the specific product being manufactured. Nearly all the emission is odorous. For instance, esterification may result in the evolution of allyl sulfide, hydrogen sulfide, butyl mercaptan, and thiophene from the sulfur in the oils. Pentaerythritol fumes are evolved when that polyol is used. Linseed-oil blowing and cooking evolve acrolein. When phthalic anhydride and maleic anhydride are used as modifiers, odorous fumes are emitted.

The compounds emitted from cooking of oleoresinous varnish include water vapor, fatty acids, glycerin, acrolein, phenols, aldehydes, ketones, terpene oils, and carbon dioxide. Of all these compounds, perhaps the one most noxious and normally associated with varnish manufacture is acrolein, because of its pungent, disagreeable odor and its eye, nose, and throat irritating characteristics.

Typical losses of materials from various cooking processes are as follows:

- Total losses from oleoresinous varnish cooks average 3–6%, with some losses as high as 12%.
- Losses from alkyd resin cooks range from 4 to 6%.

- Cooking and blowing of oils produce losses of 1–3%.
- Heat polymerization of acrylic resins produces losses of less than 1%, unless the reaction gets out of control.
- Losses of solvent during thinning operations have been estimated at 1–2% of the solvent used.

A variety of control methods have been used to curtail odorous emission from varnish manufacturing. In every case, however, the first requisite is capture of emitted fumes, so that they may be conveyed to the control equipment. Adequate hooding of the kettle and a sufficient velocity of ventilation air are basic requirements for any control system.

According to one study, a 1,000-gal closed kettle should be furnished with exhaust capacity of 300–400 cfm; an open kettle may require 800–1,000 cfm. The exhaust-system capacity for the open kettle may be reduced drastically if the kettle and system are well-designed. For both types of kettles, ductwork should be of stainless steel, to reduce corrosion. Provision should also be made for frequent cleaning (e.g., by use of rectangular ductwork with one side hinged for easy opening), particularly in alkyd resin manufacure, where phthalic or maleic anhydride or pentaerythritol is added. These materials have a tendency to sublime and deposit in exhaust ducts. Conveying velocities of 1,500–2,500 ft/min (460–760 m/min) are suitable for most installations.

Once the odorous gases from the cooker or thinning operation are diverted into suitable ductwork, a variety of means may be used to capture or destroy the odorous effluent. The most common methods involve the use of scrubbers, adsorbers, or afterburners, of the flame or catalytic type. To avoid putting too great a load on the control equipment, condensers are often installed just ahead of the control equipment. In this way, valuable materials may be recovered. Precautions against flashback must be provided. Uncondensed vapors and noncondensable gases are then led to the control unit. In the case of a scrubber, efficiency of odor removal is related to the solubility of the odorous constituent in water and to the surface area provided for gas-liquid contact. Various means for increasing contact are available. Adsorption with activated charcoal is a possibility, but is usually more expensive than other, equally efficient methods. The expense is due to the fact that the compounds may polymerize on the carbon and ruin it. The most effective means of controlling odorous emission from varnish manufacturing is combustion, usually in an afterburner preceded by either a scrubber or a condenser.

It has been stated[12] that the factors of greatest importance in afterburner design for varnish cookers are these:

- Adequate, tight, corrosion-proof hooding with a minimal face velocity of 150 ft/min (46 m/min).
- Easily cleanable noncorrosive ductwork suitable for conveying velocities of 1,500–2,000 ft/min (460–760 m/min).
- Flashback protection—e.g., constriction in duct immediately preceding the afterburner, so that gas velocity is increased to 50 ft/s (15 m/s).
- Residence time in afterburners of 0.3–0.5 s at 1,200–1,400°F (650–760°C), with a refractory-lined combustion chamber, good turbulence, and gas velocity through the chamber less than 15 ft/s (4.6 m/s).

The same authors reported on a test of two varnish-cooker afterburners, carried out by the Los Angeles County Air Pollution Control District. The tangentially fired unit cost $6,500 when it was installed in 1956; the axially fired unit cost $2,500 installed.

Catalytic-combustion units have had a mixed reception in the varnish industry. Selheimer *et al.* (cited in U.S. EPA[16]) tried catalytic combustion and found that irritating fumes passed through the unit. Despite their lack of success with catalytic combustion, the method is fairly widely used, and various manufacturers point with pride to many successful installations. However, it is generally accepted that some cooking operations are not adaptable to catalytic afterburners, because the vapor products of esterification oxidize to very irritating compounds at temperatures up to 600°F (about 315°C).

The Los Angeles County Air Pollution Control District recommends direct-flame afterburners as the most positive method of control. In many cases, however, where fuel costs are high, the advantages of catalytic combustion warrant consideration.

PHARMACEUTICALS

Pharmaceuticals embrace all substances that are used for therapeutic purposes, so the list is long. Whereas, in earlier years, medicinals were chiefly extracted from crude drugs of vegetable origin, today a wide variety of synthetic products are also produced. Others are produced by fermentation and by the refining of technical-grade products.

Pharmaceuticals, or medicinals, include antibiotics and other antiinfective agents, antihistamines, autonomic drugs, cardiovascular agents, central nervous system depressants and stimulants, hormones and synthetic substitutes, vitamins, and other therapeutic agents for human and veterinary use and for animal-feed supplements.

Two types of pharmaceutical operations produce highly odorous effluent gases: antibiotic manufacture (fermentation process) and the processing of biologic material—specifically, glands, urine, and blood—by extractive processes.

Antibiotics

Antibiotics are usually produced by fermentation. The principal odor emission from fermentation processes is of the off-gases during fermentation and the processing and disposal of waste products, such as spent mash and micella.

In one fermentation plant, odors developed from three main sources: vent gases from the fermenters, spilled "beers" and filter cake, and vapors from the waste aeration tank. Vent gases from erythromycin production were especially objectionable. The problem was solved, after pilot tests, by collecting all vent gases, scrubbing them in an evaporator-cooler equipped with an entrainment separator, and discharging the scrubbed gases through a series of activated-carbon panels.

In another plant manufacturing a fermentation antibiotic for use as an animal-feed supplement, odorous vapors were emitted when the product was dried on steam-heated rolls. The odors were abated by passing the dryer gases through a water scrubber. The different types of odorous materials emitted from fermentation and later processing are so diverse that all common odor-abatement methods are used somewhere or other in the industry.

One particularly difficult odor problem arises from the disposal of liquid waste products. Both anaerobic and aerobic processes are used, to prepare the wastes for liquid disposal. The most suitable process for digesting the liquid waste is not necessarily the best for odor reduction, so generalities are difficult to make.

Extractives

One example of an odorous process of treating biologics is described here. Exhaust air from a unit processing beef lung to produce an anticoagulant was particularly odorous. A variety of absorbing solutions, such as caustic and acid, and oxidizing and reducing substances were evaluated. Hypochlorite solutions containing 5–70% available chlorine were found suitable for destroying the odor. Activated carbon was also effective, but silica gel and lime were not. Catalytic incineration was satisfactory. The system finally selected consisted of water scrubbing followed by the feeding of chlorine gas into a stack with the scrubbed gases. It took 25 lb (about 11 kg) of chlorine gas to treat an airflow of 8,000 cfm for 24 h.

ODOR-CONTROL COSTS

The costs of odor control for various types of chemical-process odor emission vary greatly. For instance, process changes can be accomplished with no capital investment for batch-process plants, whereas accomplishing the same basic task could involve a capital cost of millions of dollars for a petroleum refinery. Operating costs for odor-control systems are difficult to estimate in the absence of data taken from similar applications. Odor-control hardware costs are a function of the application and source size.

For cost information on specific odor-control processes, refer to the sections related to those processes in Chapter 6.

REFERENCES

1. Albrecht, O. E. Odor problems at the combustion of waste water sludges. Schriftenreihe Ver. Wasser-Boden-Lufthyg. (Berlin) 35:81–84, 1971.

2. Barker, J. E., J. McN. Bruce, and R. Kemmetmüller. Symbiotic potential: The integration of preheating and dry cooling in cokemaking. J. Inst. Fuel 51:91–99, 1978.

3. Callahan, W. F., and A. B. Pincince. Physical-Chemical Treatment System at Fitchburg, Massachusetts, p. 33. Presented at the TAPPI Environmental Conference, Chicago, Illinois, April 25–27, 1977. Atlanta: TAPPI, 1977.

4. Danielson, J. A., Ed. Air Pollution Engineering Manual, County of Los Angeles Air Pollution Control District. 2nd Ed. Publ. No. 999-AP-40. Cincinnati: U.S. Department of Health, Education, and Welfare, Public Health Service, National Center for Air Pollution Control, 1973.

5. Duffee, R. A., S. Cha, F. R. Farrow, and J. P. Wahl. Evaluation of Odor Measurement Techniques. Vol. II, The Pulp and Paper Industry. EPA Contract 68-02-0662, Program Element 1A1010. TRC Project No. 62184. Wethersfield, Conn.: The Research Corporation of New England, 1974. 72 pp.

6. Fair, G. M., and E. W. Moore. Measurement of intensity and concentration of odors associated with sewage treatment processes. Sewage Works J. 7:182–199, 1935.

7. First, M. W. Control of systems, processes, and operations, pp. 4–39. In A. C. Stern, Ed. Air Pollution. 3rd ed. Vol. 4. New York: Academic Press, 1977.

8. Hicks, R. Odor control in waste-water treatment systems, pp. 867–876. In H. M. Englund and W. T. Beery, Eds. Proceedings of the Second International Clean Air Congress. New York: Academic Press, 1971.

9. Kenson, R. E., N. E. Bowne, and W. A. Cote. The cost-effectiveness of coke oven control technology: Study of ambient air impact, pp. 6–18. In Proceedings of the Western Pennsylvania Section Technical Conference on Coal Utilization and Air Pollution Control, Air Pollution Control Association. Pittsburgh: Air Pollution Control Association, 1976.

10. Kent, J. A., Ed. Riegel's Industrial Chemistry. New York: Reinhold Publishing Co., 1962.

11. Largest costs of saving environment ahead. Eng. News Res. 200(26):19–20, 1978.

12. Mills, J. L., W. F. Hammond, and R. C. Adrian. Design of afterburners for varnish cookers. J. Air Poll. Cont. Assoc. 10:161–168, 1960.

13. Perry, R. H., C. H. Chilton, and S. D. Kirkpatrick, Eds. Perry's Chemical Engineers Handbook, pp. 20–35. 4th ed. New York: McGraw-Hill Book Co., 1963.

14. Spitz, P. H. Phthalic anhydride revisited. Hydrocarbon Processing 47(11):162–168, 1968.

15. Union Carbide Corporation. Oxo process. Hydrocarbon Processing 56(11):190, 1977.

16. U.S. Environmental Protection Agency. Air Pollution Aspects of Odors. 2 vols. February 1971. (unpublished)

17. U.S. Environmental Protection Agency. Atmospheric Emissions from the Pulp and Paper Manufacturing Industry, p. 5. Publ. No. EPA-450/1-73-002. Research Triangle Park, N.C.: U.S. EPA, September 1973.

18. U.S. Environmental Protection Agency. Control Techniques for Volatile Organic Emissions from Stationary Sources. Publ. No. EPA-450/2-78-022. Research Triangle Park, N.C.: U.S. EPA, Office of Air and Waste Management, May 1978. 533 pp.

19. U.S. Environmental Protection Agency. Draft Guideline Document: Control of TRS Emissions from Existing Kraft Pulp Mills. Publ. No. EPA-450/2-78-003a. Research Triangle Park, N.C.: U.S. EPA, January 1978. 210 pp.

20. U.S. Environmental Protection Agency. Standard Support and Environmental Impact Statement. Vol. 1. Proposed Standards of Performance for Petroleum Refinery Sulfur Recovery Plants. Publ. No. EPA-450/2-76-016a. Research Triangle Park, N.C.: U.S. EPA, Emission Standards and Engineering Division, September 1976. 200 pp.

21. U.S. National Air Pollution Control Administration. Control Techniques for Hydrocarbons and Organic Solvent Emissions from Stationary Sources. PHS Publ. No. AP-68. Washington, D.C.: U.S. Department of Health, Education, and Welfare, Public Health Service, Environmental Health Service, National Air Pollution Control Administration, 1970.

22. U.S. Public Health Service. Atmospheric Emissions from Petroleum Refineries: A Guide for Measurement and Control. PHS Publ. No. 763. Washington, D.C.: U.S. Department of Health, Education, and Welfare, Public Health Service, Division of Air Pollution, 1960. 56 pp.

23. Working Party on the Suppression of Odours from Offensive and Selected Other Trades. Odours: Report of the Working Party on the Suppression of Odours from Offensive and Selected Other Trades. Part 1: Assessment of the Problem in Great Britain. Stevenage, England: Warren Springs Laboratory, 1974. 116 pp.

24. Yocom, J. E., and R. A. Duffee. Controlling industrial odors. Chem. Eng. 77(13):160–168, 1970.

9 Control of Odors from Mobile Sources

The mobile sources discussed here include various kinds of engines. The classification is by type of propulsion system and encompasses diesel engines, gasoline engines, gas turbines, and jet engines.

DIESEL ENGINES

In a 1967 survey,[13] 54% of those interviewed tended to agree with the statement that "trucks burning diesel fuel cause the most air pollution"; 20% disagreed; and 25% were not sure. The public's opinion of diesel buses and trucks was also noted in voluntary remarks observed during a survey on diesel-engine exhaust odor conducted by Southwest Research Institute for the National Air Pollution Control Administration (NAPCA).[96] Because the odor and smoke from diesel-powered vehicles is so easily associated with air pollution by the general public, the diesel receives more than its share of the blame. The readily observable odor has been termed a nuisance emission by some (Elliott *et al.*[36] and Collins, cited in U.S. EPA[125]), because a number of studies have attributed no known health hazard to the odor itself.[3,6,15,17,18,41,45,48,65,80,84,105,106,108,113,137,141] However, the presence of polynuclear aromatic compounds and recent findings that diesel-exhaust aerosol particles showed mutagenicity in the Ames test prompted the EPA[126] to issue warnings against unnecessary exposure to diesel exhaust, and more studies are in progress.[16]

Of the estimated 150 million vehicles in this country,[27,133] diesels (excluding tractors) accounted for less than 1% of vehicle regis-

TABLE 9-1 Population of Mobile Sources of Air Pollution in the United States

Vehicle	Fuel or Engine	No. in Use
Passenger cars	Gasoline	113,696,111 (1977)[130]
Trucks	Total (gasoline and diesel)	27,719,597 (1976)[130]
	Diesel, butane, and other	1,362,406 (1977)[130]
Buses	Total (gasoline and diesel)	491,674 (1977)[130]
	Diesel, butane, and other	83,397 (1977)[130]
Motorcycles	Gasoline	5,110,000 (1977 est.)[130]
Off-highway wheel-type tractors (contractors)	Gasoline and diesel	21,647[a] [98]
Wheel-type tractors (except contractors)	Gasoline and diesel	2,128,914[a] [98]
Track-laying tractors	Gasoline and diesel	245,595[a] [98]
Aircraft (nonmilitary)	Gas turbine	7,722 (1977)[129]
	Piston engine	196,224 (1977)[129]
Locomotives	Diesel	27,473 (1977)[11]
Ships	Diesel	Not available
Outboard motors	Gasoline	7,737,000 (1976 est.)[120]
Boats (registered, pleasure)	Outboard	6,619,490 (1976)[119]
	Inboard	1,051,723 (1976)[119]
Utility	Gasoline	Not available
Mines	–	4,000 (1968)[98]

[a] Total units manufactured, 1959 through September 1968.

trations[12,20,23,59,124,133] and less than 5% of the fuel consumed by on-highway vehicles in 1969.[25,59,133] Table 9-1 shows a recent breakdown, by engine type, of vehicles of all types that contribute to air pollution. Nationally, the automobile accounts for approximately half the mass of emission to the atmosphere.[4,47,68,103,112,123,135,136] The overall contribution of diesel engines to total air pollution from mobile sources is relatively minor, as illustrated in Figure 9-1.[98]

With the thrust to save petroleum fuels, the proportion of diesel-powered vehicles, including passenger cars, is expected to increase rapidly. Diesels consume less fuel per mile than gasoline engines. Because federal regulations mandate specific increasing miles-per-gallon targets to be met by the entire production mix of an automobile manufacturer of the coming years, incorporation of diesel-operated passenger cars provides an avenue for reaching such targets and is being pursued by automobile manufacturers. Diesels are more difficult to start in cold weather, but require less maintenance; technologic advances toward easier-starting diesels are under way. It has been reported that some progress has been achieved in the

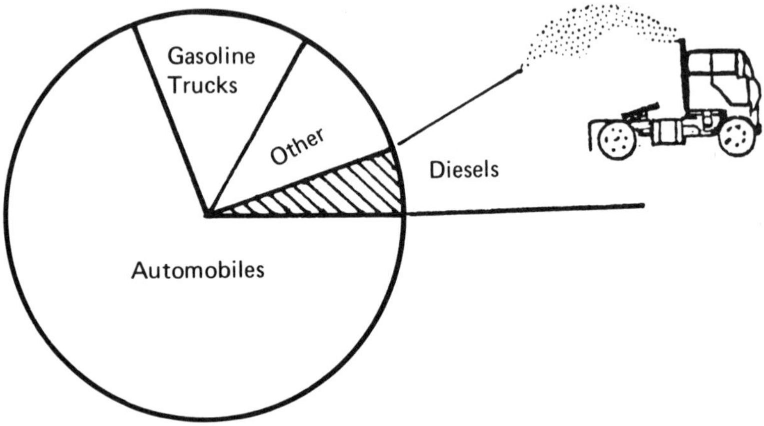

FIGURE 9-1 Diesel-engine contribution to air pollution from mobile sources. Data from U.S. Environmental Protection Agency.[128]

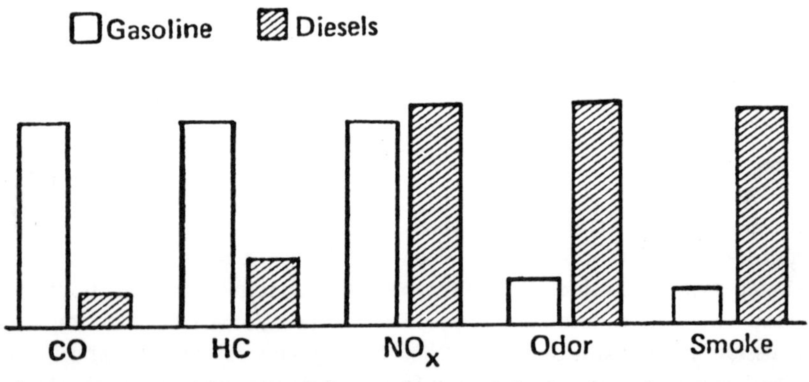

FIGURE 9-2 Comparison of emission components of diesel and gasoline engines. Data from Springer and Ludwig.[98]

reduction of diesel-exhaust odors.[19] Figure 9-2[98] compares a gasoline- and a diesel-powered vehicle with respect to the amounts of various components of their emission. Diesel engines emit far less carbon monoxide and smog-forming hydrocarbons (HC) than gasoline engines. Oxides of nitrogen (NO_x) are emitted in about the same amounts from both engines. Odor and smoke, however, are emitted in much greater quantities from a diesel than from a gasoline engine.

CAUSES OF DIESEL-EXHAUST ODOR

There are two sources of odors in diesel engines. One is unburned fuel and its thermal breakdown products, which are principally various forms of hydrocarbons and some nitrogen compounds. The other is incomplete combustion, as in zones of the combustion chamber close to its internal walls. Incomplete combustion results in the partial oxidation of hydrocarbons to oxygen-containing organic compounds termed oxygenates. Any means that promotes complete combustion and reduces the zone of partial oxidation will reduce odor.

COMPOSITION OF ODORANTS

Investigations of the chemical species responsible for diesel-exhaust odor were conducted with increasing sophistication as analytical instrumentation dramatically advanced from the late 1960's into the early 1970's.[8,32,52,-58,66,76,77,81-83,85-87,116-118,125,138] High-surface polymeric adsorbents were used for sample collection. Component separation used silica gel filtration and high-resolution gas-chromatographic (GC) columns. Mass spectrometry was used to identify the chemical species corresponding to the odorous components. Odor assays of GC effluent during analysis permitted sorting out the components that were relevant to odor.

At least 1,000 species were found at concentrations above 0.1 $\mu g/m^3$ of exhaust. Of these, about 100 were judged to be possible contributors to the exhaust odor individually, and others may have contributed collectively.

Contributors to oily-kerosene odor notes were alkylbenzenes, naphthalene derivatives, indans, indenes, and tetralins. Contributors to smoky-burnt odor notes were oxygenated compounds. Smokiness carriers were hydroxy and methoxy indanones; methylphenols and methoxyphenols also assisted in this note. Burnt character was associated with furans and alkylbenzaldehydes. Aliphatic aldehydes up to octanal were found, probably contributing some aldehydic character. Aliphatic acids up to nonanoic were found by the GC odor assay, and one foul-odor species containing sulfur, trimethylthiophene, was found by mass spectrometry.

No work is known to be in progress on further identification of odorants in diesel exhaust.

The influence of particulate matter on the exhaust odor has been investigated with a corona-free electrostatic precipitator to remove the smoke particles.[104] Smoke removal had only a marginal effect on the exhaust odor; this indicates that the major odorous components are carried in vapor form. However, in another study with a thermal

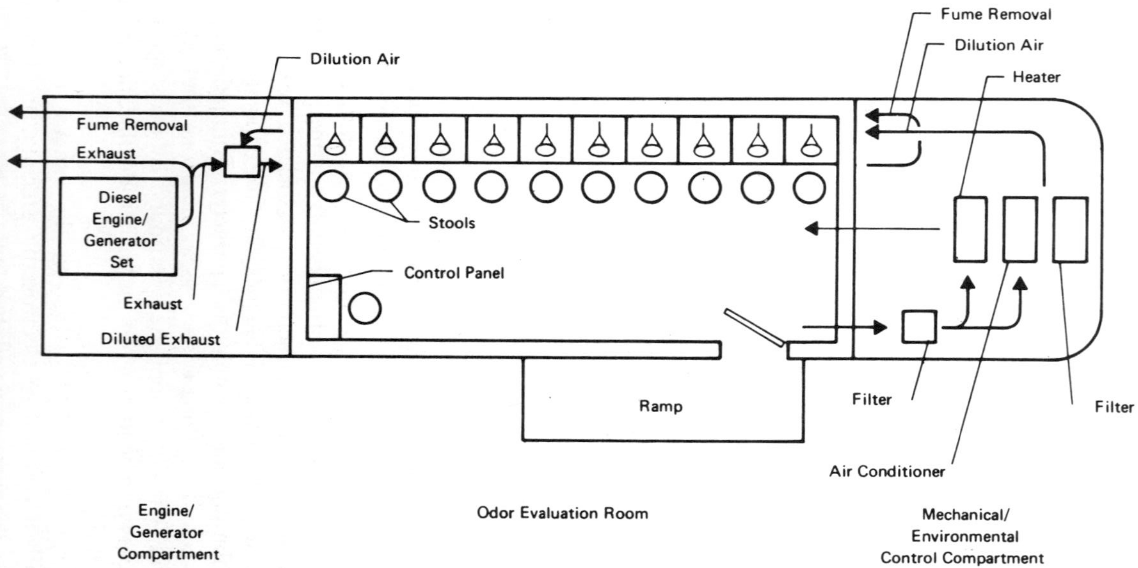

FIGURE 9-3 Schematic of mobile odor-evaluation laboratory. Reprinted with permission from Hare *et al.*[49]

precipitator, smoke removal did reduce diesel-exhaust odor (Rossano and Ott, cited in U.S. EPA[125]).

ODOR MEASUREMENT

Both sensory and analytical methods have been applied to diesel-exhaust measurements (see Chapter 4). Sensory methods use simple intensity-category scales, a reference-sample scale, or signal-detection theory; all these require the use of panelists (sensory judges) and are considered by many engineers involve in diesel design as cumbersome and not sufficiently accurate. Recent EPA data have shown diesel-exhaust particulate matter to be mutagenically active in the Ames test.[16,126] This has required that exposure of odor panels to diesel exhaust be eliminated in any work supported by EPA. An analytical system that measures the content of oxygenated species in exhaust is used in diesel development programs, principally as an index of the degree of incomplete combustion.[7,58]

Sensory Measurements

Because of the recent results that indicated that diesel exhaust may be carcinogenic, it is possible that sensory measurements of diesel-exhaust odor will have to be abandoned.

Exhaust is usually diluted with nonodorous air before sensory evaluation. The dilution is designed to obtain the odor experienced by a pedestrian at a curb close to the tailpipe of a diesel bus. Tracer measurements[90] indicated that dilution by a factor of 100 is typical for such a situation. For trucks equipped with vertical exhaust stacks, this factor is around 700.

In a typical evaluation,[97,99] exhaust is diluted by a factor of 100 before it is piped for evaluation in a specially constructed laboratory (Figure 9-3). The evaluation uses a diesel-odor reference set[116] of 28 plastic bottles, each containing a different odor sample. Four sets of four bottles are used to characterize, on a 1–4 scale, the intensities of burnt-smoky, oily, pungent-acid, and aldehydic-aromatic notes. A set of 12 bottles, termed the D-scale, contains mixtures of all four odor notes at 12 different concentrations. This set is used to characterize the overall diesel-odor intensity. The sets are used to train carefully selected panelists in the recognition of different notes and intensities. The standard deviation for a panel of six to 10 is about one point on the 12-point D-scale.

Another form of evaluation[7] uses an odor-test room in which diesel exhaust is diluted by a factor of 600 before evaluation. Odor intensity is evaluated with a so-called TIA (total intensity of aroma) scale, which has the following points:)((trace), 0.5, 1, 1.5, 2, 2.5, and 3 (strong). Sample concentration must be increased approximately by a factor of 10 to produce a one-unit increase on the TIA scale. Intensity of oily, burnt, and kerosene notes is rated separately on a similar scale. The presence or absence of eye, nose, and throat irritation is also recorded. Typically, up to four expert odor panelists are used in this method.

Scales other than the D-scale and the TIA scale have also been used, e.g., the 1-butanol scale.[79]

A quite different approach has been applied to the study of ambient traffic odors in Stockholm, Sweden.[63] Panelists were in a mobile laboratory that could be stationed in various street-traffic locations, and the detectability index *d'* (see Chapter 4) was calculated to characterize the extent to which the ambient odor could be distinguished from nonodorous air. For example, it was found that the traffic odor was not detectable 75 m from a thoroughfare.

Public-opinion surveys have been conducted using a Sniffmobile.[96] This mobile laboratory contained a concealed diesel engine that produced diesel-exhaust odor for presentation from sniffing funnels. The laboratory was placed in various cities and locations, and members of the public were invited to enter, smell the funnels, state the degree of their dislike, and answer some survey questions.

With a possible abandonment of sensory methods because of the potential health hazards from diesel exhaust, techniques may need to be developed to probe changes in public annoyance caused by diesel-exhaust odor in normal street use of diesels, without intentional experimental exposure to such odors.

Analytical Measurements

To satisfy the need for objective measurement of diesel-exhaust odors, a group-analysis system has been developed, named DOAS (diesel-odor analysis system).[7,58] A sample of exhaust is collected by adsorption in a styrene-divinyl copolymer, Chromosorb 102, eluted by cyclohexane, and analyzed in a silica gel liquid-chromatography column equipped with an ultraviolet absorption detector, with cyclohexane as the first eluting solvent. A peak proportional to the content of aromatic (benzenoid) compounds appears. Cyclohexane is then replaced by isopropanol, and a second peak, characteristic of the content of organic oxygenates (oxygen-consuming species), appears. It is claimed that the larger of these peaks, usually the oxygenates peak, is proportional to the TIA.

Further research on the usefulness of the DOAS method (Cermansky *et al.*;[24] Degobert;[28] J. H. Johnson, personal communication; R. J. Hames, personal communication; C. W. Savery, personal communication; K. J. Springer, personal communication; W. F. Marshall, personal communication; J. M. Perez, personal communication) has shown that the correlation between odor intensity and the logarithm of DOAS readings is limited to closely related exhaust samples. Major changes in fuel and diesel-engine type destroy the correlation. For example, in one study,[94] a good correlation between the D-scale values obtained from judgments of 10 panelists and TIA values calculated from DOAS data was found for light-duty diesels; but no correlation was found in another study by the same laboratory[95,100] for heavy-duty diesels. For cases with a correlation, the DOAS-predicted TIA values had a standard deviation of 0.4 TIA unit in a plot of actual versus predicted TIA.

Nevertheless, the DOAS procedure appeals to engineers engaged in diesel-technology research, because it is simple and does not depend on human judgment. It also does not require exposing humans to exhaust pollutants with a potential for long-term health effects.

The Coordinating Research Council, Inc., is an organization of engine and fuel manufacturers. A panel of its cooperative study group CAPI-64 is concerned with the development of a standardized DOAS sampling and analysis procedure and the establishment of its reproducibility in interlaboratory comparisons. Some improvements have been made in the United States[24] and Europe.[28] The reproducibility of the DOAS itself appears to be characterized by a standard deviation of 0.1 (instrumental) TIA unit.

The use of DOAS is in establishing whether some change in the design or operation of an engine reduces the concentration of the oxygenates in its exhaust and thus perhaps reduces the odor. The odor reduction itself can be verified by sensory methods.

ODOR CONTROL

Telephone interviews were conducted with researchers known to be working on various aspects of diesel odor and its control (J. H. Johnson, Michigan Technological University; R. J. Hames, General Motors Corp.; C. W. Savery, Drexel University; K. J. Springer, Southwest Research Institute; W. F. Marshall, U.S. Department of Energy; and J. M. Perez, Caterpillar Co.). Their statements and specifically cited references on particular aspects of control are the basis for the following discussion of findings and trends.

Influence of Fuel

The type of fuel used in a diesel engine has only a marginal influence on its exhaust odor.[24,53,92]

In one study, pure cetane was used as the fuel.[1] Exacting odor measurement was not conducted, but the experimenters detected the usual diesel-exhaust odor at substantial intensity. Although cetane is a chain hydrocarbon, aromatic ring compounds and various other species characteristic of diesel exhaust were analytically detected in the exhaust in the cetane experiment.

Changing from the usual diesel fuel, which contains a substantial fraction of aromatic ring compounds, to a fuel that was 99.9% paraffinic did not influence the DOAS-derived odor-intensity index.[24] No significant change in D-scale sensory intensity values was found with a change in fuel in other experiments.[7,92]

In general, engines are optimized by manufacturers to operate with either DF-1, a kerosene type of fuel used by many municipal transit systems, or DF-2, the principal fuel used in diesel-powered trucks. An engine optimized for DF-1 may result in a more objectionable odor when operated on DF-2. For example, a heavier, burnt-smoky, and sometimes more acrid odor has been reported from a municipal bus optimized for DF-1 but operating on DF-2. The exhaust odor was lower in the kerosene odor note.

Fuel Additives

Materials may be added to diesel fuel to influence the odor of its exhaust. Some additives may be intended to improve or modify the combustion process and thus reduce the concentration of odorants in the exhaust.[30] Others may be intended to survive the combustion process and modify the exhaust odor itself, making it less recognizable and less offensive[29,78,110,111] (see also various bulletins, etc., from Rhodia, Inc., Metropolitan Petroleum Petrochemicals Co., Inc., Lubrication Engineers, Inc., Whitnor Chemicals, Inc., and Ellis Chemicals and Lubricants, Inc.—all cited in U.S. EPA[125]).

NAPCA, and more recently EPA, proposed a set of requirements[38,134] that additive manufacturers and fuel refiners would have to meet regarding registration of additives used in motor gasoline. In addition to the chemical formula and concentration, the effects of the additive on other exhaust emission and on toxicity would be requested.

Most diesel manufacturers have definite fuel recommendations for satisfactory engine operation[37] and specifically warn against the use of fuel additives other than those which they have approved. In some cases, the engine warranty may be jeopardized by the use of additives.

Smoke Suppressants

A barium compound added to the fuel to suppress smoke was tested, but had little influence on odor.[92] Similar observations have been made on the use of a manganese compound (W. F. Marshall, personal communication).

Odor Modifiers

Odor modifiers are substances that, when mixed with malodorous samples, may reduce odor intensity, detectability, and recognizability or change the odor character to make it more acceptable. A rich commercial nomenclature has developed that claims various effects of modifiers on odors. A scientific classification of the intensity-related modifying effects has been proposed[22] for odors in general. Controlled studies on the effects of a variety of odor-modifying formulations and single odorants on typical malodors indicated that the modifiers need to exhibit an odor themselves and that their dominant effect is to decrease the recognizability of malodors, with or without decreasing odor intensity (A. Dravnieks, personal communication).

Additives available for controlling odor may be commercially classed as masking agents (covering an odor with another odor) or counteracting agents (interacting with an odor to form another odor or to reduce or even "destroy" odor). Some operators have found that odor additives, principally of the masking type, have helped to reduce complaints.[78] Although laboratory tests have not shown a reduction in odor intensity,[69,70,92,99] it may be that an active public-information program would make the modified odor more acceptable. Aside from cost, the major disadvantage of masking agents may be their creation of odors that are more unpleasant to some people than the original odor,[37] inasmuch as the unpleasantness is a highly individualistic property of odor. No engine difficulties or maintenance problems have been reported to result from the use of such modifiers.

In one experiment,[126] a fragrant odor was introduced to modify the odor of diesel exhaust from urban buses. Complaints about the exhaust odor decreased temporarily, then built up again as the public learned to associate the modified odor with smoke from the bus exhaust pipes.

There is increasing concern over the unknown carcinogenic and mutagenic potential of many substances. In view of the inadequacy of knowledge on the actual (nominal and impurity-related) composition of commercial modifiers, caution is advised in the use of this form of odor control, especially because its effectiveness is doubtful.

Operational Variables

A guide to smoke and odor reduction has been published.[98,127] Among the operational steps that reduce smoke and odor are an increase in the air-to-fuel ratio, engine derating, maintenance programs that include correction of faulty fuel injectors and of restricted air cleaners, turbocharging at higher elevations, avoidance of excessive engine idling and underspeeding, and careful selection of fuels and fuel additives. Other measures are replacement, in two-stroke diesels, of the original crown injectors with needle injectors. Some jurisdictions have regulations controlling, for example, idling practices, because exhaust odor is strongest during idling.[25]

The effect of the lubricating oil on the exhaust odor has not been clarified. As much as 50% of the smoke particles may originate from the lubricating oil in two-stroke engines, and up to 20% in four-stroke engines (J. H. Johnson, personal communication). However, in some exploratory tests in which odors were judged by simple observation, a change from the regular lubricating oil to a synthetic oil based on glycols did not result in an obvious odor change (W. F. Marshall, personal communication).

Correlations have been found between odor intensity and such combustion indexes as unburned hydrocarbons, carbon monoxide, aldehydes, and ethylene.[31] This supports the relation between odor and combustion conditions, which depend on engine design and operation.

Engine and Vehicle Design

Design features that reduce the incomplete combustion of diesel fuel are probably the most significant factors in diesel-exhaust odor control.

In a comparison of several makes of diesel-powered passenger cars in 1974, the exhaust odors tested at a 100-fold dilution ranged from 3.2 to 5.3 on the D-scale of odor intensities.[94]

Improvements in combustion are achieved by precombustion chambers, by stratified-charge design, by prevention of fuel from entering the combustion chambers at times other than immediately before ignition, and by changes in the geometry to improve fuel-air mixing. Thus, reduction in the uncontrolled volume of the fuel flow to the injector tip resulted in a reduction in hydrocarbon emission and odor.[40,71] Although four-stroke engines are considered less odorous than two-stroke engines, and those with precombustion chambers less odorous than those with direct injection, it appears that improved fuel-system and injector designs narrow these differences.

Other modifications to reduce diesel-exhaust odor deal with treatment of the exhaust itself. Treatment may be mechanical, such as a change in the position of the exhaust pipe, or chemical, such as scrubbing or catalytic oxidation.

Exhaust Location Although a change in the location of the exhaust tailpipe does not in itself change the odorous-emission rate, it may reduce the impact of the odor on a curbside bystander. It has been shown that such an observer receives exhaust odors at a 100-fold to 600-fold dilution when the horizontal exhaust pipe ends below the end of the bus, and at a 1,000-fold to 7,000-fold dilution if exhaust is through a vertical stack, as in trucks.[90] The change from a horizontal to a vertical stack on buses reduced the odor threshold around the buses by a factor of 8 and made odor unnoticeable at a distance only two-thirds of that with horizontal exhaust;[26] a rooftop diffusion system for the exhaust was less effective than a vertical stack.

Catalytic Oxidation Available catalytic systems for treatment of diesel exhaust require temperatures of around 370°C (700°F) or higher for proper operation. When catalysts are incorporated in the usual way in an engine exhaust system, such temperatures are maintained only at steady, relatively high-load operation, as in diesels used in mines where catalysts have been used.

Of mine diesels, almost all of which have precombustion chambers, approximately half are equipped with catalytic converters. These are limited principally to other than coal mines, because safety regulations for coal mines impose many restrictions. Catalysts differ widely in effectiveness; the best ones decrease odor intensity by 2–3 units of the D-scale and last without replacement for 6–12 months (they eventually fail by carbon accumulation).

Catalyst treatment of city-bus diesel exhaust had an inconsistent effect on exhaust odor;[92] because of long idling periods and unsteady flow conditions, exhaust from such buses does not provide a high enough temperature to maintain adequate catalyst activity. These findings were extended to a variety of catalysts, including precious metals and transition metals. Similar considerations apply to diesels in passenger cars.

It has been suggested that incorporation of the catalysts in the engine design may permit operation at adequate temperatures (J. H. Johnson, personal communication). This would require a drastic redesign of diesel engines.

Other Exhaust Treatment Adding air at the exhaust valves in a bus engine[91] or passing the exhaust through water in a wooden box in a mine diesel (W. F. Marshall, personal communication) had an insignificant effect on diesel-exhaust odor.

Experimental scrubbing of diesel exhaust with various aqueous solutions indicated that chemical additions to the water did not significantly improve the slight effect of the water itself on the odor. The low solubility of diesel-exhaust odorants was probably responsible for the negligible effect of the chemicals.

Prospects for Odor Control

Adherence to the best operating conditions[43,98,101,125,127] and changing to better fuel injectors and vertical stacks (instead of the beneath-vehicle horizontal exhaust systems) can considerably alleviate the diesel-odor problem. General Motors developed, with the support of the U.S. Department of Transportation (DOT), an environmental improvement program (EIP). A kit was developed for modifying urban buses to produce cleaner and less odorous exhaust;[109] the process included catalytic treatment of the exhaust. A number of buses have been modified in this manner, and new buses are manufactured accordingly, because DOT requires this in transit development projects that use federal funding (J. H. Johnson, personal communication).

It has been estimated[93] that 100-fold dilution decreases exhaust-odor intensity by 3 units on the D-scale. Reduction by 3 units relative to the present diesel-exhaust odor should result in major relief from the annoyance of such odor. This degree of reduction should be possible with the more recent technology, with the new injector design, and with some other design changes that improve combustion control of odorants. Odor character would then be the most significant factor (K. J. Springer, personal communication). The best present diesels appear to be able to reach D-scale values of 3.2, averaged over many modes of operation—still higher than gas-turbine exhaust, which typically has D-scale ratings around 1.

A 1977 evaluation of two production versions of diesel-powered passenger cars (Volkswagen and Oldsmobile Cutlass)[95] showed odor-intensity ratings, at a 100-fold exhaust dilution, in the range of 2.5–4.0 on the D-scale, which is judged sufficient to trigger substantial public reaction if many such units were in use in the United States.

The cost of EIP modification has been estimated at several thousand dollars per unit. The cost of catalytic treatment, such as for diesels in mines, is \$200–500 per engine.

Public-opinion surveys on diesel odors of different D-scale intensities[49] have indicated that objections are significant at a D-scale rating of 2 and increase rapidly at higher ratings. An extrapolation of such data seems to indicate that ratings around 1 may significantly reduce complaints. A prototype exhaust-treatment system based on two catalysts and a molten carbonate scrubber has been designed,[107] but is probably too cumbersome for routine use. Its objective was to remove controlled emission species, such as hydrocarbons, carbon monoxide, and nitrogen oxides. No practical technologic means are in sight to reduce diesel-exhaust odor to such low intensities.

GASOLINE ENGINES

Most of the research on odors from gasoline engines was conducted before the wide use of catalytic converters. There appears to be a considerable difference between the European experience, which includes much work on gasoline-engine odors, and U.S. experience, which before the introduction of catalytic mufflers did not reveal a substantial gasoline-engine exhaust-odor problem.

In a comparison of several diesel-powered and gasoline-powered passenger cars, the odor intensities of exhausts (at a 100-fold dilution) were 1.1–3.0 on the D-scale for the gasoline engine, averaged over several operational conditions; those for the diesels were 3.2–5.3.[94] One engine, the Texaco TCCS, could be operated on either diesel fuel or gasoline; with gasoline, the intensities were about 30% lower (on the D-scale) than with diesel fuel. A gasoline engine with a catalytic converter produced odor-intensity ratings close to 1.

Work in Europe has been conducted almost exclusively with odor dilution thresholds as the measure of intensity.[35] In Sweden, gasoline and diesel engines were compared by a magnitude-estimate method;[64] gasoline-engine emission odor was found to be more intense than that from a diesel and varied less with changes in the engine operating conditions than that from a diesel.

In a Swedish study,[64] odor thresholds for emission from gasoline engines with several types of odor-control devices were at dilutions of a factor of 5,300–6,800, and for diesels without such devices at a dilution of a factor of 1,500. Thus, under practical conditions of operation, catalytic converters do not eliminate emission odors, but may substantially change the distribution of odorant types.[114]

Reasons for the divergence between European and U.S. findings on the difference between gasoline- and diesel-engine exhaust odors are not clear.

Odorous exhaust from gasoline engines has been significantly changed by introduction of catalysts and catalytic converters,[2,60,73,74,88] which are necessary to reduce residual organic vapors and carbon monoxide that are emitted at a higher rate from gasoline engines than from diesels (see also bulletins from Oxy Catalyst, Inc., and General Motors Corp., cited in U.S. EPA[125]). Although controlled studies have not been reported, it is common knowledge that there are many complaints about odors from catalytic converters and that these complaints sometimes result in removal of such converters (K. J. Springer, personal communication). There are some indications that the odor problem may be alleviated in the newer models of automobiles (G. D. Kittredge, personal communication).

CAUSES OF GASOLINE-EXHAUST ODOR

As in diesel engines, the cause of exhaust odors from gasoline engines is incomplete combustion of fuel.[46] Without catalytic converters, hydrocarbons in the exhaust are more highly concentrated than in diesel exhaust. With catalytic converters, even if a considerable amount of organic vapors, including partially oxidized organics, is burned, new odorous materials can be generated and survive through the muffler, especially during acceleration. Thus, odors may be caused both in the engine and in the exhaust system.

COMPOSITION OF ODORANTS

The composition of odorants in gasoline-engine exhaust has been studied much less extensively than that of odorants in diesel exhaust.

Unburned hydrocarbons, partial-oxidation products, and nitrogen oxide have been found responsible for the exhaust odor.[46] Formaldehyde and other aldehydes have been found among the partial-oxidation products,[46,61,72,87] but appear at concentrations insufficient to cause odor in the vicinity of the vehicle.[131] Olefins, toluene, alcohols, ozone, and hydrogen cyanide have been found in the exhaust, irrespective of the fuel used.[34] Hydrogen cyanide at concentrations of 1–15 ppm has also been found.[51,56,139]

Catalytic converters usually operate under oxidizing conditions and promote the oxidation of unburned hydrocarbons and partial-combustion products. However, when the temperature is too low and the gas flow is too rapid, the catalysts themselves may promote the formation of additional partial-oxidation products. Detailed studies of potentially odorous emission from catalytic converters under oxidizing conditions do not seem to have been conducted, but passenger cars equipped with catalytic converters emitted aldehydes at rates (per kilometer) not grossly different from those emitted by diesel cars without such converters.[100] Because of continuing changes in catalytic-converter technology, it is difficult to estimate their odorant-removal efficiency in general terms.

Under the reducing conditions that may exist during deceleration, hydrogen sulfide and probably hydrogen cyanide may be generated in catalytic converters (by the catalytic reduction of sulfur dioxide and by a reaction between nitrogen oxides and carbon monoxide, respectively).[50,139] Both gases have highly unpleasant foul odors; hydrogen cyanide has an almond-like note.

ODOR MEASUREMENT

Sensory measurements use approaches similar to those for diesel-exhaust odors, except that odor intensities and odor notes are different.

Analytical measurements usually consist of analysis for aldehydes.[100] The DOAS method is not applicable to gasoline-engine exhaust odors (K. J. Springer, personal communication).

ODOR CONTROL

Most of the work on odor control of gasoline-engine exhaust has been conducted without the use of catalytic converters. Hence, many of the reported findings may not apply to engines equipped with such converters.

Influence of Fuel

In experiments with 17 different fuels in a single-cylinder four-stroke gasoline engine, exhaust odor did not change substantially with fuel composition;[34] the fuels used encompassed marketable compositions of paraffins, olefins, aromatics, and alcohols.

Fuel Additives

Evidence of the effect of fuel additives on gasoline-engine exhaust odor is not available.

Operational Variables

Fuel-to-air ratio, ignition delay, stroke, volume, speed, and so forth all have effects on the exhaust odor.[34,35,46] An increase in excess air results in increased odor, although the amounts of unburned hydrocarbons and carbon monoxide decrease. Deceleration enhances aldehyde production.[72]

Operational conditions that would minimize odor will increase gasoline consumption.[46] Most of the available findings probably need to be modified to apply to gasoline engines equipped with catalytic converters.

Engine Design

No systematic effort to design engine-exhaust systems to reduce odor from gasoline engines seems to be in progress. Promotion of oxidation in the converters and of better combustion in engines by improving fuel-air mixing is usually motivated by a need to reduce the fuel consumption and the emission of hydrocarbons, carbon monoxide, and nitrogen oxides.

Prospects for Odor Control

Reduction in gasoline-engine exhaust odor may be a byproduct of the progress toward fuel-saving and reduction of controlled (EPA-regulated) types of emission. There appears to be no effort to monitor concurrent changes in the emission-odor intensity and character.

An obvious means of odor control is to increase the catalytic-converter temperature and the residence time of exhaust in the converter. Such changes carry corresponding economic penalties and may affect fuel consumption. It appears that proper vehicle adjustment and specific elimination of sources of overrich fuel-air mixtures may solve the catalyst exhaust-odor problem (G. D. Kittredge, personal communication). The cost increment per engine for converter changes needed to keep exhaust odor at a tolerable point cannot be estimated unless experimental programs on odor versus converter variables are conducted. Public dissatisfaction with catalytic-converter odors, if it continues despite changes in catalytic-converter technology, might require technical changes in the converters or in the design of engine components.

GAS TURBINES[114]

Gas turbines operate at air-to-fuel ratios of 200:1 to 300:1, whereas diesels operate at 20:1 to 100:1. Consequently, they produce much less odorous exhaust than diesels and gasoline engines[94]—a D-scale rating of 1 after 100-fold dilution. Thus, they should not generate a major odor-nuisance problem, unless they are used in congested areas in large numbers.

AIRCRAFT

The density of pollutant emission by aircraft and the resulting pollutant concentrations in the vicinities of air terminals are comparable with emission densities and concentrations of the same pollutants in adjacent communities from other combustion and traffic sources.[9] Thus, the principal impact of aircraft emission is local, and it is expected to become more severe. It is also likely that aircraft emission will constitute a more important portion of community-wide pollutant loadings as new aircraft are introduced and as emission from other sources is reduced.[10]

Odorous emission from aircraft constitutes one aspect of the air-pollution problem.[75] In one study,[42,67,75] odor thresholds and the concentrations of nitrogen oxides, aldehydes, carbon monoxide, and hydrocarbons from three representative types of jet engines were measured. The odor measurements were based on dilution-to-threshold ratios. Table 9-2 shows the relationship between hydrocarbons, aldehydes, and odor. The odor-threshold dilutions correlate somewhat better with the aldehyde emission than with emission of hydrocarbons or any of the other pollutants measured in the study.

In another study,[21,57] the TIA odor intensity and the character of exhaust from a jet combustor were evaluated after dilution of the exhaust by a factor of 1,000 with nonodorous air. The DOAS method was used to measure concentrations of total aromatics (benzenoids) and oxygenated compounds in the exhaust. These methods were similar to those used in some diesel-exhaust odor studies. The TIA odor intensity correlated with the logarithm of the concentration of oxygenates and to some extent that of aromatics and total hydrocarbons. Typically, TIA values were in the range of 1–2 units. Changes in operating conditions, fuel types, and nozzle design influenced the burning efficiency; higher efficiency resulted in weaker odor. Odor character was influenced by the type of fuel. The principal character was smoky or burnt-smoky, but fuels with a low aromatic content had an odor that was more sweet and burnt, and fuels with a high aromatic content had an odor that was smoky and tarry.

The following approaches to the control of aircraft-exhaust odor have been suggested:[9]

- *Greater Dispersion*: This could be achieved by increased separation of airfields from densely populated communities or by modification of flight patterns or ground operations. The operational difficulties involved are complex and beyond the scope of this document.
- *Activated-Carbon Purification of Intake Air*: For passenger terminal

TABLE 9-2 Pollution Emission from Jet Aircraft and Odor Threshold[a]

Pollutant	Operating Mode of Aircraft								
	Takeoff			Cruise and Approach			Idle		
	T-56	J-57	TF-33	T-56	J-57	TF-33	T-56	J-57	TF-33
Total hydrocarbons, as carbon atoms, lb/h	0.3	1.2	2.4	0.1	1.1	9.3	3.0	10.5	43.2
(kg/h)	(0.1)	(0.5)	(1.1)	(0.05)	(0.5)	(4.2)	(1.4)	(4.8)	(19.6)
Total aldehydes, as formaldehyde, lb/h	0.5	0.5	0.04	0.2	0.4	0.14	0.3	0.4	3.2
(kg/h)	(0.2)	(0.2)	(0.02)	(0.1)	(0.2)	(0.06)	(0.1)	(0.2)	(1.5)
Odor-threshold dilution (factor)	100	600	75	—	600	15	—	500	1,000

[a] Data from George *et al.*,[42] Lozano *et al.*,[67] and Northern Research and Engineering Corp.[75]

buildings, this is a reasonable control method, and it has been used in some terminals.

• *More Complete Oxidation of Fuel*: Any procedures—such as modification of engines, aircraft, or fuels—used to reduce the quantity of smoke or total organic matter in aircraft exhaust will probably also reduce odor. This relationship is not always ensured, however, because partial oxidation may change odor character. The data in Table 9-2 show that partial oxidation of hydrocarbons to aldehydes in jet-engine exhaust may increase odor intensity. Increased total conversion to carbon dioxide, however, will necessarily decrease odor.

• *Odor Modification*: Odor modifiers have not been used in direct conjunction with aircraft engines, and it is indeed doubtful that such an approach would be feasible, in view of the temperature extremes in the exhaust. Odor modifiers would more reasonably be considered for application to the air intake of terminal buildings or for dispensing into the outside air near the terminal buildings or community residences.

OTHER ENGINES

Engine types other than those mentioned above have been considered.[5,14,39] Fuel-powered steam engines[44,54,55,89,102,121,122,124,125,123] and Stirling-cycle engines (in which a gaseous nonconsumable working fluid is used in the cylinders) do not operate under combustion cycles, as do internal-combustion engines. Such mobile power sources probably do not present significant odor-nuisance problems.[33,62]

OTHER MOBILE SOURCES

Mobile odor sources include vehicles that transport odorous matter under conditions that permit leakage. The types of odors will depend on the nature of the materials transported. Some typical examples can be cited briefly. Truck transport of raw materials to rendering plants may involve chicken offal, meat scraps and bones, fish scraps, feathers, and whole dead animals. When any of these materials are in an advanced state of decomposition, the odor potential during transport is great. In addition, some of the material is often liquid and may leak from the truck, causing a considerable odor nuisance. Extreme precautions in sealing and in outside cleanliness are necessary. Closures in ordinary truck bodies are usually insufficiently tight for adequate odor control in such cases.

The transport of fertilizers to farms and of farm wastes to disposal sites may also constitute odor problems.

The problem of disposal of chemical wastes has become more difficult as opportunities for local processing have diminished. Accordingly, there is an increasing use of central waste-disposal facilities, often by independent contractors, with the attendant problems of odors during transport. Typical waste materials are tarry process residues with great odor potential.

In all such cases, sealing of the contents during transport and removal of any odorous matter from exposed outside surfaces must adhere to rigid standards.

CONCLUSIONS

1. Odorous air pollution by mobile sources, such as motorized vehicles and aircraft, is caused by engine exhaust. Carrying odorous industrial and agricultural materials in vehicles may also cause odorous pollution if such materials are not properly confined.

2. The causes of the exhaust odors are incomplete combustion of fuel, thermal degradation of some fuel components, and partial oxidation of some fuel components. The role of lubricating oils in the generation of odorants emitted in exhaust is not clear.

3. Diesel-engine exhaust is more odorous than the exhaust from other types of engines used in mobile equipment.

   a. There are technical instructions for operating diesel engines to maintain low odorous-emission rates.

   b. Design changes, including modification of the fuel injector, to promote more complete combustion are potentially the most effective approach to reduction of diesel-engine exhaust odor.

   c. Catalytic converters like those used in gasoline-powered cars are not useful in reducing diesel-exhaust odor, because of the lower exhaust temperatures. Such catalysts are in use in cases where diesel engines are operated continuously at high loads, as in the mining industry. Diesel-engine designs in which converters are integrated with the engine and can operate at higher temperatures may be possible.

   d. Changes in the composition of diesel fuel appear to have little effect on exhaust odor.

   e. Chemical additives introduced into diesel fuel had little effect on exhaust odor, in the case of the few additives tried.

   f. Adding odor modifiers to change the odor character of diesel exhaust might result in temporary reduction of complaints about diesel-exhaust odor, but complaints may return as the public begins to associate the modified odor with diesel exhaust. The possible health effects in this approach are uncertain; technically, odor modifiers add to the air pollution caused by chemicals.

4. Odors caused by gasoline-engine exhaust do not thus far appear to present a substantial public nuisance in the United States. Catalytic converters have been reported to generate odorous hydrogen sulfide and hydrogen cyanide under some transient load conditions and have been a cause of odor complaints; this problem may be gradually alleviated in newer cars and with better engine adjustments.

a. Operating conditions have an important effect on gasoline-engine exhaust odors.

b. Fuel composition does not appear to have an important effect on gasoline-engine exhaust odor.

c. Fuel additives do not appear to have an important effect on gasoline-engine exhaust odor.

5. Gas-turbine exhaust odor is weak and does not have a substantial odor-complaint potential.

6. Aircraft jet-engine exhaust odor is a local problem. People in terminal buildings may be protected from such odor by carbon filtration of the air supplied to the terminal buildings. Odorous pollution in the vicinities of airports is similar to local pollution in the vicinities of industrial plants, but is more difficult to alleviate, because the source is not stationary. Redesigning of jet engines and reduction in the density of aircraft traffic do not appear to be valid approaches to reduction in the odorous pollution in the vicinities of airports. Some jet-combustor changes may promote more complete combustion and decrease the exhaust odor.

7. Engine-exhaust odor measurements by sensory methods in the laboratory may have to be abandoned, at least in the case of diesel exhaust, because some diesel-exhaust components, not necessarily odorous, are suspected mutagens.

8. Engine-exhaust odorant measurements by analytical methods do not provide values from which human responses to exhaust odor can be reliably estimated for various types of engines. Analytical methods can provide data on the completeness of fuel combustion in engines. They can also provide data on the chemical identities and concentrations of odorants responsible for the characteristic odor of exhaust.

9. Because intentional exposure of people to exhaust is becoming unacceptable, the only remaining approach to estimating the extent of reduction in the odor nuisance of exhaust, as its odor is minimized by various technologic innovations, is through methods that measure changes in the public reaction to unavoidable normal exposure to traffic odors.

RECOMMENDATIONS

1. Reduction in the odorous emission from mobile sources, especially those with diesel engines, should be sought by:

   a. adherence to specific operating procedures and maintenance (short-range approach),

   b. design changes to result in more complete combustion of fuel in the engines (long-range approach), and

   c. (only to a minimal extent) selection of fuels and use of fuel additives, in the absence of new significant findings.

2. Means should be developed for estimating the extent of decrease in odorous emission brought about by technologic innovations in engine design and operation, especially in the case of diesel engines, gasoline engines with catalytic converters, and jet engines.

3. Such means should avoid intentional systematic exposure of people to odorous exhaust, because of the presence of potentially carcinogenic but nonodorous substances in exhaust, at least in diesel exhaust.

4. Because simple analytical data, such as those developed on odor intensity of some types of diesel exhaust, are not applicable to the spectrum of diesel and other engines, alternative methods should be developed to monitor progress in emission-odor control. Such means may be as follows:

   a. developing techniques to measure the extent of reduction in public annoyance in normal exposure to odors of traffic (including airports), and

   b. identifying the significant odorants in exhaust and following the reduction in their concentrations as technology progresses, working toward an understanding of how odors of odorants combine to produce the characteristic odor and odor intensity of their mixtures (this should assist in the interpretation of the analytical data on odorants and their concentrations, with respect to the composite odor of exhaust).

5. The potential for emission of odorous effluents should be a consideration in the development of new types of engines for propulsion of vehicles.

REFERENCES

1. Aaronson, A. E., and R. A. Matula. Diesel odor and the formation of aromatic hydrocarbons during the heterogeneous combustion of pure cetane in a single-cylinder diesel engine, pp. 471–481. In Proceedings of the 13th International Symposium on Combustion. Pittsburgh: The Combustion Institute, 1971.

2. A catalytic muffler for buses. Chem. Eng. News 43(18):58, 1965.

3. Air pollution from motor vehicles is not dangerous to health. Chem. Eng. 74:56, August 28, 1967.

4. Air pollution—the problem and the risks. SAE Journal 76(5):47–52, 1968.

5. Alternative nonpolluting power sources. SAE Journal 76(12):40–80, 1968.

6. Altshuller, A. P. Air pollution. Anal. Chem. 37(5):11R–20R, 1965.

7. Arthur D. Little, Inc. Chemical Analysis of Odor Components in Diesel Exhaust. Final Report to Coordinating Research Council, Inc., and Environmental Protection Agency. No. ADL 74744-5. Cambridge, Mass.: A. D. Little, Inc., 1973. 82 pp.

8. Arthur D. Little, Inc. Chemical Identification of the Odor Components in Diesel Engine Exhaust. Final Report (Year 1) to Coordinating Research Council, Inc., and National Air Pollution Control Administration, Public Health Service, U.S. Department of Health, Education, and Welfare. Cambridge, Mass.: A. D. Little, Inc., 1969. 92 pp.

9. Arthur D. Little, Inc. Chemical Identification of the Odor Components in Diesel Engine Exhaust. Final Report (Year 2) to Coordinating Research Council, Inc., and National Air Pollution Control Administration, Public Health Service, U.S. Department of Health, Education, and Welfare. Cambridge, Mass.: A. D. Little, Inc., 1970. 112 pp.

10. Arthur D. Little, Inc. Chemical Identification of the Odor Components in Diesel Engine Exhaust. Final Report (Year 3) to Coordinating Research Council, Inc., and Environmental Protection Agency. No. ADL 62561-5. Cambridge, Mass.: A. D. Little, Inc., 1971. 139 pp.

11. Association of American Railroads. Yearbook of Railroad Statistics, 1978, p. 48. Washington, D.C.: Association of American Railroads, 1978.

12. Automobile Manufacturers Association. Motor Truck Facts. Detroit: Automobile Manufacturers Association, 1969. 61 pp.

13. Automobile Manufacturers Association, Statistical Department. Public Attitudes Toward Trucks. Detroit: American Manufacturers Association, 1968.

14. Ayres, R. U. Two Possible Alternatives to the Internal Combustion Engine, pp. 14–22. Annual Report 1968. Washington, D.C.: Resources for the Future, Inc., 1968.

15. Bailey, C. L., A. R. Javes, and J. K. Lock. Investigations into the consumption of diesel engine exhausts, Section VI, pp. 209–226. In Proceedings of the Fifth World Petroleum Congress, New York, June 1–5, 1959. New York: Fifth World Petroleum Congress, Inc., 1959.

16. Barth, D. S., and S. M. Blacker. The EPA program to assess the public health significance of diesel emissions. J. Air Poll. Cont. Assoc. 28:769–771, 1978.

17. Battigelli, M. C. Air pollution from diesel exhaust. Occup. Med. 5:54–57, 1963.

18. Battigelli, M. C. Effects of diesel exhaust. Arch. Environ. Health 10:165–167, 1965.

19. Bowditch, F. W. G.M.'s push for marketing diesels faces environmental rules hurdles. Environ. Rep. 9:1246–1247, 1978.

20. Brubacher, M. I. Role of Trucks and Buses in the Air Pollution Problem. Berkeley: State of California Motor Vehicle Pollution Control Board, 1962. 7 pp.

21. Butze, H. F., and D. A. Kendall. Odor Intensity and Characterization Studies of Exhaust from a Turbojet Engine Combustor. NASA Tech. Mem. Report No. TMX-71429. Washington, D.C.: National Aeronautics and Space Administration, 1973. 12 pp.

22. Cain, W. S., and M. Drexler. Scope and evaluation of odor counteraction and masking. Ann. N.Y. Acad. Sci. 237:427–439, 1974.

23. California Air Resources Board. Bulletin 1(13), March–April 1969. 4 pp.

24. Cernansky, N. P., C. W. Savery, I. H. Suffet, and R. S. Cohen. Diesel Odor Sampling and Analysis Using the Diesel Odor Analysis System (DOAS). SAE Paper 780223. Warrendale, Pa.: Society of Automotive Engineers, 1978.

25. City of New York, Department of Air Pollution Control. New York City Air Pollution Control Code, August 14, 1964.

26. Colucci, J. M., and G. J. Barnes. The Effect of Exhaust System Geometry on Exhaust Dilution and Odor Intensity. SAE Paper 710219. New York: Society of Automotive Engineers, 1971. 11 pp.

27. Dateline news. Fleet Owner 64(4):29, 1969.

28. Degobert, P. Diesel exhaust odor measurements, pp. 566–581. In Proceedings of the Fourth International Symposium on Automotive Propulsion Systems, Washington, D.C., April 18–22, 1977. Washington, D.C.: Automotive Propulsion Systems Pilot Study, February 1978.

29. Diesel fume perfume. Chem. Week 99(4):75, 1966.

30. Diesel smoke suppressant. Chem. Week 100(15):40, 1967.

31. Dietzmann, H. E., K. J. Springer, and R. C. Stahman. Diesel emissions as predictors of observed diesel odor. SAE Paper 720757. SAE Trans. 81:2268–2292, 1972.

32. Dravnieks, A., A. O'Donnell, R. Scholz, and J. D. Stockham. Gas-Chromatographic Study of Diesel Exhaust Using a Two-Column System. Presented at meeting of the American Chemical Society, April 1971, Los Angeles, Calif. (unpublished)

33. Dutch on the road to a pollution-free engine. Bus. Week No. 2106:52–53, January 10, 1970.

34. Eberan-Eberhorst, V. R., and D. Gruden. The influence of fuel composition on engine performance, emission, and odor of exhaust gas of four-stroke Otto cycle engines. Erdöl und Kohle 26:257–264, 1973. (in German)

35. Eberan-Eberhorst, V. R., and D. Gruden. The odor of exhaust gases from Otto engines. Motortech. 35:246–250, 1974. (in German)

36. Elliott, M. A., G. J. Nebel, and F. G. Rounds. The composition of exhaust gases from diesel, gasoline, and propane powered motor coaches. J. Air Poll. Cont. Assoc. 5(2):103–108, 1955.

37. Ellis, H., Jr. Reducing diesel odors. Chem. Week 99(10):13, September 3, 1966. (letter)

38. Federal Register: Environmental Protection Agency. Registration of fuels and fuel additives. 40(216):52009–52014, Nov. 7, 1975.

39. Feedback on alternative nonpolluting power sources. SAE Journal 77(4):62–64, 1969.

40. Ford, H. S., D. F. Merrion, and R. J. Hames. Reducing Hydrocarbons and Odor in Diesel by Fuel Injector Design. SAE Paper 700734. New York: Society of Automotive Engineers, 1970. 7 pp.

41. Frey, J. W., and M. Corn. Physical and chemical characteristics of particulates in a diesel exhaust. Amer. Ind. Hyg. Assoc. J. 28:468–478, 1967.

42. George, R. E., J. A. Verssen, and R. L. Chass. Jet aircraft: A growing pollution source. J. Air Poll. Cont. Assoc. 19:847, 1969.

43. GMC develops system to trim bus emissions. Automotive News, December 22, 1969, p. 7.

44. GM reveals hidden steam and turbine research. Ind. Res. 11(7):36–37, 1969.

45. Goldsmith, J. R. Public health hazards of air pollution, pp.20–35. In Proceedings of the Conference on Air Pollution in California. Berkeley: California State Public Health Service, 1968.

46. Gruden, D. Exhaust emission and exhaust odor from four-stroke spark ignition engines. Automobiltech. Z. 74:180–188, 1972. (in German)

47. Hanks, J. J., and H. D. Kube. Industry action to combat pollution. Harv. Bus. Rev. 44(5):49–62, 1966.

48. Hannay, W. F. Air pollution by diesel fumes, II. Petrol. Times 63:402–403, 1959.

49. Hare, C. T., K. J. Springer, J. H. Somers, and T. A. Huls. Public opinion of diesel odor. SAE Paper 740214. SAE Trans. 83:949–970, 1974.

50. Hightower, J. Consultant Report to the Committee on Motor Vehicle Emissions, Commission on Sociotechnical Systems, National Research Council, on an Evaluation of Catalytic Converters for Control of Automobile Exhaust Pollutants. Washington, D.C.: National Academy of Sciences, 1974. 116 pp.

51. Hurn, R. W., J. R. Allsup, and F. Cox. Effect of Gasoline Additives on Gaseous Emissions. Prepared by the Bureau of Mines, Bartlesville Energy Research Center, Bartlesville, Oklahoma, for Office of Research and Development, U.S. Environmental Protection Agency. Publ. No. EPA-650/2-75-014. December 1974. 64 pp.

52. Hurn, R. W., B. Dimitriades, and R. D. Fleming. How hydrocarbon types determine smog-forming potential of exhaust gases. SAE Journal 74(2):59–61, 1966.

53. Hurn, R. W., and W. F. Marshall. Techniques for diesel emissions measurement. SAE Paper 680418. SAE Trans. 77:1492–1501, 1968.

54. Inventor Lear abandons steam car efforts. Fleet Owner 65(1):31, 1970.

55. Joseph, J. Inside Bill Lear's steam motor. Diesel and Gas Turbine Progress 35(6):24–26, 1969.

56. Kaneko, F. Excess death through air pollution in Osaka City. J. Jap. Soc. Air Poll. 8:198, 1973.

57. Kendall, D. A., and P. L. Levins. Odor Intensity and Characterization of Jet Exhaust and Chemical Analytical Measurements. Prepared by A. D. Little, Inc., for National Aeronautics and Space Administration, Lewis Research Center. NASA Publ. No. CR-121159. ADL 74443. Cambridge, Mass.: A. D. Little, Inc., 1973. 58 pp.

58. Kendall, D. A., P. L. Levins, and G. Leonardos. Diesel exhaust odor analysis by sensory techniques. SAE Paper 740215. SAE Trans. 83:971–985, 1974.

59. Kinosian, J. R., J. A. Maga, and J. R. Goldsmith. The Diesel Vehicle and Its Role in Air Pollution: A Report to the California Legislature. Berkeley: California State Department of Public Health, Bureau of Air and Sanitation, 1962. 70 pp.

60. Leak, R. J., J. T. Brandenburg, and M. D. Behrens. Use of alumina-coated filaments in catalytic mufflers testing with multicylinder engine and vehicles. Environ. Sci. Technol. 2:790–794, 1968.

61. Levine, B. S. Survey of U.S.S.R. Literature on Air Pollution and Related Occupational Diseases. A Survey, Vol. 3, p. 135. Washington, D.C.: U.S. Department of Commerce, 1960.

62. Lienesch, J. H., and W. R. Wade. Stirling engine operates quietly with almost no smoke and odor and with little exhaust emission. SAE Journal 77:40–44, 1969.

63. Lindvall, T. Sensory measurement of ambient traffic odors. J. Air Poll. Cont. Assoc. 23:697–700, 1973.

64. Lindvall, T., and S. E. Mortstedt. Investigation of the Odor Intensity of Internal Combustion Engine Exhaust. Report BIL-52. Stockholm: AB Atomenergi, 1970. 13 pp. (in Swedish)

65. Linnell, R. H., and W. E. Scott. Diesel exhaust analysis. Preliminary results. Arch. Environ. Health 5:616–625, 1962.

66. Linnell, R. H., and W. E. Scott. Diesel exhaust composition and odor studies. J. Air Poll. Cont. Assoc. 12:510–515, 545, 1962.

67. Lozano, E. R., W. W. Melvin, Jr., and S. Hochheiser. Air pollution emissions from jet engines. J. Air Poll. Cont. Assoc. 18:392–394, 1968.

68. Ludwig, J. H. Progress in control of vehicle emissions. ASCE Paper 5380. Proc. Amer. Soc. Civ. Eng., J. Sanit. Eng. Div. 93(SA4):73–79, August 1967.

69. Marshall, W. F., and R. W. Hurn. Factors influencing diesel emissions. SAE Paper 680528. SAE Trans. 77:2139–2150, 1968.

70. McConnell, G., and H. E. Howells. Diesel fuel properties and exhaust gas—Distant relations? SAE Paper 670091. SAE Trans. 76:598–615, 1968.

71. Merrion, D. F. Effect of design revisions on two stroke cycle diesel engine exhaust. SAE Paper 680422. SAE Trans. 77:1534–1545, 1968.

72. Mikita, J. J., H. Levin, and H. R. Kichline. Gasoline engine exhaust odor (a study of odors produced under conditions of deceleration). SAE Trans. 51:12–19, 1943.

73. Miller, M. R., and H. J. Wilnoyte. A study of catalyst support systems for fume-abatement of hydrocarbon solvents. J. Air Poll. Cont. Assoc. 17:791–795, 1967.

74. Mooney, J. J., and K. W. Blamble. Diesel exhaust purification with fixed bed catalyst. Engelhard Industries Technical Bulletin 9(3):85–89, 1968.

75. Northern Research and Engineering Corp. Nature and Control of Aircraft Engine Exhaust Emissions. Report No. 1134-1. Cambridge, Mass.: U.S. Department of Health, Education, and Welfare, Public Health Service, National Air Pollution Control Administration, 1968. 49 pp.

76. O'Donnell, A., and A. Dravnieks. Chemical Species in Engine Exhaust and Their Contributions to Exhaust Odors. Final Report to Coordinating Research Council, Inc., and National Air Pollution Control Administration, Public Health Service, U.S. Department of Health, Education, and Welfare. Chicago: IIT Research Institute, November 1970. 99 pp.

77. Odors from industries need controls. Environ. Sci. Technol. 3:623–624, 1969.

78. Oil dispersant reduces diesel pollution. Oil Daily No. 4143:1, December 28, 1967.

79. Perez, J. M., and E. W. Landen. Exhaust emission characteristics of precombustion chamber engines. SAE Paper 680421. SAE Trans. 77:1516–1533, 1968.

80. Ray, S. K., and R. Long. Polycyclic aromatic hydrocarbons from diffusion flames and diesel engine combustion. Combust. Flame 8:139–151, 1964.

81. Reckner, L. R., W. E. Scott, and W. F. Biller. The composition and odor of diesel exhaust. Proc. Amer. Petrol. Inst. 45(III):133–147, 1965.

82. Reckner, L. R., and R. E. Squires. Diesel exhaust odor measurement using human panels. SAE Paper 680444. SAE Trans. 77:1694–1705, 1968.

83. Reckner, L. R., and R. E. Squires. Final Report on Diesel Exhaust Composition and Odor. Prepared by Scott Research Laboratories, Inc., Project No. 1076, for Division of Air Pollution, Public Health Service, U.S. Department of Health, Education, and Welfare, Contract No. PH 86-67-134; and Coordinating Research Council, Inc., Project No. CD-9-61. Perkasie, Pa.: Scott Research Laboratories, Inc., September 3, 1968. 105 pp.

84. Reed, L. E., and C. F. Barrett. Air pollution from road traffic-measurements in Archway Road, London. Intern. J. Air Water Poll. 9:357–365, 1965.

85. Rounds, F. G., and H. W. Pearsall. Diesel exhaust odor—Its evaluation and relation to exhaust gas composition. SAE Paper 560863. SAE Trans. 65:608–627, 1957.

86. Schmidt, R. C., A. W. Carey, and R. Kamo. Exhaust characteristics of the automotive diesel. SAE Paper 660550. SAE Trans. 75:102–109, 1967.

87. Schumann, C. E., and C. W. Gruber. Motorist exposures to aldehydes from diesel-powered buses. J. Air Poll. Cont. Assoc. 14:53–55, 1964.

88. Smithers, K. A. Firm has new system aimed at eliminating odor from diesel fuel. Wall Street Journal 172(62):23, September 26, 1968.

89. Specs set for 4 experimental steam buses. Fleet Owner 64(7):15, 1969.

90. Springer, K. J. An Investigation of Diesel Powered Vehicle Odor and Smoke. Part I, Final Report. A Report for the Automotive Research and Development Section, National

Center for Air Pollution Control, Public Health Service, U.S. Department of Health, Education, and Welfare. PHS Contract No. PH 86-66-93. San Antonio, Tex.: Southwest Research Institute, Vehicle Emissions Laboratory, September 1967. 647 pp.

91. Springer, K. J. An Investigation of Diesel Powered Vehicle Odor and Smoke. Part II. A Report for the Automotive Research and Development Section, National Center for Air Pollution Control, Public Health Service, U.S. Department of Health, Education, and Welfare. PHS Contract No. PH 86-67-72. San Antonio, Tex.: Southwest Research Institute, February 1968. 143 pp.

92. Springer, K. J. An Investigation of Diesel Powered Vehicle Odor and Smoke. Part III, Final Report. A Report for the Division of Motor Vehicle Pollution Control, National Air Pollution Control Administration, Public Health Service, U.S. Department of Health, Education, and Welfare. PHS Contract No. PH 22-68-23. San Antonio, Tex.: Southwest Research Institute, October 1969. 783 pp.

93. Springer, K. J. Combustion odors—A case study, pp. 227–262. In A. Turk, J. W. Johnston, Jr., and D. G. Moulton, Eds. Human Responses to Environmental Odors. New York: Academic Press, 1974.

94. Springer, K. J. Emissions from Diesel and Stratified Charge Powered Cars. EPA Report No. 460/3-75-001A. San Antonio, Tex.: Southwest Research Institute, 1974. 328 pp.

95. Springer, K. J. Investigation of Diesel-Powered Vehicle Emissions. VII. EPA No. 460/3-76-034. Ann Arbor, Mich.: U.S. Environmental Protection Agency, Office of Mobile Source Air Pollution Control, 1977. 181 pp. + 12 app.

96. Springer, K. J., and C. T. Hare. A Field Survey to Determine Public Opinion of Diesel Engine Exhaust Odor. Final Report. A Report for Division of Motor Vehicle Research and Development, National Air Pollution Control Administration, Public Health Service, U.S. Department of Health, Education, and Welfare. Contract PH 22-68-36. San Antonio, Tex.: Southwest Research Institute, February 1970. 113 pp.

97. Springer, K. J., and C. T. Hare. Four Years of Diesel Odor and Smoke Control Technology Evaluations—A Summary. ASME Paper No. 69-WA/APC-3. Presented at American Society of Mechanical Engineers Winter Annual Meeting, Los Angeles, November 16–20, 1969.

98. Springer, K. J., and A. C. Ludwig. Documentation of the Guide to Good Practice for Minimum Odor and Smoke from Diesel-Powered Vehicles. Final Report. Contract CPA22-69-71. A Report for Emission Control Engineering Branch, Division of Motor Vehicle Pollution Control, National Air Pollution Control Administration, U.S. Public Health Service, Department of Health, Education, and Welfare. San Antonio, Tex.: Southwest Research Institute, November 1969. 100 pp.

99. Springer, K. J., and R. C. Stahman. Control of diesel exhaust odors. Ann. N.Y. Acad. Sci. 237:409–426, 1974.

100. Springer, K. J., and R. C. Stahman. Unregulated emissions from diesels used in trucks and buses. SAE Paper 770258. SAE Trans. 86:1210–1234, 1977.

101. Stahman, R. C., G. D. Kittredge, and K. J. Springer. Smoke and odor control for diesel-powered trucks and buses. SAE Paper 680443. SAE Trans. 77:1657–1693, 1968.

102. Steam car gets strong push in Senate report. Oil Daily No. 4498:18–19, May 19, 1969.

103. Stern, A. C. Acute pollution problems still face automotive engineers. SAE Journal 75(11):69–71, November 1967.

104. Stockham, J., A. O'Donnell, and A. Dravnieks. Chemical Species in Engine Exhaust and Their Contribution to Exhaust Odor. Final Report. Prepared for the Coordinating

Research Council, Inc. Report No. IITRI C8150-5. Chicago: IIT Research Institute, June 30, 1969. 92 pp.

105. Styles, H. E., J. Vrebos, J., and J. Lawther. Public Health Aspects of Air Pollution from Diesel Vehicles. WHO/AP/67.28. Geneva: World Health Organization, 1967. 40 pp.

106. Styles, H. E., J. Vrebos, and P. J. Lawther. The diesel engine and atmospheric pollution. WHO Chronicle 21:201–206, 1967.

107. Sudar, S., and L. Grantham. Diesel Exhaust Emission Control Program. Final Report, No. A1-73-61. Prepared under subcontract to Southern California Rapid Transit District, Los Angeles, Calif., for U.S. Department of Transportation, Urban Mass Transportation Administration. Canoga Park, Calif.: Atomics International, January 1974. 142 pp.

108. Sullivan, J. L., and G. J. Cleary. A comparison of polycyclic aromatic hydrocarbon emissions from diesel- and petrol-powered vehicles in partially segregated traffic lanes. Brit. J. Ind. Med. 21:117–123, 1964.

109. Swetnam, G. F., and F. L. Willingham. Evaluation of City Transit Bus "EIP" Kits to Reduce Engine Smoke, Odor, Noxious Emissions and Noise. M71-49. Washington, D.C.: U.S. Department of Transportation, 1971. 35 pp.

110. Tessier, K. C., and H. E. Bachman. Fuel Additives for the Suppression of Diesel Exhaust Odor and Smoke. Part I: Proposed Mechanism for Smoke Suppression. Paper 68-WA/DGP4. Presented at American Society of Mechanical Engineers Annual Meeting, New York, 1968. 8 pp.

111. Tessier, K. C., and H. E. Bachman. Fuel Additives for Suppression of Diesel Exhaust Odor and Smoke. Part II: Field Trials. Paper 68-WA/DGP-5. Presented at American Society of Mechanical Engineers Annual Meeting, New York, 1968. 5 pp.

112. The case for clean air. Mill & Factory 80(7):41–57, 1967.

113. The pall above, the victims below. Med. World News 8(5):60–69, 1967.

114. Truck turbines: Full bore by 1971. Heavy Duty Trucking 48(3):24–32, 1969.

115. Turk, A. Evaluation of catalytic mufflers for buses of the New York City Transit Authority. Report to the New York City Department of Air Pollution Control, September 19, 1967. 36 pp. (unpublished)

116. Turk, A. Selection and Training of Judges for Sensory Evaluation of the Intensity and Character of Diesel Exhaust Odors. PHS Contract No. PH 27-66-96. Cincinnati: U.S. Department of Health, Education, and Welfare, Public Health Service, 1967. 45 pp.

117. U.S. Bureau of Mines. Composition, Smoke, and Odor of Diesel Exhausts. Summary Quarterly Progress Report from Bartlesville Petroleum Research Center to U.S. Public Health Service for the Quarter Ended March 31, 1968.

118. U.S. Bureau of Mines. Composition, Smoke, and Odor of Diesel Exhausts. Summary Quarterly Progress Report from Bartlesville Petroleum Research Center to U.S. Public Health Service for the Quarter Ended September 30, 1968. BM-125, Section XIV.

119. U.S. Coast Guard. Boating Statistics, 1977, pp. 8, 10. Publication CG-357. Washington, D.C.: U.S. Department of Transportation, Coast Guard, 1977.

120. U.S. Coast Guard. Recreational Boating in the Continental United States in 1973 and 1976, p. 29. Washington, D.C.: U.S. Department of Transportation, Coast Guard, 1977.

121. U.S. Congress. Senate. Joint Hearings before the Committee on Commerce and the Subcommittee on Air and Water Pollution of the Committee on Public Works. Automobile Steam Engine and Other External Combustion Engines. Serial No. 90-82. Ninetieth Congress, Second Session, May 27–28, 1968.

122. U.S. Congress. Senate. The Search for a Low-Emission Vehicle. Staff Report for the Committee on Commerce, United States Senate. Ninety-first Congress, First Session. 149 pp.

123. U.S. Department of Commerce. The Automobile and Air Pollution: A Program for Progress, Part I, pp. 10–11. Washington, D.C.: U.S. Department of Commerce, 1967.

124. U.S. Department of Commerce. The Automobile and Air Pollution. A Program for Progress, Part II. Washington, D.C.: U.S. Department of Commerce, 1967. 160 pp.

125. U.S. Environmental Protection Agency. Air Pollution Aspects of Odors. 2 vols. February 1971. (unpublished)

126. U.S. Environmental Protection Agency. EPA says more diesel research needed. Press Release No. 264. Environmental News, November 12, 1977.

127. U.S. Environmental Protection Agency. Guide to Reduction of Smoke and Odor from Diesel-Powered Engines. Research Triangle Park, N.C.: U.S. EPA, Office of Air Programs, 1971. 28 pp.

128. U.S. Environmental Protection Agency, Office of Air and Waste Management, Office of Air Quality Planning and Standards. 1973 National Emissions Report, p. 1. EPA No. 450/2-76-007. Research Triangle Park, N.C.: U.S. EPA, 1976.

129. U.S. Federal Aviation Administration. Census of U.S. Civil Aircraft, Table 1. Washington, D.C.: U.S. Department of Transportation, F.A.A. (in press)

130. U.S. Federal Highway Administration. Highway Statistics, 1977, Tables MV-1, MV-9, MV-10. Washington, D.C.: U.S. Department of Transportation, F.H.A., 1977.

131. U.S. National Air Pollution Control Administration. Air Quality Criteria for Hydrocarbons. Publ. No. AP-64. Washington, D.C.: U.S. Department of Health, Education, and Welfare, 1970.

132. U.S. National Air Pollution Control Administration. Award of two contracts designed to advance the development of a practically pollution-free power plant for the family car. U.S. Department of Health, Education, and Welfare, Public Health Service, News Release HEW-X35, July 15, 1969.

133. U.S. National Air Pollution Control Administration. Control Techniques for Carbon Monoxide, Nitrogen Oxide, and Hydrocarbon Emissions from Mobile Sources. Publ. No. AP-66. Washington, D.C.: U.S. Department of Health, Education, and Welfare, Public Health Service, Environment Health Service, National Air Pollution Control Administration, 1970.

134. U.S. National Air Pollution Control Administration. Federal regulations to require the registration of fuel additives were proposed today by the Commissioner. U.S. Department of Health, Education, and Welfare, Public Health Service, News Release HEW-X41, July 30, 1969.

135. U.S. Public Health Service. Proceedings of the Third National Conference on Air Pollution. PHS Publ. No. 1649. Washington, D.C.: U.S. Department of Health, Education, and Welfare, Public Health Service, National Center for Air Pollution Control, Office of Legislative and Public Affairs, 1967. 667 pp.

136. U.S. Public Health Service. The Sources of Air Pollution and Their Control. Consumer Protection and Environmental Health Service. PHS Publ. No. 1548. Washington, D.C.: U.S. Department of Health, Education, and Welfare, Public Health Service, 1966. 14 pp.

137. Vehicle exhausts in relation to public health. Chem. Ind. February 12, 1966, p. 290.

138. Vogh, J. W. Contribution of Some Carbonyl, Phenol, and Hydrocarbon Components to Diesel Exhaust Odor. RI No. 7632. Washington, D.C.: U.S. Department of the Interior, Bureau of Mines, 1972. 12 pp.

139. Voorhoeve, R. J. H., C. K. N. Patel, L. E. Trimble, and R. J. Kerl. Hydrogen cyanide production during reduction of nitric oxide over platinum catalysts. Science 190:149–151, 1975.

140. Wade, W. A., III, K. C. Tower, and S. Cha. Technical Report to the Illinois EPA on the Evaluation of Four Odor Measurement Systems. TRC Project No. 32390. Wethersfield, Conn.: The Research Corporation of New England, 1974. 61 pp. + app.

141. World Health Organization. Urban Air Pollution with Particular Reference to Motor Vehicles. Report of a WHO Expert Committee. Technical Report Series No. 410. Geneva: World Health Organization, 1969. 53 pp.

10 Techniques for Assessing the Economic Value of Odor Control

INTRODUCTION

Everything said in this chapter unequivocally presumes that man is the measure of all things. Whatever a person does must be the best thing for him to do, given his knowledge of his circumstances at the moment—otherwise, he would not do it: the person's autonomous preferences are revealed by his behavior. This is the perspective of value that pervades economic analysis. Contrary, however, to much common usage, "economics" and "pecuniary" are not viewed as synonymous. In fact, human function and the health or aesthetic effects of an odor are directly "economic." The effects of an odor on real property are "economic" only insofar as that property contributes to human health and happiness.

The preceding perhaps conveys the stance of economics with respect to the basis of values. It fails, however, to state the units in which values are to be measured or the context that bestows meaning on these units. Assume, for example, that a person derives satisfaction from an aesthetic phenomenon, such as the absence of malodor. If there is a local increase in malodor, the person will possibly feel that he has been made worse off; but, if there are other worldly things capable of providing him satisfaction, then there must be some additional provision of these other things causing him to feel as well off as he would without the increased malodor. Finally, if these things can be secured by the expenditure of income, or time that can be used to earn income, then there is some additional income that, with the malodor increase, would make the person feel no worse off. The unit, therefore, in which economics would have us measure value is money stated in terms of income. Implicit in the acceptance of this unit is the presumption that, even if the thing being valued cannot be secured in the marketplace, there are in the marketplace collections of other things from which the person receives equal satisfaction. These other things, which have money prices attached, can, under a fairly wide range of well-specified conditions, serve as vehicles to infer the "values" of entities and services for which no directly observable pecuniary prices exist.

In spite of the reasonableness of the approach to valuation sketched above, it will often yield, depending on the conditions adopted for the analysis, different values for the same quantity variation in the entity being valued. For example, if one is interested in the control of an odor that is aesthetically offensive, the value that a person will attach to its reduction can depend on whether one is measuring what the person is willing to pay for the reduction or what the person would have to be paid in order not to have the reduction. In the latter case, because the person is viewed as holding the legal right to stop the pollution, his revelation of his preferences is not limited by his income. However, his income does limit what he can do when he must buy a cessation of pollution from someone else. As goods become scarce, he becomes reluctant to trade money for goods. Thus, the two measures would be identical only when variations in income play a trivial role in determining the quantity of the good that the person will choose to hold.

Other sources of variations in values for identical variations in the quantity of a particular good include whether, in an original and in a new state, the original quantity of the good is the most preferred or the least preferred quantity; whether the valuation in the new state is independent of adjustments in overall patterns of consumption in moving from the original quantity of the good to the new quantity; and whether the person can by his own actions adjust his consumption of the good in question or, as with some malodors, must become resigned to an externally imposed fate. In short, to be meaningful and communicable, the exact context of a particular economic valuation measure must be explicitly and fully stated. The criteria for judging which of the several analytically correct valuation measures to apply to a particular real problem must often come from outside economics.

At a thoroughly abstract level, aesthetic phenomena in general, and malodors in particular, play no role in economic analysis, other than as just another class of goods to which the consumer may attach value. This class of goods, however, possesses some rather special attributes of relevance to economic valuation in the context of public policy.

The first question to consider is why malodors might be thought relevant for public policy. One obvious reason is that malodors are frequently externalities. The particulate and gaseous emission from a feedlot may reduce the ability of a neighbor to use his own backyard. No part is played by the neighbor in the feedlot manager's decision of whether or not to produce malodor at any given time, even though the neighbor's welfare is affected by these decisions.

From the perspective of economic efficiency, the initial existence of an externality is no more than a necessary condition for public-policy relevance. In many situations in which, at first glance, there appear to be externalities, a person has in fact voluntarily exposed himself to the harmful agent only on the condition that he receive what he regards as adequate compensation. Thus, superficially, it might seem that the feedlot neighbor is involuntarily exposed to malodor. Nevertheless, if the neighbor was fully aware of the existence of malodors before he bought his property and therefore demanded and received a reduction in the property price sufficient to make him indifferent as to a choice between this property and another more expensive property without malodor, then from the perspective of economic efficiency no public-policy problem arises.

In contrast, if the person were imperfectly knowledgeable about the presence of the malodor, or if substitute properties or other goods were not readily available, or if no channels were available for him to communicate without distortion his wish for greater compensation, then no adequate private means to ameliorate the externality would exist. With respect to economic efficiency, the problem of neighborhood exposures would then be a relevant one for public policy to resolve. The measurement techniques sketched in the following sections would be useful as a means to approximate the magnitude of the aforementioned compensation, or the expense that the source of malodors might reasonably be required to bear for bringing them under control.

A thorough search of the technical economics literature has failed to unearth a single recent instance of the application of common techniques of economic valuation to a malodor problem. (Copley International Corporation did publish a report[8] on the valuation of malodors, but much of it is now somewhat outdated.) The objective of this discussion must therefore be limited to providing enough information on the possibilities and the limitations of the uses of these techniques to permit the reader to decide whether they might advantageously be applied to malodor problems.

COMPLAINTS AS VOTES

The great bulk of extant attempts to evaluate the responses of people to odors were reviewed in Jonsson.[14] No explicit attempts were made in these studies to assign pecuniary values to the presence of malodors, although the frequent references to the counting of complaints to surveyors or public officials make it appear as though such counting is the preferred method of valuation.

Complaints may be viewed as the result of a binary decision: one either objects to the presence of a malodor or does not object. Presumably, complainants make the choice or abstain by comparing what their welfare would be under the alternative outcomes. A complaint reveals information about the direction of the welfare change. However, without an extended chain of stringent assumptions, a complaint says nothing about the magnitude of the change, i.e., it provides no information about the *intensity* of objections to the presence of a malodor. If the income equivalent of the change for one person is $1 and for another is $1,000, comparisons of relative frequencies of complaints can be grossly misleading as a basis for public-policy decisions. A fine example of the long series of sometimes shaky steps that the investigator must traverse to proceed from information on binary choices to inferences on preference intensities was ably presented by Deacon and Shapiro.[9] A specific example of the way in which simple complaint frequencies can be misleading as to the seriousness of the effects of malodors was presented by Diemer and McKean.[11] Complaints about odors from cattle feedlots in the vicinity of Fort Collins, Colorado, were too few to appear in a rank-ordering of outstanding community problems. But when the authors provided for intensity of feeling, feedlot odors were only a place or two in the rank-ordering below inadequate trash collection, inadequate hospital facilities, and the absence of a public auditorium.

PROPERTY-VALUE STUDIES

Recently, many studies have tried to infer the influence of environmental-quality improvements on property values. Particular attention has been devoted to air quality in urban areas. Among the better known of these studies are those by Ridker and Henning,[18] Anderson and Crocker,[1] Deyak and Smith,[10] and Harrison and Rubinfeld.[12] The studies have concluded that air pollution—generally measured as annual geometric mean concentrations of total suspended particles, sulfur dioxide, and nitrogen dioxide, individually or in combinations—has a detrimental impact on residential-property values: a 1% change in air pollution, however measured, tends, on the average, to induce a 0.1–0.3% change of the opposite sign in residential-property values.

The analytical development of the hypotheses tested in the aforementioned studies proceeds from the following fundamental scenario. Suppose that people find air pollution or malodors to be offensive and that the extent of pollution differs from one location to another. Suppose further that in some initial state property prices and the prices of all other goods are the same everywhere. Thus, people in the most polluted locations would be worse off then others: they would be paying the same unit prices for property and other goods, but they would suffer more pollution. It would therefore be in the interest of those who are subject to worse pollution to offer higher prices for locations subject to less pollution. The market prices of the lower-pollution sites would then rise, and the welfare of persons wishing to use the cleaner sites would fall, relative to the welfare of those in the higher-pollution sites. The rise in the prices of the cleaner sites, relative to the prices of the dirtier sites, would continue until no person could gain by moving from a dirty to a clean site or vice versa: the relative increase in the prices of the cleaner sites would then have been sufficient to make the representative person indifferent to the choice of living in a cleaner or dirtier location.

Empirical tests of the hypotheses derived from the analytical constructs built on the above scenario have used linear and nonlinear statistical regression techniques. Typically, residential-property prices have been regressed on housing and neighborhood characteristics, including air pollution. The regression coefficient of the air-pollution variable is then used as a datum to infer the value of cleaner air. There has, however, been an extended debate in the technical economics literature as to the validity of many such inferences. Mäler[15] discussed at length the "unrealistic" assumptions (e.g., perfect mobility and nonsegmented markets) required to make models constructed from the preceding scenario fully operational. Unfortunately, no attempts have yet appeared in the literature dealing with property values and environmental quality in which the sensitivity of empirical results to the set of assumptions adopted has been tested with the same data set.

SUBSTITUTION STUDIES

If the uses of some goods that have directly observable market prices are functionally related to the level of a nonmarketed environmental-quality good, the value of the latter can be inferred from changes in the quantities of the former that are used.

Two variants of substitution studies have thus far been used to estimate welfare changes due to changes in one or another facet of environmental quality. Both variants typically adopt the household-production function approach to consumer behavior. The essential elements in this framework are a utility function (set of preferences), a user technology, budget and time constraints, and the prices of marketed goods. The user is presumed to derive utility or satisfaction from a set of characteristics that he produces. Inputs into the production process are the market goods that the person purchases, given his income and market prices.

The first variant is fundamentally similar to property-value studies. It has been applied often in studies of the willingness to pay for outdoor recreational activities. (For a review of this work, see Smith.[19]) The first step in the variant is the estimation of a cost function. This is accomplished by relating the expenditures of individuals on particular activities to individual observations on the characteristics of these activities, including environmental quality and aesthetic characteristics. As with the property-value studies, the estimated coefficient of whatever environmental quality or aesthetic variable is of interest is used as a datum to infer the value of additional units of the variable.

The second variant has the benefit of increased generality, in that it allows the investigator to remove from the value estimate the effects of real-income *changes* caused by changes in the environmental-quality parameter. In this procedure, one uses information on changes in activity patterns to estimate the household technology. This results in a representative household technology. A specific form for the utility function is then elected, the expenditure data are used to ascertain prices of the alternative activity patterns, and the representative individual's decision problem is then solved. Horst and Crocker,[13] who studied the value of atmospheric visibility in the Four Corners area of the United States, presented an example of this approach.

BIDDING-GAME STUDIES

Since 1973, an untraditional (to economists) technique for assessing the demand for unmarketed goods, such as environmental quality, has been widely used by a small group. The technique, known as bidding games, uses structured interviews to establish directly a measure of welfare change—the willingness to pay or to be compensated for a change in environmental quality.

Initially, alternative quantities of the environmental good of interest are described to the respondent. Careful attention is devoted to detailing the relevant quantity, quality, location, and time dimensions of the good in question. In addition, the hypothetical market (including the property-rights structure) in which the respondent is to bid is described, and the currency in which the bid is to be stated is specified. The currency is typically current income. The setting having been established for the respondent, the respondent is asked to react to prices for changing environmental quality. The prices are posed by the interviewer. For example, the interviewer might ask the respondent whether he is willing to pay $15 for a particular improvement in the environmental variable. This price is then varied iteratively until the price at which the respondent is indifferent to the specified improvement is identified. Another degree of improvement is then stated and the bidding process restarted. The collection of bids is then used to construct Bradford[4] bid curves from which various measures of welfare change may be obtained. (Brookshire and Randall[6] have explained the transition from bid curves to measures of consumer surplus.)

This procedure has recently been applied to a wide variety of aesthetic environmental phenomena. Randall *et al.*[17] and Blank *et al.*[3] used a bidding game to acquire data to value man-induced atmospheric-visibility changes in the Four Corners region. Brookshire *et al.*[5] used a bidding game in their study of the potential visibility reductions that could be caused by the proposed Kaiparowits coal-fired power plant in the Glen Canyon recreation area. Bidding games have also been used by Ben-David *et al.*[2] and Randall *et al.*[16] to gather data on the aesthetic effects of geothermal development in the Jemez Mountains of New Mexico and the aesthetic effects of strip-mining in Appalachia, respectively. A recent report submitted by Brookshire *et al.*[7] to the EPA noted substantial accord between the values developed by a bidding-game study and by a property-value study for air-pollution reductions in the South Coast Air Basin of southern California.

The bidding-game procedure obviously bears a close resemblance to the panel procedures often used to assess the presence and severity of malodors. In fact, it appears that traditional panel practices in odor measurement can be easily transformed into bidding-game formats. Having obtained a panel member's statements about odor quality, intensity, extension, etc., the enumerator may simply go through the iterative bidding procedure. However, in spite of the seeming ease with which bidding games might be appended to an important and widely used odor-measurement method, the validity of the value measures obtained will currently be viewed with skepticism by the vast majority of economists. Concern is often expressed about strategic behavior, where the respondent biases his bid to attempt to shift the aggregate bid of all respondents in a direction he favors. In addition, it is feared that the dollar value at which the interviewer initiates the bidding process may help to determine the respondent's final bid. Finally, questions are raised about whether the circumstances posited in bidding games have sufficient correspondence to reality to be meaningful to respondents.

SUMMARY AND CONCLUSIONS

This chapter has outlined three ways to establish the economic benefits of malodor control. Nonmarketed aesthetic phenomena other than those related to odors have been assessed by the property-value and substitution techniques. At a minimum, the studies have provided evidence that many benefits of pollution control traditionally viewed as intangible and thereby nonmeasurable can, in fact, be measured and made comparable with economic values expressed in markets. Similar results have been obtained with a third technique, the bidding game, that has much less acceptance among economists. Nevertheless, bidding games may often be the preferred technique when assessing malodors, because they seem firmly in the tradition of the panel procedures that have been used to measure the character of odors. The technique can thus probably be readily implemented.

REFERENCES

1. Anderson, R. J., Jr., and T. D. Crocker. Air pollution and residential property values. Urban Studies 8:171–180, 1971.

2. Ben-David, S., S. Noll, W. Schulze, and M. Thayer. The Economics of Geothermal Energy: A Case Study. Second-year report to the National Science Foundation for Project #APR 75–18249, 1977. 220 pp.

3. Blank, F. M., D. S. Brookshire, T. D. Crocker, R. C. d'Arge, R. L. Horst, Jr., and R. D. Rowe. Valuation of Aesthetic Preferences: A Case Study of the Economic Value of Visibility. A research report to the Electric Power Research Institute for Contract #RP 7855-2. Washington, D.C.: E.P.R.I., October 1978. 337 pp.

4. Bradford, D. F. Benefit-cost analysis and demand curves for public goods. Kyklos 23(4):775–791, 1970.

5. Brookshire, D. S., B. C. Ives, and W. D. Schulze. The valuation of aesthetic preferences. J. Environ. Econ. Manag. No. 4, 3:325-346, 1976.

6. Brookshire, D. S., and A. J. Randall. Public Policy Alternatives, Public Goods and Contingent Valuation Mechanisms. A paper presented at the Annual Meetings of the Western Economics Association, Honolulu, Hawaii, June 22, 1978.

7. Brookshire, D. S., R. C. d'Arge, W. Schulze, and M. Thayer. Experiments in Valuing Non-Market Goods: A Case Study of Alternative Benefit Measures of Air Pollution Control in the South Coast Air Basin of Southern California. A report to the U.S. Environmental Protection Agency for Grant No. R805059010. Washington, D.C.: U.S. EPA, November 1978. 287 pp.

8. Copley International Corporation. A Study of the Social and Economic Impact of Odors. Phase II. Report to the U.S. Environmental Protection Agency. Contract No. CPA 70-116. La Jolla, Calif.: Copley International Corporation, November 1971. 308 pp.

9. Deacon, R., and P. Shapiro. Private preference for collective goods revealed through voting on referenda. Amer. Econ. Rev. 65:943–955, 1975.

10. Deyak, T. A., and V. K. Smith. Residential property values and air pollution: Some new evidence. Quart. Rev. Econ. Bus. 14:93–100, 1974.

11. Diemer, J. A., and J. R. McKean. The assessment of community preference: A methodology and case study. Land Econ. 54:244–252, 1978.

12. Harrison, D., Jr., and D. L. Rubinfeld. Hedonic housing prices and the demand for clean air. J. Environ. Econ. Manag. 5:81–102, 1978.

13. Horst, R. L., Jr., and T. D. Crocker. The Value of an Aesthetic Phenomenon: Atmospheric Visibility in the "Four Corners" Area of the United States. A paper presented at the Annual Meetings of the Western Economic Association, Honolulu, Hawaii, June 22, 1978. 30 pp.

14. Jonsson, E. Annoyance reactions to environmental odors, pp. 329–333. In A. Turk, J. W. Johnston, Jr., and D. G. Moulton, Eds. Human Responses to Environmental Odors. New York: Academic Press, 1974.

15. Mäler, K-G. A note on the use of property values in estimating marginal willingness to pay for environmental Quality. J. Environ. Econ. Manag. 4:355–369, 1977.

16. Randall, A., D. Grunewald, A. Pagoulatos, R. Ausness, and S. Johnson. Estimating Environmental Damages from Surface Mining of Coal in Appalachia: A Case Study. Environmental Protection Technology Series 600/2-78-003. Cincinnati: U.S. Environmental Protection Agency, January 1978. 141 pp.

17. Randall, A., B. Ives, and C. Eastman. Bidding games for valuation of aesthetic environmental improvements. J. Environ. Econ. Mang. 1:132–149, 1974.

18. Ridker, R. G., and J. A. Henning. The determinants of residential property values with special reference to air pollution. Rev. Econ. Stat. 49:246–257, 1967.

19. Smith, V. K. The estimation and use of models of the demand for outdoor recreation. Appendix B, pp. 89–123. In National Research Council, Committee on Assessment of Demand for Outdoor Recreation Resources. Assessing Demand for Outdoor Recreation. Washington, D.C.: National Academy of Sciences, 1975.

11 Regulatory Aspects of Odor Control

INTRODUCTION

Preceding chapters of this report have described the effects of odors on people who perceive them. This chapter considers whether and under what circumstances odorant sources should be controlled in order to reduce or eliminate those effects. For regulatory purposes, one can hardly avoid trying to reach judgments that evaluate odors in terms of their effects: adverse, beneficial, or neutral; toxic or nontoxic; injurious to health or injurious to welfare. Obviously, the case for regulating odors is more compelling when and if they can fairly be characterized as hazardous to human health than when they are merely experienced as an unpleasantness by some sectors of the population. However, because of differences in perspective, insufficiencies of data, or ambiguities in the data, not everyone will agree on which evaluative label is appropriate to which set of circumstances. It is therefore advisable to take a practical, open-ended view of the matter and to avoid getting bogged down in semantic distinctions.

There is, first of all, a group of effects that everyone would acknowledge as undesirable, ranging from mild displeasure to extreme upset. If perception of an odor provokes, for example, reactions of nausea, respiratory disturbance, or loss of sleep in persons of relatively normal sensibilities, there should not be much difficulty in recognizing these as health-related effects from which the public ought to be protected. Odors that produce lesser degrees of annoyance or discomfort may be accorded lower priorities for control in the interest of public welfare. It should be emphasized, however, that the same type of odor may be toxic in some exposures and nontoxic in others. The production and the evaluation of effects—as well as the decision of whether, when, and how to regulate their causes—will depend on a number of variables, including the character, intensity, duration, and frequency of odor and the sensibilities of exposed populations.

A second group of possible effects consists of morphologic changes in the human body, such as a shrinking of cells in the olfactory bulbs or a lowering of heart rate, which may or may not be accompanied by unpleasant sensations. In the present state of knowledge, it is not yet clear whether such effects should be viewed as evidence of damage or risk to human health. On the one hand, they may represent significant departures from normal or expected physiologic function; on the other, they may be readily and completely reversed on cessation of exposure to the odorant. The matter is confused, moreover, by uncertainty over whether such effects would occur on exposure to the pollutant even if the subject did not perceive it as an odor. Further studies of these phenomena, including their relationships and consequences, may have to be undertaken before their implications for regulatory decision-making can be spelled out.

This report is concerned with effects on health or welfare that are associated, at least in part, with perception of odors. Odorous substances may have adverse effects both because they are perceived as malodorous and because they are harmful substances even apart from their perceived characteristics. In such cases, controlling malodor may also serve, in whole or in part, to reduce the harm that is caused directly by the substance itself. Whether or not that is so, the control of odors *per se* calls for regulatory techniques (to be discussed in this chapter) that differ appreciably from regulation of substantive air pollutants without regard to any odorant potential they may have.

Section 403(b) of the Clean Air Act poses the question to which EPA must ultimately respond: whether national ambient-air quality standards should be established for odorous substances. Let the reader be advised at the outset, however, that no clear-cut answer is offered in the following pages. In the present state of the art, the need for odor control is still highly controversial. The subject of odor control is riddled with uncertainties of methodology, of measurement, of perception, and of social preference. Nor can legal questions in this field be disentangled from technical ones. The nature of a regulatory program, including choice of standards and enforcement tools, depends largely on what is known about the properties of odor and on the technology available for measuring and abating it. Most significant of all, from a regulatory point of view, is the difficulty of determining the acceptability of an odor by some qualitative or quantitative test that can be administered or otherwise understood apart from the community setting in which odors from various sources are originally experienced.

Although a cautious approach to the subject is therefore warranted, there can be no doubt that malodors should be regulated by one means or another, as a matter of public environmental policy. "It has been amply documented that in the recent past, as many as up to 50% of all citizen complaints to local (air quality) agencies are associated with odors."[10] Numerous judicial opinions have reported the dire effects of odor nuisances, as described in no uncertain terms by complaining witnesses. In one case, for example, a rendering plant was responsible for "a putrid smell which brings tears to the eyes of nearby residents, drives them from their yards to the protection of their homes and robs them of their sleep."[6] In another case, odors from the defendant's chicken-processing plant made the plaintiffs unable "to eat their meals without nausea."[15] Prolonged and repeated exposure to offensive odors in a third case made people "very irritable, upset and nervous" and even prompted one witness to move to another town "because of the stink."[1] Malodors have received judicial recognition as "crimes against the environment."[4]

The foregoing cases were all actions brought under the law of public or private nuisance, as developed through a long line of judicial precedents and as codified in some instances by local ordinances. Nuisance law is the oldest and most pervasive source of law for controlling odors. But it no longer monopolizes the subject. Increasingly, state and local regulations are adopting more scientific approaches to odor control that depart substantially from the jurisprudence of the common-law nuisance action. The U.S. EPA may consider promulgating national ambient-air standards for odors or new-source performance standards for odorant sources under the federal Clean Air Act.

This chapter attempts to outline regulatory alternatives, mainly in terms of the three dimensions of law indicated above: nuisance as a judicially enforceable cause of action under state law, regulation under various state or local ordinances, and regulation under the federal Clean Air Act. Jurisdiction and regulatory approach are the two major variables to be considered. The purpose here is not to endorse any particular regulatory strategy, but to indicate advantages and drawbacks of each alternative and to discuss circumstances that may argue for or against its use. The result of such an exercise should be to shed some light on whether it is both feasible and desirable to bring odor pollution within the regulatory ambit of the Clean Air Act.

PUBLIC AND PRIVATE NUISANCES

Nuisance law (which varies little from state to state) encompasses two distinct fields of liability, termed "public nuisance" and "private nuisance," both of which deserve close study in the context of odor regulation. A public nuisance is created when an act, usually criminal, invades a right common to all members of the public, such as the right to free and safe use of a highway or to enjoyment of a park. A private nuisance involves an invasion of a private party's interest in the use and enjoyment of his land. Because the Clean Air Act does not explicitly preempt the field of odor regulation, the two types of nuisance action would remain valid avenues for seeking abatement of unpleasant odors, even if standards were established under that Act and the defendant were in compliance with them.

PRIVATE NUISANCE

Private-nuisance actions to settle grievances about odors have occupied a prominent place in the common law since complaints about a hog-sty gave rise to *William Aldred's Case*[14] in 1611. The facts that the plaintiff must establish in such cases are related to interference with his use of land by disagreeable odors that result in a nuisance condition for which the defendant can be held responsible. The plaintiff must show causality, but need not show any negligence or recklessness on the part of the defendant in allowing the offensive odorants to be released from sources under his control.[3]

To rise to the level of a nuisance, the odors complained of must be judged a substantial annoyance by the standards of the ordinary, reasonable person living in that locality. An unusually sensitive person may find it impossible to establish a nuisance on the basis of odor pollution in an industrialized neighborhood, both because the odor is characteristic of the locality and because it is considered harmless by most residents of the area. However, even the odors from a well-maintained horse barn may be deemed a nuisance in a residential community where the average homeowner is not used to the smell of manure.[9]

The fact that an odorous activity substantially impairs another's right to the use and enjoyment of land does not automatically subject the actor to liability. The plaintiff who is troubled by odors will usually find himself without a remedy unless he can show that the defendant's odor-causing conduct is unreasonable. In other words, the plaintiff will have to establish, by balancing the defendant's interests against his own, that the harm to him is greater than he should be required to bear without compensation. A private plaintiff may have difficulty making such a showing, especially against an industrial source on which the local economy depends. Moreover, even if he successfully demonstrates that he is entitled to recompense, his usual remedy will be an award of damages, rather than an injunction forcing the defendant to abate the odor.

As the classic case of *Boomer* v. *Atlantic Cement Co.*[19] illustrates, courts are reluctant to enjoin an industrial nuisance on the basis of a balance of equities. The plaintiff landowners in *Boomer* alleged that their property was injured by dust, smoke, and vibrations from a neighboring cement plant and established at trial that a nuisance existed in fact. The court, however, refused to issue an injunction, largely because of the tremendous disparity between the economic consequences of the nuisance to the plaintiff and the value of the defendant's operations. The trial court found that the total property damage amounted to $185,000, whereas the cement company's investment in its plant was valued in excess of $45,000,000. The opinion suggests that the court might have ordered the defendant to take corrective action if this had been technologically feasible, but the nuisance could not have been abated without major technologic innovations.

Legal and practical considerations other than the difficulty of obtaining injunctions also limit the role of private litigation as a technique for regulating odorants. Such defenses as laches* or legislative authorization for the defendant's conduct may bar the nuisance action altogether. The theory of "coming to the nuisance" will usually prevent recovery by a plaintiff who knowingly moves into the vicinity of an odor-emitting source. For example, plaintiffs who had bought property near an old railroad could not obtain an injunction after it recommenced active operation, which produced noise and smoke that interfered with the plaintiffs' use and enjoyment of their land.[11] Because the private nuisance is tied to interests in land, a nuisance action cannot be maintained by an employee, invitee, or other person who has no property right in the affected land. Private litigation, moreover, is a costly and uncertain route to vindicating one's legal rights, and plaintiffs only rarely have the resources to pursue remedies available to them under nuisance law.

PUBLIC NUISANCE

The public-nuisance action avoids many of these difficulties, but has serious limitations of its own as a tool for odor regulation. Because laches, prescriptive or prior rights, and legislative authority are no defenses in a public-nuisance action, this type of lawsuit is easier to initiate than a

*This familiar defense is available to a defendant in an equitable action who can show that the plaintiff's undue delay in asserting his legal rights caused a detrimental change in the defendant's position.

private-nuisance suit. Furthermore, the action can be maintained by public authorities, because the conduct that gives rise to a public nuisance is usually a crime at common law or under an appropriate statute. Private resources need not be expended to protect rights that are common to the general public. Nevertheless, odor regulation through the public-nuisance action can be a frustrating task, because of evidentiary problems, and inertia or bias on the part of the enforcing authority can stand in the way of forceful public litigation.

For a public nuisance to exist, the rights invaded must be genuinely public rights, and not merely the rights of a large number of individuals. Although it need not be shown that the whole community has suffered, the prosecutor must establish that the alleged nuisance interferes with some exercise of public rights. The strength of the case depends mainly on the number and reliability of witnesses who testify to the existence of the nuisance. For example, in *Fort Smith* v. *Western Hide and Fur Co.*,[8] the evidence of eight businessmen was sufficient to establish that an odor-producing fur business in the heart of the city was a public nuisance. The witnesses testified to the offensiveness of the odors and also gave evidence of damage to their own businesses.

NUISANCE ON BOTH COUNTS

When disagreeable odors are dispersed widely enough to interfere seriously with public comfort, they frequently also infringe on the right of private individuals to use and enjoy land near the emission source. This makes the nuisance a private as well as public one, and the private landowner may sue in tort on his own behalf, although the odorous activity is also a crime against public welfare. Failing some interference with his property rights, however, a private plaintiff cannot sue for abatement of a public nuisance unless he suffers damage different in kind from any other members of the general public. Injury to health, such as persistent nausea and headaches, is usually considered sufficiently particularized harm to form the basis for such a suit. But it is doubtful whether an individual could establish standing to sue simply because he found an odor more disagreeable than did his neighbors. Unless he could prove the existence of a private nuisance, he would therefore have to rely on public enforcement to correct the condition.

EVALUATION

As the foregoing discussion indicates, traditional nuisance law has some serious shortcomings as a source of authority for securing abatement of malodors. Lawsuits are costly, time-consuming, and risky to undertake.

Witnesses must be lined up to testify, and convincing evidence of the defendant's responsibility must be presented at the plaintiff's expense. If he is a person of unusual sensitivity to malodors, he may fail altogether to secure relief. Courts may be willing to award damages, but not necessarily to enjoin the nuisance itself or to order its abatement. Where no suit is brought, because prospective plaintiffs lack the means or the initiative to assert their rights or because the public prosecutor has other priorities in mind, malodorous nuisances will continue to escape regulation.

However, the disadvantages of this traditional route to relief should not be exaggerated, nor should its merits be lightly dismissed. The very fact of a nuisance suit, supported by complaining witnesses, may be more convincing evidence of community annoyance with malodors than all the scientific attempts to determine acceptability of an odor through public-attitude surveys, numbers of complaints, dilution of odor samples, or comparisons of samples with reference standards. Of course, nuisance law involves subjective judgments that may vary considerably from case to case; but malodors themselves are subjective perceptions, and reactions to them also vary greatly from one person or circumstance to another. Courts are well equipped, moreover, to evaluate the evidence of each witness and to reject testimony from witnesses who appear to be biased or influenced by improper motives. A court can fashion relief that is appropriate to the particular case, leaving other sources free to emit odorants so long as it has not been shown that they, too, are causing nuisances. This is perhaps as it should be.

At the same time, arguments can certainly be advanced that nuisance law should be supplemented by more scientific and more comprehensive regulatory approaches to odor control.

MODERN STATE AND LOCAL CONTROLS

In recent years, many state and local authorities have adopted specific odor-control regulations, usually as part of a broader program for air-quality maintenance. These regulatory efforts represent varied ways of dealing with the same fundamental question: How can we determine what constitutes an "acceptable" level of odorous emission? The most interesting solutions borrow from the old nuisance action the idea that community consensus should be an important factor in defining acceptable limits for odor. At the same time, they avoid the cost and uncertainty of litigation by providing for a simpler administrative determination that a nuisance exists and for an administrative enforcement order to bring the violator into compliance.

The earliest and still most common type of local regulation is the provision that makes emission of unpleasant odors a public nuisance. These can be worded in different ways, but a fairly typical example is the following regulation of the State of Texas:[18]

> No person shall discharge . . . one or more air contaminants [including odors] or combinations thereof in such concentrations and of such duration as are . . . injurious to . . . human health or welfare, animal life, vegetation or property. . . .

A violation of such a law is proved primarily on the basis of testimony from affected residents of the community. This approach enables official investigation and public prosecution of cases that would otherwise have been left to the law of private nuisance, but otherwise does little more than codify traditional nuisance standards.

To avoid the evidentiary problems that often arise under vaguely worded laws like Texas's, a newer generation of regulations has begun experimenting with more precisely defined methods for measuring odor.

On the basis of research done by Arthur D. Little, Inc., on 53 odorous substances, the State of Connecticut has adopted numerical ambient standards as a basis for regulation. An objectionable odor is found to exist when it is caused in whole or in part by one of these 53 pollutants present in the ambient air at concentrations higher than the specified threshold limits. But the enforceability of this regulation has been called into question,[17] especially for substances that are detectable by odor at concentrations lower than 1 ppb. It is unlikely that these odorants can be measured accurately by analytical techniques in use today.

Limitations on emission sources, although not in general use, may be a more promising regulatory approach than maximal ambient-air concentrations. The San Francisco Bay area, for example, adopted an odor regulation in 1973 that sets emission limits at 100 times the odor threshold (to be determined by chemical analysis or dilution to threshold of stack samples) for five compounds identified as the most likely causes of industrial odor in that region. Similar limits have been adopted by a number of state and local agencies for sulfur compounds emitted by kraft pulp mills, and by the States of Connecticut, Illinois, and Minnesota. This approach has the merit of going directly to the sources of the problem, but its effectiveness may be limited by lack of any proven correlation between emission standards and annoyance thresholds in the community.

A different type of source-specific regulation was developed in Los Angeles County for the rendering industry and has since been adopted by a number of other jurisdictions. They require effluent gases from rendering plants and a few other highly odorous industrial processes to be incinerated under specified operating conditions (temperature and duration) that will bring about substantially complete oxidation. Incineration standards, however, are viewed as unduly discriminatory by the rendering industry;[16] and, unless processes of "equivalent" efficacy are also permitted, there is some danger that standards of this sort will deter technologic innovation by any industry to which they may be addressed.

Another regulatory approach that does not refer to measurements or performance standards at the source depends on sensory evaluation of odors in the ambient air. For example, a violation may be deemed to exist if the odor in an ambient-air sample is still detectable after 7 or 31 dilutions with equal volumes of clean air. The ordinance may specify use of the Scentometer for this purpose. How often and within what periods of time observations must be made, and by how many inspectors, may also be specified in the ordinance. This approach has the merit of attempting to define the objective—elimination of odorous nuisances—in quantitatively measurable terms, but it can also be criticized on a number of technical grounds, including the problematic link between dilution factors and community annoyance, the difficulty of obtaining ambient-air samples, and doubts concerning the reliability of observers or odor panels (see Chapter 4) that must make the necessary perceptual judgments.

Of course, dilution readings need not be the sole basis for determining that odors have reached nuisance levels. Nevada, for example, uses the Scentometer to find a violation only after 30% of a sample population of at least 20 persons have complained of objectionable odors. If fewer than 20 persons were exposed to the odors, 75% of the affected group must agree that the odor is unacceptable in order to trigger a Scentometer test. Nevada thus combines the Scentometer approach with the odor-panel approach adopted by several other jurisdictions. Iowa defines an odor as objectionable when 30% of a random sample of at least 30 persons (75% if fewer than 30 are affected) so testify, and South Dakota convenes a special five-member panel, including members from the Department of Health, to judge whether an odor eliciting at least five complaints is unreasonably injurious to human health and welfare or to animal and plant life.

Although each of the regulatory approaches described above incorporates innovative thinking on the subject of odor control, each displays fairly clear deficiencies and can be criticized as arbitrary, unfair, or unreasonably expensive. Moreover, state or local standards are sometimes set by nontechnical personnel and without a sufficient basis in experience to warrant the standards, and the reliability of the Scentometer as an enforcement tool may be questioned.

Dissatisfaction with existing odor-control strategies has led to the development of at least two innovative model ordinances that deserve analysis here. These are founded on the important insight that the acceptability of an odor depends on a wide variety of factors, all of which should be weighed and considered by the authorized regulatory agency.

The ordinance proposed by M. W. First[7] takes account of the following factors: odor intensity, odor quality, odor duration and frequency, time of day and week, and wind direction. These factors are proportionally weighted and combined according to a suggested formula to yield an overall measure (called the "odor perception index") for the magnitude of the annoyance. The weighting of each factor involves a relative ranking of all the possible variables and assignment of a numerical value to each. For example, odor quality is determined from a schedule listing commonly experienced odors (from acrylic plastics, foods, fishmeal-processing, etc.) with assigned values for each ("pleasantness quality factors") ranging from 2 for the least pleasant to 1 for the least objectionable. On the basis of such weightings, an odor with an odor perception index at or above some magnitude of annoyance would be characterized as a nuisance requiring abatement.

This approach raises a number of technical and legal issues that cannot be lightly dismissed. Further empirical research is necessary on how to translate the variables of olfactory experience into numerical values and weights and on whether weightings of particular variables or marginal effects of adding odor units may vary under different circumstances, instead of remaining constant.

The initial choice of factors, the numerical values assigned within each set of variables, and even the combining formula would all be open to challenge as arbitrary and capricious, unless the regulatory agency offered generous opportunities for public participation before adopting them in the form of a regulation. A series of public hearings would have to be held, possibly at considerable expense, in order to establish enforceable proportional weightings. A procedure should also be provided for revising the numerical schedules from time to time to ensure that they continue to reflect changing community attitudes. Finally, the ordinance should respect the due-process rights of alleged offenders by allowing them to dispute a finding that a nuisance exists. For example, an odor perception index above the normally acceptable maximum might function only to establish a rebuttable presumption that a violation has occurred, instead of serving as conclusive proof of a violation. Conversely, in the face of numerous complaints of malodor, the administering agency should be allowed an opportunity to show that a violation has occurred even though the prescribed index may not have been exceeded.

The ordinance proposed by Copley International Corporation[5] (CIC) would rely essentially on public-attitude surveys to establish both the existence of a community odor problem and the success of efforts to cure it. Complaints from residents, verified by sensory measuring techniques, would serve to trigger such surveys, but would not be used to define any ambient-air odor standard. Sophisticated survey techniques would be used, with both an experimental group (odor panel) consisting of randomly selected residents from the affected community and a control group drawn from another nearby odor-free community of similar socioeconomic characteristics. In this way, it could be determined whether a significantly greater proportion of residents in the affected community than of residents in the control community say that they are bothered by odors. If so, then a community odor problem has truly been identified, inasmuch as odor is the controlled variable. Complaints and sensory measurements would then be used again to pinpoint the odorant sources and to prescribe abatement measures. Further surveys might be necessary, after those measures have been undertaken, to test their sufficiency.

This approach can be commended for recognizing that the most important—and perhaps even the only—measurable impact associated with a vast majority of odor problems is community annoyance; that measurement of annoyance is central to the success of a regulatory strategy for odor control; and that public-attitude surveys, when carefully conducted, are more reliable measures than the use of only sensory standards or only citizen complaints. However, controlled public-attitude surveys are costly, time-consuming, and unwieldy as regulatory devices. Nor do they serve to inform polluters of what they must do to eliminate the problem. Survey data might well be challenged as hearsay in a court of law. The ordinance proposed by CIC would therefore have to be supplemented by carefully defined procedures and techniques for translating survey results into a workable program of regulatory controls.

The CIC model has one further advantage that should be mentioned here. Repeated surveys will respond to shifts in community attitudes over time, especially as abatement of some odorant sources brings others to the forefront of attention. In contrast, ambient-air standards for odor express fixed levels of acceptability, which can be changed only through the formalities of the rule-making process. But ambient-air and emission standards still need to be explored in some detail, especially in the context of the Clean Air Act.

FEDERAL REGULATION UNDER THE CLEAN AIR ACT

AMBIENT-AIR QUALITY STANDARDS

Section 108(a)(1) of the Clean Air Act sets forth three conditions for listing pollutants that will be made subject to federal ambient-air standards. First, a pollutant shall be listed by the Administrator if, in his judgment, its emissions "cause or contribute to air pollution which may reasonably be anticipated to endanger public health or welfare. . . . " Second, the presence of the pollutant in the ambient air must result "from numerous or diverse mobile or stationary sources." Finally, even if the Administrator determines that the substantive requirements of the two foregoing conditions are met, it remains within his discretion whether to plan to issue air-quality criteria for the particular pollutant. As summarized in other parts of this report, some odors do have adverse effects on "public health or welfare."

Air-quality criteria must be issued for any pollutant within 12 months of its inclusion on the Administrator's list. These criteria shall reflect "the latest scientific knowledge useful in indicating the kind and extent of all identifiable effects on public health or welfare which may be expected from the presence of such pollutant in the ambient air, in varying quantities" and under variable circumstances [S 108(a)(2)]. The mode of quantification is not specified; some form of sensory measurement could presumably be used for odors. The harder question (discussed further below) is whether, in the current state of scientific knowledge, particular points on any odor-measurement scale can be correlated with identifiable effects on public health or welfare. This difficulty should not, however, be overemphasized. Reasonable approximations may be sufficient for the practical purpose of standard-setting for odors.

Under S 109(a)(2), issuance of criteria must be accompanied by publication of proposed "national primary and secondary ambient air quality standards for any such pollutant." On the basis of the overall intention of the Clean Air Act and its legislative history, EPA has interpreted this provision as requiring these standards to be uniform nationwide. Primary standards must be established at levels that "in the judgment of the Administrator, based on such criteria and allowing an adequate margin of safety, are requisite to protect the public health" [S 109(b)(1)]. Secondary standards are to be set at levels that are "requisite to protect the public welfare from any known or anticipated adverse effects associated with the presence of such air pollutant in the ambient air" [S 109(b)(2)]. Thus, the Act contemplates establishment of federal standards for protecting everyone's health and welfare. At the same time, states are free to adopt stricter ambient standards if they so choose. In this manner, regional variations might well appear, but not by allowing relaxation or selective application of any federally established standard.

Section 302(h) of the Act defines "effects on welfare" as including effects "on personal comfort and well-being." Thus, it appears that not every form of discomfort or degree of annoyance caused by an odor must be classified as a detriment to health. If this is the case, the Act takes a narrower view of "health" than the definition of this term by the World Health Organization as "a state of complete physical, mental and social well-being and not merely the absence of disease and infirmity." It would not be necessary, however, to designate different levels at which health and welfare impairments are experienced from exposure to an odorant. Primary and secondary standards are in fact identical for a number of currently regulated pollutants, and it may also be that, at whatever level an odorant is deemed detrimental, it injures both health and welfare.

The Act makes the states primarily responsible for achieving federal ambient standards, through execution of state implementation plans (SIP's). A primary odor standard would have to be attained "as expeditiously as practicable"—within 3 years, but with a possible extension to 5 years, from the date of approval of such a plan [S 110(a)(2)(A) and S 110(e)]. Secondary standards must be attained within "a reasonable time" [S 110(a)(2)(A)].

With the foregoing provisions of the Act in mind, we may now ask whether ambient-air standards for odor should be established on a national scale. To what extent is federal intervention justified at all in a field where nuisance law and local or state ordinances have been the traditional modes of regulation?

It will be difficult enough for states to revise and execute SIP's for controlling the more dangerous pollutants, without also taking on the burden of regulating odors through the SIP process. What is needed here, but apparently still lacking, is a reliable assessment of how well states and localities are already performing in the field of odor control, or are likely hereafter to perform, in the absence of any mandatory federal guidance. As already noted in this chapter, the nonfederal approaches have their shortcomings. But these may be largely overcome as states and municipalities, spurred on by a rising tide of citizen complaints, adopt modern odor-control ordinances that are tailored to their own perceptions of the problems they face.

Uniform national standards, moreover, may leave too little room for variable community preferences. Reactions to odor, like attitudes toward pornography, depend closely on local values and local aesthetic judgments. Odors that can pass as acceptable in industrialized or urban surroundings may seem intolerable in rural or recreational settings. A small community depending on a paper mill or rendering plant for its livelihood may be willing to tolerate substantially higher levels of odor than a larger, mainly residential community that supports very little industry. It can be argued that an odor regulation program should be flexible enough to indulge these community sentiments if it is possible to do so without endangering public health and welfare.

However, federal standards would protect people who are not free to move away from malodorous environments, as well as interstate travelers in states and localities that fail to adopt reasonable odor controls. In the absence of federal standards, states may fail to establish their own, because they are competing with one another for industry at the expense of environmental quality. (The argument has been advanced that uniform federal standards are needed to discourage polluters from shopping for locations where they will be largely free of environmental controls, and thus to avoid penalizing localities that take the initiative in establishing such controls.[13]) Federal standards would also simplify the task of resolving interjurisdictional disputes when odorants travel across local and state boundaries and are especially appropriate for regulating mobile sources of odor.

Above all, federal involvement in odor regulation appears justified if the health of the public is at stake. In the last decade, the federal government has assumed a greatly expanded role in protecting public health from a variety of environmental hazards, recognizing that state and local governments lack the expertise or the will to perform this function effectively. However, as indicated in other chapters of this report, whether and when an odor problem constitutes a public-health problem is a difficult question of judgment. Data on human physiologic responses to different levels of odor are incomplete and often inconclusive. No test has yet been widely adopted to assess the health effect of exposure to odor, although it has been suggested that such a test could be based, in part, on the rapidity and completeness with which physiologic reactions to odor can be reversed. It is unclear whether psychosomatic or other indirect adverse reactions to odor, such as loss of appetite or of sleep, should be treated as health effects for purposes of regulation, along with direct physical responses. Further research on the effects of odors would help to clarify these issues. Judgment in deciding whether federal intervention is justified on public-health grounds cannot be made merely on the basis of how people typically react to odors. Rather, "included among those persons whose health should be protected by the ambient-air standard are particularly sensitive citizens. . . . In establishing an ambient-air standard necessary to protect the health of these persons, reference should be made to a representative sample of persons comprising the sensitive group."[20]

Apart from the central question of federal policy, there are also technical and perceptual difficulties that need to be resolved on the way to adopting ambient-air odor standards at *any* level of government—federal, state, or local.

The first difficulty is how to classify odorants for purposes of standard-setting. Odorants and their odors vary widely in character and in acceptability at different intensities, frequencies, and durations. Perceptions of odor will deviate in character, not only in intensity, from any common reference standard with which different odors may be compared. It follows that there cannot be a single standard for all offensive odors. But how many different odorants or classes of odorant should there be for purposes of differential standard-setting, and how shall they be identified? Too many or too few will result in an unworkable regulatory classification. Meeting this challenge will not be simple.

How will ambient-air standards for odors be formulated? For other pollutants, the prevailing form is a maximal concentration of the polluting substance expressed in micrograms per cubic meter, but this approach is feasible for only a few known odorants. Many odors may be unacceptable at odorant concentrations too small to be measured by instruments; at present, concentrations of odorants cannot readily be used to define thresholds of acceptability.

What is done instead is to attempt the formulation of ambient-air odor standards in sensory terms that reflect the perceptual basis of the problem. At some risk of oversimplification, it can be said that there are two major competing techniques for performing this task: dilution of an odorous sample to the threshold of perception and direct comparison of an odorous sample with a reference standard that represents the maximal acceptable intensity (see Moskowitz *et al.*[12]). A standard of the first sort might provide for a maximal dilution ratio of 7:1, which means that a violation exists if it takes more than seven equal volumes of clean air to dilute an ambient sample down to the point where the odor is no longer perceived. A standard of the second sort might provide that a violation exists if the intensity of an ambient-air sample is perceived as being stronger than a sample of butanol at some specified concentration, such as 250 ppm. In both cases, reliance is ultimately placed on the human nose, or on the collective perceptions of a panel of noses, both to define and to adjudge compliance with whatever standards are posited.

The complexities and refinements of these techniques are described in other chapters of this report (such as Chapter 4). Suffice it to say here that they both rest on a critical presupposition: the ability to define thresholds of annoyance in terms of measurable odor levels or standardized odor intensities. It is fair to say that this ability has not yet been practically and convincingly demonstrated. Inquiry into dose-response relationships, which would serve to correlate community annoyance with sensory measurements, is at the top of the agenda for further research.

These relationships and the means of testing them are complicated by a number of variables. Reactions to malodors frequently involve varying judgments by individuals as to what is or is not tolerable. The volume of complaints is not necessarily a reliable guide. Nor is acceptability of an odor solely a function of its intensity. Duration, frequency, and time of occurrence are also important factors, which the standard-setter must take into account. Attempts to define acceptability in terms of dilution to threshold must reckon with the problem of drawing inferences from dilution ratios to community acceptance; with highly variable ratios, even for odors of equal intensity; and with the difficulty of determining thresholds of perception in the presence of possible odor fatigue or extraneous distorting influences. Comparison with reference standards would appear to be a simpler approach, but it depends on empirical consensus as to what the standard should be, on having a sufficient variety of reference samples at hand to permit meaningful comparisons, and on assurances of uniformity from one sample to the next. Both approaches call for observers whose perceptual judgment, trained or untrained, will be accepted as a basis for enforcement, and this may be a tricky or a costly requirement to meet. How to obtain and preserve representative samples of ambient air for sensory measurement is also recognized as a basic problem (see Chapter 4 and Prokop[16]).

The choice of ambient-air standards will also involve some conclusions as to how the balance should be struck between benefits and costs. In applying nuisance law, courts balance the equities as between the parties and decide how much of the loss, if any, will be shifted from defendant to plaintiff. But national standards for odors could result in raising the prices of goods and services, so ideally standards ought to be set in the light of what the citizenry at large would be willing to pay for abatement or would have to be paid to forego abatement. The former of these tests would be appropriate if one starts with the assumption that odorant sources have the privilege of discharging odorants to the environment; the latter, if it is initially assumed that citizens have the right to breathe air not contaminated by objectionable odors. This distinction is critical in striking the balance between costs and benefits in pollution control. (See Chapter 10.)

However, the Clean Air Act requires that primary ambient-air quality standards be set solely on the basis of public health and apparently does not allow a tradeoff for costs. Thus, to the extent that it is concluded that odors do produce adverse health effects (as opposed to welfare effects), the Act might preclude this balancing approach.

Because of space limitations, this chapter can deal no further with the multiplicity of issues involved in ambient-air standard-setting for odors. It should be clear enough from the foregoing discussion that problems of political philosophy, methods (especially of measurement), resources, and social judgment remain to be resolved on the way to deciding whether ambient-air standards should be adopted for odors under the Clean Air Act.

PERFORMANCE STANDARDS FOR NEW AND EXISTING SOURCES

Even if ambient-air standards for odor are not adopted, new malodorous sources can be controlled by emission or operating standards under the Clean Air Act. The possibility remains of establishing federal new-source performance standards for any category of odorant sources under S 111 of the Act, on a finding by EPA that the category "causes, or contributes significantly to air pollution which may reasonably be anticipated to endanger public health or welfare" [S 111(b)(1)(A)]. The standard applicable to a new odorant source (i.e., to a source hereafter constructed or modified) could be a set of emission limitations that reflect the degree of odorant control "achievable through application of the best technological system of continuous emission reduction which . . . the Administrator determines has been adequately demonstrated," after taking account of cost, energy requirements, and other impacts. Alternatively, if it is not feasible to prescribe or enforce emission limitations, EPA may instead promulgate "a design, equipment, work practice, or operational standard" for one or more categories of odorant sources. It seems reasonable to require that new sources, wherever they may be, include in their plans the use of advanced technology for controlling malodors associated with their operations.

If a new-source standard is established for an odorant that is not also covered by an ambient-air quality standard, corresponding standards of performance must also be developed for similar *existing* sources [S 111(d)(1)]. EPA would furnish general guidance to the states on what degrees of control can be achieved, and the states would be responsible for applying that guidance to particular sources after taking into account their remaining useful life and other practical considerations. The advantage of this approach is that it could impose direct requirements on emitters to install new-source and retrofit technologies for odorant control, without having first to go through the problematic process of setting primary and secondary ambient-air standards for odors.

To date, EPA has promulgated new-source performance (NSP) standards for only two categories with a view to controlling their malodors: refinery fuel-gas systems and total reduced sulfur (TRS) from kraft pulp mills. Significantly, however, these standards are couched in objective, nonsensory terms. EPA identified a best available technology, estimated how much emission reduction it could achieve, and wrote the NSP standards accordingly, in the hope that compliance with them would result in reducing the associated malodors to an acceptable level. This is a familiar method, but there is some doubt as to whether it attains the objective of odor control. National ambient-air standards for odors have not been promulgated. Part of the reason is that odor-measurement technologies have been inadequately tested and poorly validated against community responses.

In fact, it is no easy matter to predict ambient odor intensities on the basis of either sensory or objective measurements of stack emission (see Chapter 5 and Prokop[16]). Correlations of source emission with ambient-air quality, constructed with the aid of dispersion models, are a necessity for regulating any air pollutant under the Clean Air Act; but the task is complicated for odors, because of the need to rely on subjective sensory measurements for both standard-setting and compliance-monitoring. Moreover, fugitive sources are hard to identify and assess; multiple sources of odors may combine to produce unacceptable effects; and changes in the quality of an odorant may occur from stack to dosage point or from sampling to exposure time.

More broadly, it should be possible to correlate source emission levels (concentrations or dilution ratios) with community acceptability (as measured by empirically developed ambient-air standards, volume of complaints, or public-attitude surveys). But this is another topic on which work remains to be done. The interests at stake here include but go beyond whether odors shall be regulated in any sense under the Clean Air Act. The ability to interrelate the source, the dose, and the acceptability of an odor and to weave these variables into a workable regulatory scheme is critical to advancement in general beyond the traditional framework and limitations of nuisance law.

CONTROL THROUGH ECONOMIC INCENTIVES

This chapter has stressed direct regulatory approaches to odor control: the law tells polluters what degrees of control they are legally obliged to achieve in order to meet legally defined objectives. This is by far the most common mode of regulation adopted by governments in the field of environmental quality.

It is frequently observed, however, that economic incentives, such as emission charges or marketable emission permits, used in lieu of or in conjunction with direct regulatory controls, might result in more efficient solutions. There are several good reasons for considering the use of such incentives:

1. Emission charges set at levels equal to known pollution damages will, in competitive industries, tend to bring marginal costs of pollution control into equality with marginal benefits of control.[2]
2. Even if an ambient standard exists, a system of economic incentives will allocate pollutant units among emitters in a way that minimizes the total control costs of meeting the standard. Regardless of the competitive structure of the odor-causing industry, a system of pollution charges can achieve any given reduction in emission at minimal cost to society. Polluters who can reduce emission most cheaply will go furthest toward doing so, because they are the ones to whom the largest savings will accrue if they avoid the charge.
3. Because marketable emission permits can be voluntarily traded among emitters, among sufferers, between emitters and sufferers, and across space and time, they allow emission patterns to adjust quickly to changing circumstances. The regulatory agency need not intervene except insofar as it wishes to influence the market.

The foregoing advantages must, however, be weighed against the following disadvantages and relative unknowns:

1. As the discussion of standard-setting in this chapter suggests, it is exceedingly difficult to calculate damage (or benefit) functions for various odor types, intensities, and exposures.
2. Odorant emission is often difficult to monitor and measure. In many cases, direct regulation requiring installation of specified control devices or processes may be the only effective strategy.
3. Substantial portions of the U.S. economy are not highly competitive. For example, firms in many industries possess substantial monopolistic advantages, public agencies operate malodor-producing facilities, and the net revenues of public utilities are constrained by law. The behavior of polluting firms in these imperfectly competitive sectors when they are faced with economic incentives for pollution control is, as yet, not completely known.
4. Revision of effluent charges from time to time, especially in the face of inflation or economic growth, may be politically difficult to achieve.
5. The ranges of response of polluting firms to variations in effluent charges is not well understood. Similarly, the behavior of permit prices

with respect to variations in the number of permits in any given set of circumstances will often be hard to predict.

6. The costs of administering a market in discharge permits could be very substantial.

Even for pollutants that are better known and more easily measured than odors, the practicality of proposals to institute systems of pollution charges or marketable discharge permits in this country has never been tested in an actual pollution setting. It is most unlikely that odorants will be subjected to control by economic incentives before these have been tested successfully in more familiar contexts.

CONCLUSION

The legal treatment of odor problems has come a long way from judicial applications of time-honored nuisance doctrine to laws increasingly based on quantitative measurement. There is no getting beyond the fact, however, that odors are perceptions whose intensity, impact, and significance are difficult to assess. If we are to get beyond the subjectivities and vagaries of classical nuisance law, it will be necessary to determine acceptability of odors through quantitative measures or qualitative comparisons that can be repeated, without loss of validity or reliability, at different times and in different places. These capabilities are essential to establishing a sound technical basis on which to set ambient or emission standards for odors under federal, state, or local law.

REFERENCES

1. *Atlanta Processing* v. *Brown*, 227 Ga. 203, 179 S.E. 2d 752, 3 ERC 1319 (1971).

2. Baumol, W. J., and W. E. Oates. The Theory of Environmental Policy. Englewood Cliffs, N.J.: Prentice-Hall, Inc., 1975. 272 pp.

3. *Bohan* v. *Port Jervis Gas-Light Co.*, 122 N.Y. 18, 25 N.E. 246 (1890).

4. *City of Miami* v. *City of Coral Gables*, 240 So. 2d 499, 1 ERC 1184 (Fla. Dist. Ct. App. 1970).

5. Copley International Corporation. A Study of the Social and Economic Impact of Odors. Phase III. Development and Evaluation of a Model Odor Control Ordinance. Report to the U.S. Environmental Protection Agency. EPA Publ. No. 650/5-73-001. Contract No. 68-02-0095. Washington, D.C.: U.S. EPA, February 1973. 203 pp.

6. *Cox* v. *Schlachter*, 147 Ind. App. 530, 262 N.E. 2d 550, 1 ERC 1681 (1970).

7. First, M. W. A model odor control ordinance, pp. 1255–1259. In H. M. Englund and W. T. Berry, Eds. Second International Clean Air Congress. Proceedings. New York: Academic Press, 1971.

8. *Fort Smith* v. *Western Hide and Fur Co.*, 153 Ark. 99, 239 S.W. 724 (1922).

9. *Johnson* v. *Drysdale*, 66 S.D. 436, 285 N.W. 301 (1939).

10. Leonardos, G. A critical review of regulations for the control of odors. J. Air Poll. Cont. Assoc. 24:456–468, 1974.

11. *McClung* v. *Louisville and N.R. Co.*, 255 Ala. 302, 51 So. 2d 371 (1951).

12. Moskowitz, H. R., A. Dravnieks, W. S. Cain, and A. Turk. Standardized procedure for expressing odor intensity. Chem. Sens. Flav. 1:235–237, 1974.

13. National Research Council, Committee on Appraisal of Societal Consequences of Transportation Noise Abatement. Noise Abatement: Policy Alternatives for Transportation, p. 22. Report to the U.S. Environmental Protection Agency. Washington, D.C.: National Academy of Sciences, 1977.

14. 9 Co. Rep. 57b, 77 Eng. Rep. 816 (1611).

15. *Ozark Poultry* v. *Garman*, 251 Ark. 389, 472 S.W. 2d 714, 3 ERC 1545 (1971).

16. Prokop, W. H. Developing odor control regulations: Guidelines and considerations. J. Air Poll. Cont. Assoc. 28:9–22, 1978.

17. Status of regulations for source emission and ambient odors. Ann. N.Y. Acad. Sci. 237:288–308, 1974.

18. Texas Air Control Board, Rule 5.

19. 26 N.Y. 2d 219, 257 N.E. 2d 870, 309 N.Y.S. 2d 312 (1970).

20. U.S. Congress. Senate. Public Works Committee. Senate Report No. 91-1196, p. 10. National Air Quality Standards Act of 1970. Sept. 17, 1970.

12 Summary of Responses of Air-Pollution Control Agencies to Committee Questionnaire on Odors

The Committee sent a questionnaire to relevant state and selected local agencies to obtain whatever information was available from them on odor complaints, sensory data on odors, and related court cases. This action also alerted each state that the Environmental Protection Agency had asked the National Academy of Sciences to conduct a study on odors. The questionnaire afforded all the states the opportunity to submit information on the selected subjects or any others they cared to cover. In the letter to the agencies, they were asked to respond to three questions:

1. Does the agency have summaries of complaints about odors that identify the odor sources and/or the number of people affected?
2. Does the agency have sensory data and/or measurements that were used to determine the source, intensity, and quality of the odors? Who does the measurements?
3. Does the agency have summaries of court cases regarding public exposure to odorous substances?

RESPONSES TO THE QUESTIONNAIRE

A list of all agencies contacted and an overall summary of the results are presented in Table 12-1. Detailed sheets prepared from the responses of some agencies appear in Appendix C. These data sheets were selected to provide an overview of the types of odor problems that state and local agencies had encountered.

TABLE 12-1 Abridged Responses to Questionnaire[a]

Jurisdiction	Complaints and Odor Sources	Sensory Measurement	Who Measures	Court Cases	Remarks
Alabama	NR				
Alaska	See Table C-1; 1977, 12 for sulfite pulp mill	No equipment	—	No; one compliance order: fish processing	Odors are generally not a major control problem; state regulation prohibits any injurious emission; statute covers odor nuisance
Arizona: Pima County	See Table C-2; 1977, 83; 1976, 113; chemical processing, cattle feedlot, tallow plant, restaurant, sewage treatment, paint	—	Field inspectors (sensory judgment)	Hearing board order, plea agreement, fines; list provided, including action	—
Arkansas	—	—	—	—	—
California: Air Resources Board (CARB)	See Table C-3	See remarks	—	Nuisance violation and abatement orders	State regulation covers odors under general nuisance rule; each district has primary authority; some have specific regulations for controlling odors
California: Bay Area	See Table C-4 (see remarks)	—	—	—	Summary is not available; computer printout submitted showed numerous complaints for 1976-1978
California: San Diego	See Table C-5; rendering, 377; chicken-manure-processing, 380; fish-processing, 19	Dynamic olfactometer and syringe method	Consultant	Variance petitions	—

California: Santa Barbara	NR				
California: South Coast Air Quality Management District	See Table C-6; 1977, power plant, 9; refineries and chemical plants, 259; rendering, 1; industrial, 262; vehicles, 3; sanitation, 19; miscellaneous and unknown, 890 (total, 1,443)	Modified ASTM panel	Trained agency personnel	Yes; summaries not available	Odor rule based on public nuisance requires many complaints
Colorado	Summaries not available	Scentometer	Trained agency personnel	Yes; summaries not available	Regulation issued specifically for odorous air contaminants
Connecticut	1977, 313, sources not available	Person (subjective)	Agency personnel	Injunction against chemical company, 1972	State regulation for odor control forbids odor beyond property line
Delaware	NR				
District of Columbia	NR				
Florida	—	—	—	—	—
Florida: Dade County	See Table C-7; 1977, 380 (17 categories: volatile solvents, oil-burner fumes, wastewater, landfills, garbage, industrial chemicals, asphalt, diesel exhaust, etc.); 1976, 550	—	—	Fiberglass molding, fabric manufacture, coffee-roasting, rendering, galvanizing	—

TABLE 12-1 (*Continued*)

Jurisdiction	Complaints and Odor Sources	Sensory Measurement	Who Measures	Court Cases	Remarks
Georgia	NR				
Hawaii	—	—	—	—	—
Idaho	—	—	—	—	—
Illinois	See Table C-8; see local agencies for summaries; see remarks	Scentometer	2 trained inspectors; see remarks	Yes; not catalogued; see remarks	Citizen complaint procedure used; all complaints receive a response; odors are handled as non-health-related; use of Scentometer measurement procedure is specific for different sources; concurrent determination by 3 inspectors needs 2 positive determinations within 15 minutes to confirm that nuisance exists; odor violations can be prosecuted under general air-pollution regulation (Section 9a)
Illinois: Bedford Park	—	—	—	—	—
Illinois: Chicago	See Table C-9	Person (subjective); see remarks	Enforcement inspector	Yes; summaries not available	Routinely, 16 locations are checked 5 days each week; monthly summaries are prepared by engineer

Illinois: Will County	See Table C-10; 43 in 2½ years; see remarks	—	—	None; see remarks	Summaries of complaints are not available; sources include sewage, perfume, and such chemicals as ammonia, hydrogen sulfide, and sulfur; no court cases have been filed, owing to lack of measurement equipment
Indiana	25% of complaints are about odors; 3 major sources are commercial solvents, rendering, slaughterhouse	None	—	3 case histories, 1975	Study to determine need for odor regulations in state is planned
Iowa	NR				
Kansas	—	—	—	—	—
Kentucky	Summaries not available	Scentometer; summaries of measurements not available	Inspector	None; one compliance order pending	Agency deals with odors on case-by-case basis initiated by citizen complaints
Louisiana	—	—	—	—	—
Maine	See Table C-11; 1977, 63; 1976, 32; pulp and paper manufacture, chicken, fish, leather, rendering, and surface coating; summaries for 5 years available	—	—	Rendering, criminal neglect, and protein-meal-processing	There are no statutes for odor control; odor has historically been viewed as nuisance; a standard is lacking

TABLE 12-1 (*Continued*)

Jurisdiction	Complaints and Odor Sources	Sensory Measurement	Who Measures	Court Cases	Remarks
Maryland	See Table C-12; one company, 685; see summary list for 1961-1976 in table	Odor surveys	—	One; mainly chemical hazard, which began as an odor problem	—
Maryland: Towson, Baltimore Co.	See Table C-13	Person (subjective)	Field inspector	None; one hearing resulted in compliance with fine	—
Massachusetts	Yes; summaries not available	—	—	—	—
Massachusetts: Boston	Restaurants; car, truck, and bus exhaust; garbage; sewers; paint; most common type is open burning and boiler smoke	See remarks	—	Yes	Odor problems are brought to attention by citizen complaints; once source of emission is identified, control usually can be easily effected
Michigan: Wayne County	See Table C-14; rendering, foundry, cooking, paint	Modified ASTM method	Agency laboratory staff	One civil case went to state supreme court	Agency's odor guidelines stipulate emission standards of 150 odor units/ft^3 and emission rate of 1,000,000 odor units/min in stack
Minnesota	1972, 492 complaints; 35%	Modified ASTM method; odor	Certified laboratories	2 cases (roofing and paint odors); 1	2 state regulations limit emission of odor; one is specific for rendering

	due to odors; 23 sources listed, including coffee, food, fish, rendering, oil-refining, chemicals, poultry, asphalt, wood-impregnating, paint, sewage, and pulp	panel	and registered professionals	insecticide case settled out of court	plants
Minnesota: Minneapolis	—	—	—	—	—
Mississippi	—	—	—	—	—
Missouri	Summaries not available	Scentometer	—	1 court case and several commission hearings	—
Missouri: Springfield	—	—	—	—	—
Missouri: St. Louis	Summaries not available; sources listed are rendering, coil coating, charbroilers, and canning plants	ASTM method	Agency staff	—	Legal odor restrictions are strictly enforced
Montana	10-20 per year; majority are about livestock wastes; others about oil refineries and pulp mill	No equipment	—	None	Pulp mill is most noticeable

TABLE 12-1 (*Continued*)

Jurisdiction	Complaints and Odor Sources	Sensory Measurement	Who Measures	Court Cases	Remarks
Nebraska	See Table C-15	—	—	—	—
Nevada	About 3 per year	None	—	Currently not actively pursuing enforcement	There is a state odor regulation, but it lacks reliable method of measuring odor for enforcement purposes
New Hampshire	See Table C-16; tannery, underground fire, fiberglass manufacture, fuel oil and gasoline	Person (subjective)	Agency staff	2 referred to attorney general	For regulatory purposes, odor is considered public nuisance
New Jersey	2,500 complaints per year; 80% due to odors; summaries not available	Person (subjective); no instrument or panels	Trained inspectors	In 10 years, about 600 department orders and 125 court orders issued on odors	Regulations prohibit odors beyond property line; majority of cases are handled under administrative code prohibiting air pollution

New Mexico	See Table C-17	No measurements; odor complaints identified problem	—	No health-hazard legal action required to date	Air-quality regulations are not specific for odors, but odors are handled as hazardous pollutants or as public nuisance
New Mexico: Albuquerque	—	—	—	—	—
New York: Region 6	Dairy-chicken manure spreading operation	Complaints identify odor problem	—	None; voluntary abatement	Complaints usually are of short duration during manure spreading operations
New York: Region 7	Several during last 3 years; summary not available	—	—	2 significant cases: foul chemical odors and brewers grain drying; others not listed	—
North Carolina	See Table C-18	—	—	No court cases, civil or criminal	—
North Dakota	See Table C-19	Scentometer	Agency	None to date	—
Ohio	See Table C-20; 132 complaints	Information not available	—	Not available	Complete picture about odor complaints must be sought from regional offices

TABLE 12-1 (*Continued*)

Jurisdiction	Complaints and Odor Sources	Sensory Measurement	Who Measures	Court Cases	Remarks
Ohio: Cincinnati	June 1976-January 1978, about 300 complaints from 132 sources; summaries not available	Scentometer	Agency inspectors	A few resulting in remedial action	Scentometer is used only to document evidence for potential legal action in nuisance-complaint situations
Ohio: Cleveland	Summaries not available	Scentometer and odor panel	Agency inspectors	None in recent years; see remarks	Ambient odors are difficult to document
Oklahoma	—	—	—	—	—
Oregon: Portland	Summaries not available	Scentometer; see remarks	One staff observer	None; public hearings and administrative actions used for compliance	Regional standard stipulates how Scentometer is to be used
Oregon: Willamette Valley Region, Salem	See Table C-21	Scentometer	Staff	Odor standard not widely used to promote compliance	Willamette Valley Region has an ambient-odor standard that stipulates use of Scentometer to determine emission of odorous matter
Pennsylvania	Summaries not available	Person (subjective) and community survey	Department inspector	1969-1970, one involved solvent surface coating operation—successful action	Community survey is conducted when inspector cannot establish existence of a malodor problem and complaints persist; department regulation stipulates malodor as any odor that department finds objectionable to public

Pennsylvania: Philadelphia	See Table C-22	None; see remarks	—	1978, 16 cases; 1972-1977, 25 cases	Specific odor regulations were found unworkable; general-nuisance provision of Air Management Code is used now; Board determines emission standards
Puerto Rico	—	—	—	—	—
Rhode Island	See Table C-23	Persons (subjective)	Neighbors and agency personnel	2 successful	—
South Carolina	See Table C-24	Person (subjective)	Field inspector	See remarks	State has no regulations specifically for odors or odorous emission and has had little success in dealing with odors
South Dakota	—	—	—	—	—
Tennessee: Nashville	—	—	—	—	—
Tennessee: Chattanooga	See Table C-25	—	—	—	Complaints about odors are as numerous as or more numerous than complaints about any other air-pollution problem
Texas	See Table C-26 for 1976 4-month summary of sources; 1976, 842 odor, 2,038 total; 1977, 966 odor, 2,134 total	See remarks	—	1975, reported histories on 19 cases	General Rules of Texas Air Control Board regulate odor sources that cause a nuisance, relying heavily on complaints (most received via telephone); community is canvassed to trace odor; followup is by investigator to suspect pollutant source; public hearing is held when odor problem is not resolved voluntarily
Utah	—	—	—	—	—

TABLE 12-1 (*Continued*)

Jurisdiction	Complaints and Odor Sources	Sensory Measurement	Who Measures	Court Cases	Remarks
Vermont	See Table C-27	See remarks	—	None	Agency does not measure odor intensity or quality; complaints identify problem
Virginia	Major ones related to fishmeal and pulp mill	Person (subjective); see remarks	Agency staff	None; see remarks	Under Virginia regulations, public hearing is held to determine numbers of people who feel odor is objectionable; agency staff conducts thorough investigation
Washington	—	—	—	—	—
West Virginia	NR				
Wisconsin	See Table C-28; 53% due to odor	Odor survey; see remarks	Agency investigator	None	Regulations are specific for malodorous emission; department investigates frequency, intensity, etc., and premises of source; when 60% of random sample of persons feel odor is objectionable, corrective measures must be sought; public hearing may be held when order is contested
Wyoming	—	—	—	—	—

[a] Tables referred to here appear in Appendix C; NR = no response to questionnaire by time of preparation of this table; dash = response, but no data.

CONCLUSIONS AND GENERAL COMMENTS

COMPLAINTS

Among the agencies responding, there was no uniform method of reporting, summarizing, or responding to odor complaints. Most of the odor-complaint data in state files are not easily retrievable, because odor complaints are usually included with all other types of complaints related to the environment. Therefore, summaries of the number of complaints are not available from some agencies. In some instances, it was unclear how the agencies used the complaint data in their decision-making. Recorded complaints ranged in number from three per year in Nevada to about 2,500 in California's South Coast Air Quality Management District (SCAQMD) and the bay area. Of the total number of complaints received by the agencies that responded to the questionnaire, odor appears to have been responsible for 25–50%. The major sources of odor include agricultural operations and processing, animal- and fish-processing, diesel exhaust, incinerators, livestock- and poultry-raising (feedlots), manure-processing, petroleum-related processes, pulp and paper mills, and sewage treatment.

SENSORY DATA

Very little sensory-measurement information was available from state agencies. In a predominant number of the agencies, the sensory measurements were made by a single inspector whose sensory odor evaluation consisted merely of sniffing. Several agencies indicated the use of a Scentometer, and a few used the modified Mills/ASTM syringe method or more sophisticated instruments, such as the dynamic olfactometer, for the sensory measurement of odor pollution. Information obtained from the questionnaire that was based on sensory data that appear to have been well documented was collected by the California SCAQMD and by Wayne County, Michigan. Examples of the data are in Appendix C.

COURT CASES (ENFORCEMENT ACTION)

The majority of state and local agencies have used existing state regulatory authority in the courts or through compliance orders, hearings, or administrative actions to confront odor problems. Table 12-2 shows the jurisdictions that reported enforcement actions, including the use of the courts. Some state agencies reported effective procedures based on existing authority to deal with odor-nuisance problems—e.g., Texas; Dade County,

TABLE 12-2 Legal Approaches That Use Enforcement Actions to Confront Odor Problems[a]

Reporting Jurisdiction	Type of Approach Used
Alaska	Compliance orders
Arizona (Pima County)	Court cases
California (Air Resources Board)	Abatement orders
California (San Diego)	Variance petitions
California (South Coast Air Quality Management District)	Court cases
Colorado	Court cases
Connecticut	Injunctions
Florida (Dade County)	Court cases
Illinois	Pollution-control-board actions
Illinois (Chicago)	Court cases
Indiana	Court cases
Kentucky	Delayed compliance order
Maine	2 court cases (but not effective)
Maryland	Court cases
Maryland (Baltimore)	Administrative hearing
Massachusetts (Boston)	Administrative actions
Michigan (Wayne County)	Court cases tested in state supreme court
Minnesota	Court cases
Missouri	Court case and commission hearings
Missouri (St. Louis)	Strict enforcement
New Hampshire	Referred to attorney general
New Mexico (Albuquerque)	Court case
New York	Court cases
Ohio (Cincinnati)	Court case started; remedial action resulted
Oregon	Public hearings; administrative actions
Pennsylvania	Court case
Pennsylvania (Philadelphia)	16 court cases reported
Rhode Island	Court cases
Texas	Court cases; 19 case histories reported
Virginia	Public hearings

[a] Data from responses to Committee questionnaire. Table does not include all legal approaches and enforcement actions used, but only those reported.

Florida; and the California Air Resources Board. There is a lack of reliable information on the extent to which seemingly successful enforcement actions resulted in abatement of community odor problems.

GENERAL COMMENTS

In all the agencies that responded, the individual complaint is the mechanism whereby attention is first drawn to an odor problem. Some states have specific ordinances regarding odor pollution, but enforcement has been and still is largely through nuisance regulations. Some states do *not* have odor-pollution ordinances that provide any legal authority beyond the common nuisance law requiring evidence of personal damage.

The responses to the questionnaire indicated that there is no uniform mechanism among the agencies for evaluation of the magnitude of public nuisances caused by odor pollution. There is no uniform system of measurement of the odorant either at the source or in the ambient air. In most cases, the available sensory measurement data leave much to be desired.

Some factors were common to most of the agencies regarding problems in adequate followup and documentation of odor complaints, including available manpower, funds, and the degree of community "acceptance of," "objection to," or "habituation to" the odors that caused the complaints.

The responses to the questionnaire indicated that, in the absence of other regulations, all states have the alternative of using odor-nuisance laws to confront community odor problems. This approach has had various degrees of success.

Appendix A

Public-Health Aspects: Management of Environmental Odors

MELVIN W. FIRST

It has been pointed out in Chapter 2 and in parts of Chapter 3 that humans are capable of experiencing odor sensations associated with a vast array of organisms and inanimate substances, many encountered in everyday life, and a sizable fraction of them are recognizable at extraordinarily low concentrations. A catalog of possible sensory responses covers a broad spectrum, from extreme pleasure to intolerable disgust, but most odors evoke little or no response beyond recognition. Although an ability to stimulate the odor sense is a characteristic common to all manner of things, many odorous substances have additional properties that affect humans. These include:

- Odorous substances that have been well established as toxic to humans.
- Odorous substances that have produced well-defined pathologic changes in animals that have not been identified in humans.
- Odorous substances that have not been identified as toxic to humans, but that evoke violent and alarming physical symptoms in a substantial fraction of an exposed population whenever odor intensity is high and exposure more than fleeting.
- Odorous substances that have not been identified as toxic to humans, but that are capable of evoking violent and alarming physical symptoms in a small number of people even when exposure is moderate and fleeting.
- Odorous substances that have not been identified as toxic to humans (or are present at concentrations substantially below a well-established

toxic threshold), but that produce more than passing vexation by the continuing or frequent presence of their unpleasant odor.

- Odorous substances that have no known toxic properties and are universally recognized as pleasant or neutral, but that produce vexation in a substantial fraction of an exposed population because of unusual intensity or persistence.
- Odorous substances of no known toxicity that are sensed to the point of conscious recognition, but that evoke only pleasant or indifferent sensations.

These categories are summarized in Table A-1 and each is discussed in detail in following sections. Their importance stems from a need to define as precisely as possible concepts that are meaningful for the rational management of environmental odorants. Different categories and other definitions may ultimately prove more acceptable and more useful. These are proposed as a beginning. There are no clear separations between the categories, and it would be possible to insert additional descriptors with intermediate shadings between many. Nevertheless, these seven categories are believed to be sufficiently distinctive and comprehensible to serve as an adequate basis for discussion if one is willing to put aside the obvious difficulty that each, at the limit, tends to become indistinguishable from those that adjoin it.

The need for an unambiguous classification, identification, and definition of responses to odors and their effects on human populations is inherent in public-health concerns, because they are ultimately resolved through laws, regulations, and standards. There is a clear need to make them sufficiently unequivocal to avoid litigation and provide guidance to those who wish to come into and remain in compliance. Clinicians who diagnose disease and treat patients who suffer ill effects from odorous substances are also concerned with classification, identification, and definition for scientific purposes. This makes it desirable to attempt to deal with the effects of odorous substances on humans by means of a series of precisely defined categories, rather than with a continuum of effects that stretches from highly pleasing to violently repugnant and then merges into the clearly toxic, with the sole unifying factor being the ability of each substance to stimulate or act on the olfactory system in some fasion.

Therefore, the following sections will discuss the separate categories that have been identified and examine the public-health implications that flow from each. This discussion of human effects deals principally with substances that are sensed as odorous and touches only very briefly on odorous substances that affect the olfactory system in other ways. (The latter subject is reviewed in Chapter 3.) For each category that has a clear potential for creating harm, appropriate public-health responses are discussed. It is intended that this discussion provide a firm basis for reaching conclusions regarding which categories of human reponse to odorous substances merit regulation to protect the public health and, when a clear need has been identified, how best to accomplish this objective. In this process, we must recognize the unequivocal directives from Congress to protect the most sensitive segments of the population from harm and, also, to be mindful of cost-benefit considerations. As the concentration and frequency of community odors decrease, and as the fraction of the population that is affected adversely in any fashion becomes vanishingly small, the determination of when both Congressional directives are adequately fulfilled becomes increasingly judgmental. The following discussion is intended to inform that judgment because clear understanding of the diverse ways identical odorant exposures can affect people as individuals and, in the aggregate, as population groups, is an essential first step in the rational management of this most common of community air pollution problems.

ODOROUS SUBSTANCES THAT ARE TOXIC TO HUMANS

By definition, chemicals hazardous to human health are considered to be toxic; hence, their control is subject to existing laws or regulations, e.g., the Toxic Substances Act and the Clean Air Act. No new laws or regulations are needed to control substances recognized as toxic. Therefore, odor is not a decisive factor for toxic substances in the atmosphere; they are automatically subject to standard-setting under the Clean Air Act of 1970 and its amendments, and abatement to below a toxic threshold is mandatory.

Some toxic odorous substances, such as hydrogen sulfide, are such powerful odorants that they are detectable and disagreeable well below the toxic threshold. Therefore, hydrogen sulfide and other airborne substances that are malodorous below their toxic threshold must be considered in two distinct public-health categories: at higher concentrations, as toxic substances that are also odorous, and at lower concentrations, as odorous substances that are not toxic (this category is discussed later). In this categorization system, hydrogen sulfide may be contrasted with sulfur dioxide that has already been assigned an air-quality standard by EPA that is below its odor threshold. Therefore, whenever one identifies the odor of sulfur dioxide in the atmosphere over a period of an hour or more, it is clear that the concentration of this substance in air has exceeded the National Air Quality Standard. As such a degree of pollution by sulfur dioxide is already well covered by regulations designed to control its toxic

TABLE A-1 Classification of Substances Found in the Atmosphere, by Odor, Toxicity, and Recommended Action

Characterization of Odorants	Example	Recommended Action
1. Toxic to humans—nonodorous[a]	Carbon monoxide	Control by current toxic-substances regulations
2. Toxic and malodorous—toxic threshold less than odor threshold	Sulfur dioxide	Control by current toxic-substances regulations
3. Toxic and malodorous—toxic threshold greater than odor threshold	Hydrogen sulfide	Control concentrations above toxic threshold by current toxic-substances regulations; control malodorous concentrations below toxic threshold as public nuisances
4. Odorants that produce pathologic changes in laboratory animals that have not been demonstrated in humans at usually encountered concentrations	2-Nitro-2-butene	Additional study of exposed humans
5. Nontoxic malodorants universally considered obnoxious and known to produce undesirable physiologic symptoms in a substantial fraction of the exposed population—physiologic symptoms are transitory and permanently reversible when the stimulus is removed but emotional disturbance may linger	Garbage-fed pig farms; nonedible rendering; fishmeal; varnish	Control as community air pollutant

6. Nontoxic malodorants universally considered to be unpleasant and known to produce undesirable physiologic symptoms in a small number of the exposed population—physiologic symptoms are transitory and permanently reversible when the stimulus is removed but emotional disturbance may linger	Toilet soap factory	Control as public nuisance
7. Nontoxic odorants universally considered to be pleasant that cause vexation when present in unusually high concentration for long periods	Chocolate factory	Control as public nuisance
8. Nontoxic odorants universally considered pleasant that evoke pleasant or indifferent sensations	Baking bread	None

[a] Not discussed in the text but listed here for completeness.

effects, it is not necessary to regulate sulfur dioxide as a *malodorous* substance; i.e., it is adequately regulated as a toxic substance at concentrations at which it is detectable by smell.

PATHOLOGIC CHANGES IN ANIMALS FROM EXPOSURE TO ODOROUS SUBSTANCES

The evidence of pathologic changes caused by exposure to odorous substances comes primarily from experiments involving animals, as noted in Chapter 3. If part or all of the undesirable symptoms and tissue changes that have been noted in animals resulted from acute toxic or irritating properties of these chemicals, they should first be regulated as toxic substances that are odorous and dealt with accordingly, and, second, any residual health effects should be reexamined at sensory levels below those that produce acute toxic or irritating effects.

It is not possible to point to studies that have demonstrated morphologic changes in humans exposed to substances in the atmosphere that are odorous but not considered toxic. Because the significant animal data have been collected by procedures whose application to humans is impossible or highly unacceptable, it is unknown whether humans have tissue responses that are similar to those found in animals, whether human pathologic changes of a different nature may occur after similar odor exposures, or whether man suffers any discernible physiologic damage at all after identical exposures. Animal studies described in Chapter 3 have suggested that similar undesirable changes are at least possible in exposed humans, and this possibility must remain a cause for concern until more information is available. It may be that a record of harmful human effects is lacking only because suitable human studies have not yet been undertaken. Until more definitive information becomes available on this class of odorous substances, it seems prudent to assign all such chemicals, solely on the basis of judgment, either to the category of toxic substances that are odorous or to one of the categories of substances that, although not known to be toxic, evoke the undesirable human reactions discussed in following sections.

MALODOROUS SUBSTANCES THAT EVOKE VIOLENT AND ALARMING SYMPTOMS IN A SUBSTANTIAL FRACTION OF AN EXPOSED POPULATION

Odors that produce violent and alarming symptoms in a substantial fraction of an exposed population are, without exception, *malodorous.* There are many odors to which the terms "foul," "ill-smelling," "rank," "fetid," or "stinking" would be applied universally. Those commonly encountered include odors from poorly designed and badly operated sewage-treatment facilities, nonedible rendering and fishmeal plants, cattle feedlots, garbage-fed pig farms, and a variety of rubber, petroleum, and chemical manufacturing operations, including wood-pulping. These are discussed in Chapters 7 and 8. In Swedish and American surveys described by Jonsson,[12] 5–10% of respondents exposed to odor pollution caused by pulp mills reported shortness of breath, nasal irritation, and runny nose, as well as coughing, which could be elicited by stimulation of vagal receptors in the throat; 10% reported eye irritation and headache; and 15% reported nausea and sinus congestion.

Any of these physiologic reactions to foul odors may result from the direct action of the odorous substance on receptors that affect the autonomic nervous system in the same manner that food odors cause a hungry person to salivate. *The Physician's Guide to Odor Pollution* indicates that such distressing symptoms as "nausea, headache, loss of sleep, loss of appetite, impaired breathing, and even allergic reactions" may result from exposure to foul odors, especially when they occur during the night.[1] All these reactions clearly represent a matter for public-health concern and attention under the World Health Organization's definition (1958) of health—"a state of complete physical, mental and social well-being and not merely the absence of disease and infirmity"—even though the chemical components individually or collectively responsible for the foul odors have not been specifically identified as toxic substances by the usual criteria.

One criterion that may be used to distinguish discomfort, annoyance, repugnance, and similar feelings of unhappiness from true toxic symptoms is their duration. Whenever the withdrawal of the foul odor results in *prompt*, *complete*, and *permanent* remission of all the physiologic symptoms, this might be accepted as a criterion for the absence of toxicity as classically defined. This "viewpoint distinguishes between 'sensation' which ceases with the incoming signal that produces it and 'emotion' which continues afterward."[14 (p. 532)] Substances may shift from one category to another as new information becomes available, but this should be no barrier to the acceptance of presently perceived objective criteria for classifying malodorous substances with regard to their toxicity. It seems reasonable to conclude that, if it becomes possible by acceptable criteria to demonstrate pathologic change in man as a result of exposure to an odorous substance that has previously been considered nontoxic, we would want to reclassify that substance promptly, so as to make its regulation mandatory under one or another of the applicable toxic-substances acts. In spite of inherent difficulties, the general question of toxicity of malodorous substances requires resolution as a precondition for rational rule-making for the control of foul odors.

MALODOROUS SUBSTANCES THAT EVOKE VIOLENT AND ALARMING SYMPTOMS IN A SMALL NUMBER OF INDIVIDUALS EVEN WHEN EXPOSURE IS MODERATE AND FLEETING

Although it is admittedly difficult in the extreme to arrive at a precise numerical standard to distinguish unequivocally when foul odors evoke violent symptoms in "a small number of individuals" rather than "a substantial fraction of an exposed population," at least the outer bounds of each designation are recognizable. In consideration of our present uncertainty, it seems useful and prudent to focus on the extremes, rather than to attempt, at this stage, to draw a fine line to separate the two. The following discussion seeks to define an exclusive and recognizable category of severe effects of odors that are experienced by only a handful of exposed individuals.

Violent and alarming symptoms caused by fleeting exposure to foul odors of moderate, or even low, intensity that, at most, provoke distaste or annoyance reactions in the remainder of persons similarly exposed may represent physical or physiologic manifestations that result principally or solely from mental or emotional conditioning. Individual associations defy prediction. This type of response may also result from disease or genetic constitution.*

These manifestations are frequently referred to as "psychosomatic" or "psychogenic" effects, and these terms are often interpreted by the public to mean "imaginary symptoms." But nausea, vomiting, increased heart rate and blood pressure, faintness, etc., are real reactions to foul odors, although some stimuli that provoke changes in heart rate or respiration need not be considered harmful. For purposes of clearly defining some of the many ways in which people react to foul odors, it is highly desirable to reserve the terms "psychosomatic" and "psychogenic" for marked physiologic manifestations that result directly from a severe emotional upset provoked solely by perception of odors. Identical symptoms may result from the direct impact of odor perception on the same organ systems discussed near the beginning of Chapter 3, but the distinction between psychologic and physiologic symptoms is important because reactions to malodors frequently involve personal judgments of what is

*Patients with Addison's disease have been shown to have a marked increase in their ability to detect odors, some 100,000 times more acute than healthy controls.[10]

tolerable and what is intolerable. Many of these personal judgments have probably been influenced by conditioning of a seemingly unrelated nature. It is not uncommon to encounter persons who react extremely unfavorably to what are generally considered to be pleasant odors, and vice versa. In addition, human expectations that are related to cultural conditioning play a powerful role in personal judgments of which odors are tolerable and which intolerable. Presumably, these judgments are motivated by learned patterns of what constitute "familiar and acceptable" or "unfamiliar and unacceptable" environmental conditions. Furthermore, events of a horrifying nature, such as unexpected observation of severe traumatic injury to others, can evoke the same symptoms that have been ascribed to foul odors (e.g., nausea, vomiting, loss of appetite, and sleeplessness) without any suggestion that the symptoms result from overt, covert, or latent physiologic damage.

The importance of a clear distinction between physiologic symptoms that represent psychogenic effects and those which represent a toxic response has been commented upon in the proceedings of the Symposium on Measurement of Annoyance as follows: "A critical question is whether there is any evidence that a high sensory input will harm the central nervous system provided it does not cause injury to the sensory organs. Although there do not appear to exist any specific studies of this problem, the evidence indicates that such effects do not take place."[18] In fact, "there is . . . no evidence that the experience of malodor *per se* produces disease. . . . But . . . poor health may . . . increase the displeasure or at least the frequency of complaints about odor."[20]

Although this category includes only small numbers of people—and, perhaps, idiosyncratic people, on the basis of their extreme reactions to foul odors that most find tolerable—public-health practice is dedicated to attempting to protect from harm even the most sensitive fraction of the population, and this group may very well qualify as the most sensitive. The feasibility of rule-making to control odors for this small but identifiable population group remains an open policy question.

MALODOROUS SUBSTANCES THAT PRODUCE ANNOYANCE

Some nontoxic substances are of public-health concern solely because they have odors that cause annoyance to some members of the exposed population. For susceptible persons, annoyance (vexation, irritation, etc.) may increase to the point of nuisance (harm or injury usually with reference to a continuing or repeated annoyance). In general, the odors in question here are those which can be described as bothersome, unpleasant, offensive, disgusting, noxious, loathsome, or irritating—i.e., they are generally malodors, or foul odors. However, even generally pleasant odors may become unpleasant to many people whenever they are exposed to unusually high concentrations for long periods. This category is discussed in the following section.

Annoyance reactions are emotional reactions and involve all major organ systems of the body. "From the practical point of view of prevention of disease it seems to be useful to consider emotion as consisting of three elements: emotional feeling, social action or behavior, and physiological changes. These elements are interrelated to each other and also markedly affected by a fourth element 'intellectual symbolization.'"[14 (p. 531)] In support, Ingvar has pointed out that emotions influence not only the central nervous system, but also respiration, systemic circulation, gastrointestinal functions, hormonal systems, and blood chemistry.[11] Hence, nontoxic odors that produce annoyance affect different people in markedly different ways. Under identical exposure conditions, some will remain indifferent, whereas others will experience various degrees of resentment and annoyance that, in the extreme, can develop into anger or even rage.

Most odorous substances in the atmosphere that evoke complaints to air-pollution control agencies belong in this category; they do not produce dire physical symptoms, but a sizable fraction of the exposed population cannot live with them in comfort. Undoubtedly, many complaints regarding environmental odors fall neatly into this category: they involve odors that have "an effect which may not be demonstrably pathogenic but which involves a negative factor for an individual's comfort and well-being."[18]

The effects of annoyance, irritation, and inconvenience "are difficult to measure but [are,] nonetheless, real and important. They include sensory perceptions, such as . . . odors, and irritation of the eyes, nose, and throat which are not accompanied by demonstrable organic injury or disease. Such reactions . . . can be serious nuisances and interfere with performance without causing physical illness or shortening of life. Sensory perceptions and various physiologic responses . . . can be precisely measured, but their clinical significance is unknown."[21] Therefore, a fundamental question arises in an attempt to decide whether foul odors that lack demonstrable toxicity by the usual criteria and do not produce dire symptoms in a substantial fraction of an exposed population constitute a critical public-health problem, i.e., how is "unreasonable annoyance" to be defined in unequivocal scientific, medical, and legal terms? Even more difficult to define and identify is the aggravating effect of foul odors on existing disease in persons with a lowered tolerance for nonspecific stress from any source.

Measurement of annoyance is not solely a matter of odor-intensity measurement, although odor-intensity measurement is certainly vital in evaluating annoyance. Odor character, duration of exposure, time of day, day of the week, and frequency of exposure also play an important role, but the relative importance of each has not been determined objectively.[7] Studies, beginning in the early 1960's,[4,8,12] were helpful in refining techniques for the measurement of annoyance in communities and suggested the use of such measurement in the establishment of legal standards. The results pointed to the risk of relying on voluntary complaints for enforcement purposes as the volume of complaints received may reflect not only the amount of discomfort experienced by the exposed population, but also its social-class composition and degree of community organization.[22] This has become a truism of community odor control, but it is not known whether it reflects a lower annoyance threshold, a greater degree of motivation to achieve a goal, or better knowledge of how to register complaints among citizens who are better educated and in a higher social class.

Although much remains unexplored, these and other studies[6,9,13] have indicated that the impact of odorous substances in the atmosphere on exposed populations can be discovered and measured by traditional methods of epidemiology.* Future studies are expected to establish the population fraction that is likely to exhibit each type of reaction in response to an odor exposure that is well defined with respect to intensity, character, duration, frequency, and time of appearance.

NORMALLY PLEASANT AND NEUTRAL ODORS THAT PRODUCE ANNOYANCE BY REASON OF EXTREME INTENSITY OR PERSISTENCE

A great variety of manufacturing and agricultural odors are almost universally considered to be pleasant. These include the characteristic smells of baking and cooking, perfume-blending, candy-making, and haying. In spite of the pleasant feelings usually associated with these products and activities, air-pollution control agencies receive complaints about emissions of odors from restaurant kitchens, bakeries, and similar establishments that make products that we take pleasure in eating, wearing, or using. Thus, even odors that are universally thought of as pleasant may become unacceptable when they are present at an unusually high intensity for an unusually long period. Air-pollution control officers often experience great difficulty in trying to decide whether such a

*The science of studying disease in groups of people.

situation merits official control action in the face of an odor exposure that many consider trivial.

ODORS THAT EVOKE PLEASANT OR INDIFFERENT SENSATIONS

Odors that evoke pleasant or indifferent sensations are recognized as belonging to a definable category in the spectrum of human responses to odors. It is fair to assume that a pleasant and weak odor is of relatively minor physiologic and psychologic significance and is not likely to be associated with annoyance, rejection, or other negative effects. Lacking public-health significance, these odors are not a matter of concern and merit no further consideration here.

PUBLIC-HEALTH ISSUES

A clear distinction must be made between odor perception and the ability of an odor to cause a disturbance to public health. Odor perception itself does not necessarily constitute a threat to public health or welfare, even though some especially offensive malodors, such as fecal odors, might create an intolerable condition that requires prompt abatement. Withdrawal of all olfactory stimulation is considered neither desirable nor attainable as annoyance may occur as the result of understimulation (e.g., boredom). Nevertheless, overstimulation is the usual route to annoyance from odors in the atmospheric environment. Normally, the intensity, duration, character, and degree of unpleasantness of an odor in the mind of the public are all important for judging when exposures to odors constitute a matter of public concern.

A resolution of the public-health aspects of odorous substances in the atmosphere is made difficult by the many roles that odor itself plays in public awareness of, and individual responses to, this form of air pollution. (These matters are discussed in the concluding sections of Chapter 3.) Therefore, it is essential to develop a rational framework that will permit distinctions between the trivial and the serious and between unfavorable physiologic responses and responses that are triggered by personal psychologic associations before a satisfactory regulatory program can be planned and put into effect.

The designation of "foul odors" can have a variety of meanings that bear directly and importantly on rule-making, on the feasibility of enforcement, and on voluntary compliance. To help to clarify these issues, the several designations and definitions are summarized in Table A-1. It is proposed in the table that all malodorous toxic substances be dealt with in the same manner as all other toxic substances and that malodorous substances known to be nontoxic or below the threshold of toxicity be treated as undesirable air pollutants to the degree that they adversely affect the welfare of populations exposed to unreasonably high intensities for long periods and at times of day at which they are especially likely to provoke feelings of unease or discomfort. It is also proposed in the table that special rule-making be considered for handling nontoxic malodors to which a large fraction of the exposed population reacts with undesirable physiologic or emotional symptoms, even if these symptoms seem to disappear promptly and permanently when the stimulus is removed. Odor exposures that produce substantially fewer unfavorable effects, and all of low intensity, are the most difficult to deal with in practice as the cost of abatement is generally high whereas the benefits may range from marginal to moderate. The choice of abatement or tolerance certainly implies a value judgment concerning the quality of life that the affected public will wish to participate in to the fullest degree. Inasmuch as one of the purposes of expert committees is to contribute to informed decisions, it is appropriate to express opinions on the importance of foul odors relative to other public-health imperatives and to make judgments regarding preferred methods for coping with them.

It is regrettable that only a few exploratory epidemiologic studies of foul odors have been attempted, and whatever information has been developed thereby has not been widely applied. As a consequence, most of the information needed to develop practical methods for abating and avoiding the creation of foul odors is still missing. A number of studies have examined the dose-response relationships of odors in the atmospheric environment.[5,8,9,15,19] It comes as no surprise to learn from these studies that unfavorable responses were found to be most numerous when odor intensity was highest; that single-point sources of large discharge volume and high odor intensity (as uncontrolled kraft pulp mills) disturbed residents 20 km away; and that there are sex-, health-, and age-related differences in annoyance reactions to odors. Unfortunately, the correlation between degree of chemical exposure and degree of annoyance reported may be very weak, because individual annoyance perceptions have been accepted as reported by the surveyed population and have not been measured objectively. To overcome this deficiency, a number of performance measurements have been proposed as indices of annoyance. These are based on the assumption that, as the degree of annoyance increases, the performance of tasks that require skill and close attention (such as problem-solving) is affected adversely. However, these tests tend to measure annoyance reactions, rather than the annoyance itself, and it has been suggested that some somatic responses mediated by the central nervous system are more closely related to the extent of perceived annoyance. Measurements that have been proposed to monitor central-nervous-system activity include those of endocrine function, electroencephalography, and electrodermal activity. The latter type of measurement is preferred; it is noninvasive, and it has been used in other studies of discomfort and annoyance. However, it must be kept in mind that pleasant and unpleasant stimuli may induce identical changes in some of the entities proposed for measurement. The great need in using physiologic indicators of annoyance is to document their validity.[18] Examples of methods that have been used to characterize discomfort (a prominent initiator of annoyance) include: the use of psychophysical measurements of cross-modality matching and galvanic skin potential, which increases with unpleasantness of stimuli (a physiologic correlate of comfort),[16] and the use of skin-temperature rise, and sweating rate, for testing experimental respirator designs for comfort.[23] Much can be learned from studies of annoyance caused by noise and the methodology can be transferred to investigations of annoyance caused by malodors. Epidemiologic studies of annoyance from noise have shown that "it is important to search for directly measurable effects that are associated with reports of annoyance. An association was demonstrated between reported feelings of annoyance, performance efficiency, and the subjects' experience of the influence of noise on their performance. The exposure to noise was found to affect performance more negatively in the more annoyed individuals. The results indicate that the annoyance-inclined individuals in a community may constitute a special risk group that will suffer more from the adverse effects of community noise."[3] Until similar objective measurement methods are developed and used in the laboratory and in the field to evaluate the physiologic and psychologic responses to well-defined odor stresses, it is difficult to foresee that rapid strides can be made in the knowledge of how odors affect populations, what constitutes an odor nuisance, and what can be accepted as a tolerable odor exposure. None of these techniques is well established as yet because of human variability and other factors, but they may point the way to more satisfactory test methods. In addition, independent objective measurements are urgently needed to verify the results of the more conventional socioepidemiologic survey techniques for measuring annoyance described in Chapter 3.

When pathologic changes in tissue or function cannot be demonstrated in humans (although they may be observed in animals), the undesirable effects of malodorous substances in the atmosphere usually manifest themselves as annoyance reactions, although some human physiologic changes may not be accompanied by a subjective perception of annoyance.

"Annoyance" has been defined as "a feeling of displeasure associated with any object or condition believed to affect adversely an individual or group. From the medical point of view the term 'annoyance' implies an effect which may not be demonstrably pathogenic but which involves a negative factor for the individual's comfort and well being. However, the demarcation between pathogenic processes and annoyance is not distinct since the line between health and disease is partly established by current attitudes in the community."[18] Thus, although a useful distinction can be made between situations in which pathogenic changes resulting from exposure to odorous substances can be demonstrated objectively and those in which such changes are unequivocally absent, there is a middle ground, in which the criteria now used for classification tend to abut and overlap. This leads to controversies among experts that obscure a pressing need to establish more rational definitions and firmer classification criteria before the precise fate of individual cases is settled. There is no evidence that the experience of malodor itself produces diseases,[17] but the evidence is incomplete. For example, no increase in respiratory illness was found among schoolchildren in communities in which the air was polluted by the effluents from a pulp mill.[2] But in another study, higher frequencies of headache and chronic respiratory symptoms were reported in areas with greater exposure to odor.[5] In the latter case, the increase in symptoms could have been related to other atmospheric pollutants or to differences in the composition of study populations. Therefore, it seems clear that almost any conclusion may be reached regarding the nature of foul odors. How to control them is subject to challenge on the basis of inadequacy of information. Nevertheless, the notion that each odor problem must be approached on an *ad hoc* basis because it is assumed to be a unique occurrence hobbles the rational control of odor sources; most odor complaints have more in common than otherwise, and the development of unifying principles would go a long way toward resolving many of the seeming inconsistencies in odor-control efforts. In short, what is needed is a synthesizing and integrating effort in science and technology, to assure the public that adequate knowledge will become available to safeguard its health and comfort and to keep the environment in a condition close to that which exists naturally, i.e., in the absence of man-made odorants.

As a public-health concern, it is highly desirable that atmospheric odors be analyzed in conventional epidemiologic terms—i.e.:

- Identification of all the physical and emotional symptoms that odors cause.
- Development of firm dose-effect and dose-response relationships that

reach zero in the absence of stimulus but are unlikely to reach 100%, even under the strongest provocation (i.e., some degree of immunity always exists).

- Definition of susceptible populations with respect to age, sex, occupation, geographic distribution, etc., and quantification, in acceptable statistical terms, of the nature, incidence, and prevalence of unfavorable responses among identified susceptible groups.
- Development of a theoretical rationale that makes it possible to use the foregoing information for preventive purposes—the ultimate public-health goal.

As noted previously, a modest number of studies with limited objectives have been undertaken to learn more about community-wide attitudes toward foul odors. Most have taken place in areas with persistent odor problems, usually related to a single identified source, such as a large kraft pulp mill, and have used polling techniques and questionnaires. In one of these, it has been found that the majority of persons living within 1 km of a tar-oil plant in Duisburg and an insulation plant in Cologne experienced occasional-to-frequent episodes of "odor-induced" nausea and headache,[24] although such episodes may not be registered as increased visits to physicians or admissions to hospitals. To resolve the questions associated with possible physiologic damage from odors that are not considered to produce toxic effects, epidemiologic studies of populations exposed to both pleasant and offensive atmospheric odors should be undertaken to determine whether those populations contain a higher proportion of people suffering from alterations of the sense of smell and taste, including reduced sensitivity to odors, than do control populations. If increased rates of these conditions are found, an attempt should be made to establish whether the people in question have a higher incidence of work-related accidents associated with the failure to detect odors or suffer reduced enjoyment of food and lower nutritional levels that might be associated with altered appreciation of flavor. Particular attention should be paid to workers who live near a source of odor to which they are also exposed during working hours, to the very young, and to the very old.

"Malodors" constitute the subject of a large but variable percentage of complaints to air-pollution control agencies, and odor complaints are usually the principal driving force for official actions. Critical examination of official complaint records suggests that the volume of complaints is more likely to reflect the social class composition and level of community organization than the intensity and persistence of the malodor or the degree of discomfort experienced by the exposed population. The general thrust of current air-pollution regulations applicable to objectionable odors is avoidance of a public nuisance, although "public nuisance" and "objectionable odor" remain undefined in all respects other than odor intensity and fraction of the affected population that expresses dissatisfaction with existing conditions (see Chapter 11).

Perhaps the greatest need in resolving air-pollution nuisances, including odor nuisances, is for a precise quantitative definition of a nuisance and a similar definition of when such a nuisance has been abated. Although much has been made of the similarity between the physical and psychogenic effects of malodors and noise, this has not yet greatly advanced our understanding of how odors produce annoyance reactions because in both cases, the threshold of annoyance is indeterminate, the fraction of an exposed population that will remain indifferent is unknown, the intensity of annoyance will generally increase with exposure even if the stimulus remains the same, and feelings of hostility and annoyance often persist after the problem disappears. It is clear that psychologic factors play an important role in the evaluation of nuisances, including those caused by malodors. The special difficulties that lie ahead in making the necessary judgments for some compounds that are now equivocal do not invalidate a search for a rational system of differentiation. On the bright side, no special difficulties are anticipated in assigning most commonly encountered foul odors to one category or the other.

RECOMMENDATIONS

- As a general rule, substances that are toxic at the same concentrations at which they are malodorous should continue to be regulated as toxic substances; there is already excellent legislation for such regulation, and toxicity will usually be the most dire effect on populations.
- Substances released to the atmosphere that are of public concern for no other reason than their characteristic foul odor should continue to be regulated because of their detrimental public-health impact. This follows from the recognition that prolonged exposure to foul odors at high intensities usually evokes undesirable reactions in people that can range from unease, discomfort, irritation, and anger to violent physiologic manifestations, including circulatory and respiratory effects, nausea, vomiting, and headache severe enough to lead to prostration.
- The impact of persistent unpleasant odors in the atmosphere that do not produce marked physiologic symptoms in the exposed population is a matter of special concern in the context of the World Health Organization's definition of "health." This is a pervasive and important aspect of atmospheric odors that bears on the long-term mental health of exposed

populations, but it has received little attention. It is recommended that emphasis be given to this subject in future research efforts.

• Enforcement of regulations for prevention of malodors in the atmosphere is difficult in the extreme because objective criteria of annoyance and nuisance are difficult to formulate in precise and quantitative terms that can be understood and acted on in a rational manner by enforcement officers and emitters alike. Psychophysical and other studies designed to develop an epidemiology of malodors are needed to dispel this ignorance and, ultimately, to make it possible to predict, before a potential odor-emission source is constructed, the degree of odor control that will be needed at the source, to avoid community displeasure. It will be especially important for planning and enforcement purposes to learn what intensities of odor above a recognition threshold are tolerable for what periods and exposure frequencies.

REFERENCES

1. American Medical Association. The Physician's Guide to Odor Pollution. Prepared by J. A. Bell. Chicago: American Medical Association, 1973. 19 pp.
2. Anderson, D. O., and A. A. Larsen. The incidence of illness among young children in two communities of different air quality: A pilot study. Can. Med. Assoc. J. 95:893–904, 1966.
3. Arvidsson, O., and T. Lindvall. Subjective annoyance from noise compared with some directly measurable effects. Arch. Environ. Health 33:159–166, 1978.
4. Cederlöf, R., L. Friberg, E. Jonsson, L. Kaij, and T. Lindvall. Studies of annoyance connected with offensive smell from a sulphate cellulose factory. Nord. Hyg. Tidskr. 45(1):39–48, 1964.
5. Deane, M., and J. R. Goldsmith. Health Effects on Pulp Mill Odor in Anderson, California. APTIC No. 39242. Research Triangle Park, N.C.: U.S. Environmental Protection Agency, Air Pollution Technology Information Center, 1971.
6. Deane, M., G. Sanders, and E. Jonsson. Trends and community annoyance reactions to odors from pulp mills. Eureka, California, 1969–1971. Environ. Res. 14:232–244, 1977.
7. First, M. W. A model odor control ordinance, pp. 1255–1259. In H. M. Englund and W. T. Beery, Eds. Proceedings of the Second International Clean Air Congress. New York: Academic Press, 1971.
8. Friberg, L., E. Jonsson, and R. Cederlof. Studies on sanitary nuisance of smoke from a sulphate cellulose factory. Parts I and II. Nord. Hyg. Tidskr. 41:41–62, 1960. (in Swedish)
9. Goldsmith, J. R. Health and Annoyance Impact of Odor Pollution. First Report. EPA Publ. No. 650/1-75-001, Contract No. 68-02-0083. Washington, D.C.: U.S. Environmental Protection Agency, Office of Research and Development, 1973. 524 pp.
10. Henkin, R. I., and F. C. Bartter. Studies on olfactory thresholds in normal man and in patients with adrenal cortical insufficiency. J. Clin. Invest. 45:1631, 1966.
11. Ingvar, D. H. Annoyance Reactions—Neurophysiological Aspects. Paper presented at Symposium on Environmental Agents and Annoyance Reactions. Karolinska Inst., Stockholm, Aug. 30–Sept. 4, 1971. 9 pp.

12. Jonsson, E. Annoyance reactions to environmental odors, pp. 329–333. In A. Turk, J. W. Johnston, Jr., and D. G. Moulton, Eds. Human Responses to Environmental Odors. New York: Academic Press, 1974.

13. Jonsson, E., M. Deane, and G. Sanders. Community reactions to odors from pulp mills: A pilot study in Eureka, California. Environ. Res. 10:249–270, 1975.

14. Kagan, A. Epidemiology, disease, and emotion. In L. Levi, Ed. Emotions: Their Parameters and Measurement. New York: Raven Press, 1975. 800 pp.

15. Kendall, D. A., and T. Lindvall, Eds. Evaluation of Community Odor Exposure. Report of a Symposium sponsored by the U.S. Environmental Protection Agency. Cambridge, Mass.: Arthur D. Little, Inc., April 1971. 31 pp.

16. Leiderman, P. H., and D. Shapiro. Galvanic Skin Potential Level as an Indicator of Behavioral Activation. Office of Naval Research Contract Nonr 1866(4), Tech. Rep. No. 7, NR 170-518, March 1963.

17. Lindvall, T. Skeleton-Report (Aug. 25). The Karolinska Institute Symposia on Environmental Health. Stockholm, Aug. 30–Sept. 4, 1971.

18. Lindvall, T., and E. P. Radford. Measurement of annoyance due to exposure to environmental factors. Environ. Res. 6:1–36, 1973.

19. Medalia, N. Z. Community perception of air quality: An opinion survey in Clarkston, Washington. PHS Publ. No. 999-AP-10. Cincinnati: U.S. Department of Health, Education, and Welfare, Public Health Service, Robert A. Taft Sanitary Engineering Center, 1965.

20. National Research Council, Committee on Medical and Biologic Effects of Environmental Pollutants, Subcommittee on Hydrogen Sulfide. Hydrogen Sulfide, pp. 93–108. Baltimore: University Park Press, 1978.

21. President's Science Advisory Committee. Restoring the Quality of Our Environment, p. 94. Report of the Environmental Pollution Panel. Washington, D.C.: The White House, November 1965.

22. Smith, W. S., J. J. Schueneman, and L. D. Zeidberg. Public reaction to air pollution in Nashville, Tennessee. J. Air Poll. Cont. Assoc. 14:418–423, 1964.

23. Snook, S. H., W. C. Hinds, and W. A. Burgess. Respirator comfort: Subjective response to force applied to the face. JAIHA 27:93–97, 1966.

24. Winneke, G., and J. Kastka. Odor pollution and odor annoyance reactions in industrial areas of the Rhine-Ruhr region, pp. 471–479. In J. Le Magnen and P. MacLeod, Eds. Olfaction and Taste VI. London: Information Retrieval, Ltd., 1977.

Appendix B

Social-Survey Methods

JUSTIFICATION FOR USING SURVEY METHODS

In California and in many other state and local air-pollution control jurisdictions, the legal weapons most widely used to attack odor problems have been public-nuisance laws.[8] The application of such laws is usually decided on the basis of the number of persons who have complained, because it must be shown that a "considerable number of persons" have been aggrieved. An example of such a regulation is Rule 51 of the Los Angeles County Air Pollution Control District:

Rule 51. Nuisance

A person shall not discharge from any source whatsoever such quantities of air contaminants or other material which cause injury, detriment, nuisance or annoyance to any considerable number of persons or to the public or which endanger the comfort, repose, health or safety of any such persons or the public or which cause or have a natural tendency to cause injury or damage to business or property.

It is generally acknowledged by air-pollution control officials that public-nuisance laws are helpful in widespread, chronic odor situations, but not in localized odor problems in which few persons are affected or in cases where the odors are transient and the complaints cannot be validated. Without validation of complaint information by the enforcement agency, the words of complainants, even in courtroom testimony, run a risk of being impeached as biased or eccentric, as not convincingly related to the alleged source, or as based on rare and transitory episodes.

To avoid the difficulties of using complaints, other than to indicate where odor problems may exist, a few enforcement agencies have proposed limiting the emission of odorous substances to quantities related to the odor-detection threshold of these substances. This approach tends to equate the perception of odor with the existence of an odor problem. In reality, it is the personal evaluations of those who perceive odor, not the mere activation of the sense of smell, that give rise to the problem and that determine its severity. Some odors are considered pleasant; other odors, or the same odors experienced under different conditions, are considered unpleasant, sometimes to the extent of causing undesirable effects, such as annoyance and interference with activities. Thus, it seems more appropriate to base the existence of odor problems on a method that takes personal evaluations into consideration. The social survey (also called a public-opinion survey, public-attitude survey, or socioepidemiologic survey) is such a method.

BASIC ELEMENTS OF SOCIAL SURVEYS

The social-survey method involves asking questions about whether odors have been noticed, whether odors have caused bother or irritation, the circumstances at the time the odors caused bother, and in what ways the odors have been bothersome. Questions about the backgrounds of the respondents are added to characterize those who state that they have been bothered. Questions about other forms of pollution or about other aspects of the environment in the community are added, if the main interest in odors on the part of the survey organization is to be concealed. How one goes about conducting a social survey is outlined in the following paragraphs, in which the emphasis is on a survey of community reaction to odors.

As in any investigation, the objective should be carefully formulated. The key elements of population sampling, questionnaire design, and analysis depend on a clear understanding of the survey objective. Once this is known, the remaining elements can be considered.

The most difficult task to be faced is the delineation of the area affected by odor (henceforth called the "test area"). This can be done by examining recent patterns of complaints or by driving an automobile in a grid pattern throughout the general area and noting the boundaries within which the odor is perceived. The test area must be large enough to contain at least the number of people needed to permit adequate analysis of the data. This requires that at least 20 and preferably 30 families live in the community.[4]

Next, the socioeconomic characteristics of the test area should be examined by referring to published census information. This will determine whether the people living or working in the area are homogeneous or, if not, whether the area can be divided into two or more homogeneous sections. If homogeneity cannot be determined, additional questions must be added to the questionnaire, which can be used to characterize the social class of the people after the survey.

When the test area has been established, the names and street addresses of all families living in the test area or the names and workplaces of all people working in the test area should be obtained. Reverse-order (street-address) telephone directories are convenient sources of the names of the families. Employers are the sources of the names and locations of workers. A random sample of at least 20 and preferably 30 families or workers should be drawn for use in the survey.[4]

A questionnaire is then designed to obtain the information needed to reach the survey objective, given the type of people (residents or workers) to be contacted in the test area. First, it should be decided how the questionnaire is to be administered. Second, each question must be considered with respect to its ability to communicate and to elicit accurate information. Third, instructions for interviewers must be prepared or, if the questionnaire is to be self-administered, instructions for the respondents must be written. Finally, the questionnaire should be pretested in a community similar to the one in which it is to be used. Changes in the questionnaire and in the instructions for administering it should be made, if necessary, on the basis of the results of the pretest. (If the changes are extensive, it may be advisable to have another pretest.)

In preparation for the actual survey, the interviewers must be taught how the questionnaire is to be administered. The survey is then conducted in accordance with a predetermined schedule. A sample of completed questionnaires should be validated by noninterviewing personnel. This is done by reinterviewing the respondents. In addition, a sample of nonrespondents should be contacted and, if possible, the questionnaire should be administered to them. After the results are analyzed and the outcome known, appropriate action can be taken.

Throughout the survey, precautions should be taken to control for bias. Bias can be introduced in any element of the survey. Random selection of the residents or workers to be contacted, careful design and pretest of the questionnaire, thorough training of interviewers, validation of completed interviews, and interviews of nonrespondents are some of the precautions that can be taken to avoid bias.

If the objective of the survey is to determine whether a local law governing odors has been violated or to obtain an accurate measurement of community reaction to odors, simultaneous surveys should be conducted in the test area and in a matching odor-free area (henceforth called the "control area"). Comparison of the results serves either or both of two purposes:

- Under the assumption that people in both areas have equal right to odorfree air, comparison provides an equitable basis for a legal decision.
- Because what people in the test area say about odors may not reflect what they actually feel about odors, comparison of results from the test and control areas permits an estimation of true attitudes (e.g., by application of the signal-detection technique described by Swets *et al.*).[13]

The control area should be selected according to the following criteria:[4]

- The control area should be as free of odor as possible. To ensure this, the local air-pollution control agency should not have received any odor complaints from people living or working in the area during the previous 12 months.
- The control area should be near the test area—preferably within 10 miles (16 km) of it.
- The control area and the test area should have similar access to heavily traveled roadways.
- The control area and the test area should be within approximately the same distance from commercial or industrial establishments.
- Median income, home value, and gross rent for the control area should not differ from those of the test area by more than 20%.
- Median number of rooms per housing unit in the control area should not differ from that in the test area by more than 10%.

The last two criteria are pertinent to surveys of residential areas. Surveys of workers must be handled differently. Other statistics—such as percentages of blue-collar and white-collar workers, average wages or salaries for each group, average age for each group, and percentages of each group working indoors and outdoors—should be obtained from employers. The differences in these statistics between the test area and the control area should not be more than 20%.

This outline is intended only to indicate the basic elements of the social-survey method. For explanations of the theoretical concepts and essential details of conducting social surveys in general, the reader is referred to the many textbooks and survey manuals in the fields of social psychology and market research (e.g., Cannell and Kahn,[1] Kish,[7] and the Survey Research Center[12]). For the application of survey principles to measurements of community reaction to odors in the United States, the reader is referred to Medalia,[10] Goldsmith,[6] and Flesh and Burns.[4]

MEASUREMENT OF HUMAN PERFORMANCE

On the horizon is a method of measuring the social effects of odors that does not require the collection of opinions and subjective information. It includes analysis of such items as worker and student absenteeism and reduction in productivity. Two obstacles stand in the way of attempting to use this method today: adequate records are not being kept by most employers and schools; and, even where the necessary information is being collected, employers and schools are reluctant to release it to outsiders. It is possible that some sort of agreement to keep and supply records can be reached with employers and schools in the private sector, but records kept by public employers and by the public-school system are subject to federal and state privacy-act restrictions on the release of information, as well as to their own internal policies on such matters.

USE OF AVAILABLE METHODS

Of the two available methods of measuring the social effects of odors, the social-survey method is recognized by scientific groups as better. The following was a conclusion of the Fourth Karolinska Institute Symposium on Environmental Health:[9 (pp. 29–30)]

> Spontaneous complaints, such as letters to newspapers or health authorities may indicate the existence of annoyance, but this crude measure should be replaced by better techniques. Methods applied to date have emphasized population survey techniques in determining annoyance, and the possibility of errors in application of survey methods has been strongly emphasized by members of the symposium. When these survey techniques have been properly and expertly developed and applied they have yielded valid and significant results related to annoyance, and these techniques have provided the data that are the basis of current knowledge of this subject in quantitative terms. Modern psychological and sociological research offers the possibility of application of many new and potentially more quantitative methods pertinent to research in annoyance. Their practical applicability and usefulness under field conditions has to be proved, however.

A CONCEPTUAL FRAMEWORK FOR DEVELOPING DOSE-RESPONSE RELATIONSHIPS

The use of social surveys to measure community reaction to odors is attractive in theory, but it leaves much to be desired in practice. The major difficulty reported by enforcement agencies that have used the method is that it is cumbersome.[3] Some success has been achieved in minimizing the effort through the design of a brief questionnaire and a plan requiring as few as 60 interviews to reach a decision.[4] Yet, even with this design, at least a week and approximately $1,500 must be spent each time the method is used.

After experiencing many failures in seeking solutions under existing laws, state and local air-pollution control agencies are frustrated with the use of odor complaints in dealing with odor problems. The requirements of social surveys for use in routine investigations are unattractive. Consequently, air-pollution control officials have voiced an urgent need for the development of more straightforward measures aimed at dosage, instead of response. A dosage approach would take the form of a performance standard that could be evaluated at the stack or an ambient-air standard that could be evaluated in the community. Such an approach would offer the practical advantages of convenience, reproducibility, and proof of violation or compliance with regulations. However, because there are no means of relating concentrations of odorous emission or odor intensity to community reaction, dose-response relationships would have to be developed.

Human responses to odors have often been expressed in terms of thresholds, and many procedures and devices have been described for measuring thresholds.[11] The absolute or detection threshold is the minimal concentration of odorant at which a response is elicited, and the recognition threshold is the minimal concentration at which an odorant can be identified by a subject. Because people differ in their responses, it has become customary to define the "50% thresholds" as the minimal concentrations at which half the subjects in a population respond to or identify an odor. This definition is parallel to the "median effective dose," ED_{50}, used in toxicology.

In recent years, it has become understood that the concept of thresholds ignores some fundamental aspects of human responses to stimuli. Such understanding has come about from the development of the theory of signal detection,[13] which recognizes human responses to be decisions as to whether a given sensation (which always includes some random background interference, or noise) consists of noise alone or contains some signal (odor) mixed with the noise. The responses, which take the form "Yes, I detect something" or "No, I don't detect anything," are not simply expressions of sensation, but are influenced by decisions based on other factors, such as expectation, the wish not to miss an odor that is present, or the wish not to sound a "false alarm" by incorrectly reporting an odor that is absent. Thus, in this more recent view, threshold values are not absolute properties of odorant substances, but depend on the social and other

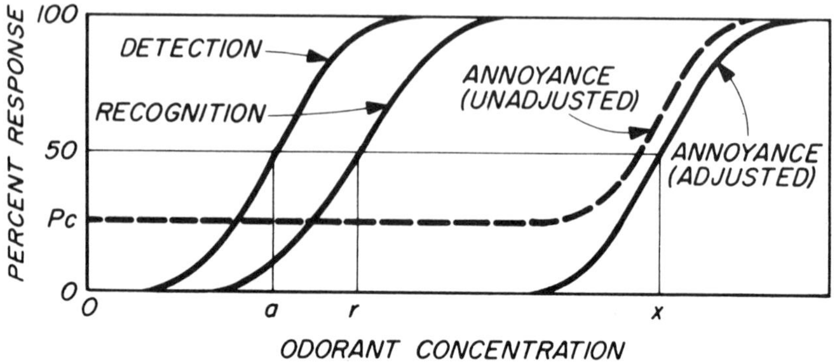

FIGURE B-1 Absolute (*a*), recognition (*r*), and annoyance (*x*) thresholds of a given odor. P_c, percentage of control-area residents expressing annoyance. Reprinted with permission from Flesh and Turk.[5]

human contexts in which they are experienced. Therefore, reported values must be regarded as functionally determined; i.e., they depend on the procedure, the subjects, and the contexts of the investigation, and thus are not necessarily quantitatively transferable from one situation to another.

With these demurrers in mind, let us examine the first two sigmoid curves of Figure B-1, of which the first represents the cumulative frequency of detection, and the second, the cumulative frequency of recognition. The absolute (or detection) and recognition thresholds corresponding to the odorant concentrations of ED_{50} are shown as concentrations *a* and *r*, respectively.

Ideally, the detection and recognition curves of Figure B-1 should be established by exposing a large group of people simultaneously to odors of various concentrations, under such conditions that each subject is insulated against any influence from other subjects or other responses. For many reasons related to the difficulty of establishing uniform vapor concentrations over a large volume or to the expense of providing an extensive network of piping and sniffing ports to service each subject, such an ideal procedure is impractical.

Instead, the sigmoid detection and recognition curves of Figure B-1 can be approximated by exposing many small groups to different concentrations of the odorous substance under investigation.

Detection and recognition thresholds can be used to monitor the effectiveness of various kinds of odor-control equipment and procedures. Surveillance of ambient or emission samples by a few trained people is often adequate to determine whether the application of odor controls has successfully reduced emission below some predetermined point. However, detection and recognition thresholds provide no information on community reaction to perceived odors.

In Figure B-1, the dashed curve represents unadjusted cumulative frequency of expressed annoyance. The long horizontal portion that precedes the sigmoid section of this curve represents a constant degree of expressed annoyance that is independent of odorant concentration and that therefore must represent positive responses provided by variables other than odor.

To correct for such irrelevant responses, the following procedure is used:

P_t = % of test-area residents who express annoyance with odors,
P_c = % of control-area residents who express annoyance with odors,
and
A = % of annoyance responses (adjusted).

The adjusted percentage of responses must be $P_t - P_c$, and the percentage must be based only on the subjects who are not responding inappropriately, i.e., on $100 - P_c$. Therefore,

$$A = \frac{P_t - P_c}{100 - P_c}. \tag{1}$$

The adjusted annoyance curve of Figure B-1 is based on Equation 1, and the 50% annoyance threshold, x, is determined graphically from the curve. This correction is far from trivial. It is noted that P_c, the percentage of control-area residents who express annoyance with odors, has been estimated at 26% for the southern quarter of Los Angeles County.[2] Even higher percentages have been encountered in control areas near a Swedish sulfate pulping plant.

Once the annoyance threshold for a given odor has been established, enforcement agencies would be able to use ambient-odor measurements to predict whether people living or working in a community will be annoyed by the smell. If the prediction is to be based on an analysis of stack samples, further steps have to be taken to relate ambient-odor intensity in the community to emission concentrations at the stack.

For enforcement agencies to have adequate control over odor problems in their jurisdictions, an annoyance threshold has to be established for each type of odor likely to be encountered in sufficient strengths to cause problems. Differences in community reaction, in meteorologic and topographic conditions, and in the distances of populated communities from the sources of odor would require that annoyance thresholds for a given type of odor be established separately in different parts of the United States. Tedious as they may seem, these steps would be far superior, from social and economic standpoints, to the imposition of identical ambient or stack limitations on all odors.

REFERENCES

1. Cannell, C. F., and R. L. Kahn. Interviewing, pp. 526–595. In G. Lindzey and E. Aronson, Eds. Handbook of Social Psychology. 2nd ed. Volume 2. New York: Addison-Wesley Publishing Company, Inc., 1968.

2. Copley International Corporation. A Study of the Social and Economic Impact of Odors. Phase II. Report to the U.S. Environmental Protection Agency. Contract No. EPA 70-116. La Jolla, Calif.: Copley International Corporation, November 1971. 308 pp.

3. Copley International Corporation. A Study of the Social and Economic Impact of Odors. Phase III. Development and Evaluation of a Model Odor Control Ordinance. Report to the U.S. Environmental Protection Agency. Publ. No. EPA-650/5-73-001. Contract No. 68-02-0095. Washington, D.C.: U.S. EPA, February 1973. 203 pp.

4. Flesh, R. D., and J. C. Burns. Procedures for the Identification and Assessment of Community Odor Problems. A Report to the U.S. Environmental Protection Agency. La Jolla, Calif.: Copley International Corporation, 1978. 72 pp.

5. Flesh, R. D., and A. Turk. Social and economic effects of odors, pp. 57–74. In P. N. Cheremisinoff and R. A. Young, Eds. Industrial Odor Technology Assessment. Ann Arbor: Ann Arbor Science Publishers, Inc., 1975.

6. Goldsmith, J. R. Health and Annoyance Impact of Odor Pollution. Publ.No. EPA-650/1-75-001. Contract No. 68-02-0083. Research Triangle Park, N.C.: U.S. Environmental Protection Agency, Human Studies Laboratory, National Environmental Research Center, 1973.

7. Kish, L. Survey Sampling. New York: John Wiley & Sons, Inc., 1965. 643 pp.

8. Leonardos, G. A critical review of regulations for the control of odors. J. Air Poll. Cont. Assoc. 24:456–468, 1974.

9. Lindvall, T., and E. P. Radford, Eds. Measurement of annoyance due to exposure to environmental factors. A report from the Fourth Karolinska Institute Symposium on Environmental Health. Environ. Res. 6:1–36, 1973.

10. Medalia, N. Z. Community Perception of Air Quality: An Opinion Survey in Clarkston, Washington. PHS Publ. No. 999-AP-10. Cincinnati: U.S. Department of Health, Education, and Welfare, Public Health Service, Division of Air Pollution, 1965.

11. Moulton, D. G., A. Turk, and J. W. Johnston, Jr., Eds. Methods in Olfactory Research. London: Academic Press, 1975. 497 pp.

12. Survey Research Center. Interviewer's Manual. Rev. ed. Ann Arbor: Institute for Social Research, 1976. 144 pp.

13. Swets, J. A., W. P. Tanner, Jr., and T. G. Birdsall. Decision processes in perception, pp. 139–155. In W. S. Cain and L. E. Marks, Eds. Stimulus and Sensation: Readings in Sensory Psychology. Boston: Little, Brown, and Co., 1971.

Appendix C

Data Sheets from State and Local Air-Pollution Control Agencies: Examples of Responses to the Committee's Questionnnaire Regarding Odor Complaints

TABLE C-1 Alaska—Complaint Summary

	City	Source Category	Number
1977	Sitka	Pulp mill	12
1976	Juneau	Open burning	4-5
		Fire place	several
		Wood smoke	
		Cars	several
	Seldovia	Crab wastes	several
1975	Kodiak	Fish process	numerous
	Juneau	Cars	several
	Anchorage	Cars	several
	Petersburgh	Fish process	several
1974	Kodiak	Fish process	several
	Wrangell	Wood smoke	several
	Fairbanks	Teepee	several
	Juneau	Asphalt	several
	Ketchikan	City landfall	several
		Teepee	several
1973	Juneau	Asphalt roofing	several
		Diesel	governor
		Demolition	several
		Pulp mill	from Ketchikan
1972	Sitka	Pulp mill	not recorded
	Ketchikan	Pulp mill	not recorded
	Kenai	Urea and ammonia plant	not recorded
1971	Juneau	Asphalt plant	not recorded

NOTE: The Department of Environmental Conservation, in response to the survey, stated: "This is not intended to be an all inclusive list because many of the complaints received are of a nuisance variety and often are not officially recorded. Further, note that this list does not include complaints received by the local air pollution control programs in Fairbanks and Anchorage, nor by the department's regional offices through the state."

TABLE C-2 Pima County, Arizona

Examples of Various Odor Sources

Source	Approximate No. of People Affected
Chemical Processing	125
Cattle Feedlot	1500 (part of residential community)
Tallow	75-100 (trailer court)
Restaurants	Varies
Sewage Treatment	Varies, but could be close to 1000
Paint Operations	Varies

Court Cases

Type of Source	Summary of Court Case or Hearing Board	Action Taken by Source to Solve Problem
Restaurant (1975)	Order of abatement by hearing board	Improved clean-up procedures
Restaurant (1975)	Plea agreement, fine suspended	Electrostatic precipitator control unit installed
Restaurant (1978)	Plea agreement, fine suspended	Different fuel use
Tallow Company (1975)	Trial, fine suspended	Improved clean-up procedures
Tallow Company (1976)	Plea agreement, fine paid	Improved clean-up procedures
Tallow Company (1977)	Plea agreement, fine paid	Alarm notification system installed
Paint Spray (1976)	Plea agreement, fine suspended	Carbon adsorption control unit installed
Chemical Company (1976)	Plea agreement, fine paid	Improved incinerator conditions
Chemical Company (1976)	Order of abatement by hearing board	Plant completely rebuilt
Feedlot (1975)	Order of abatement by hearing board	Operation closed

TABLE C-3 California Air Resources Board—Summary, Significant Odor Complaints January 1975 to July 1978

Source	Type of Odor	Number of Complaints
Auto painting operations	Paint solvent odors	57
Petroleum refinery Odors	Sulfuric odors Oil mists Sour crude oil odors Liquified petroleum Gas leaks—mercaptans	18
Chemical manufacturing plants	All kinds of chemical odors depending on products being manufactured	57
Plastic manufacturing plants	Styrene	99
Kraft pulp mills	Total reduced sulfur compounds Terpenes	236
Agriculture (except by law)	Cattle feedlots Turkey farms Chicken farms	Significant

Auto painting operations

Number of complaints

Case 1: Thirty-eight persons signed a petition against this source. Twelve additional persons complained to the California Air Resources Board.

Miscellaneous cases: Seven persons complained to CARB.

Types of odors

Paint solvent odors.

Comments

Case 1: A nuisance violation was issued by local district.

Miscellaneous cases: These sources, after investigation by the local districts, were not found to be in violation of the nuisance law.

Petroleum refinery odors

Number of complaints

Case 1: One person complained to CARB.

Case 2: One person complained to CARB

Case 3: One person complained to CARB.

Miscellaneous cases: Fifteen persons complained to CARB.

Types of Odors

Sulfuric odors.

Oil mists.

Sour crude oil odors.

Liquified petroleum gas leaks—mercaptans.

TABLE C-3 (*Continued*)

Comments

Case 1: A nuisance violation was issued by local district.

Case 2: A nuisance violation was issued by local district.

Case 3: A nuisance violation was issued by local district.

Miscellaneous cases: These sources, after investigation by the local districts, were not found to be in violation of the nuisance law.

Chemical manufacturing plants

Number of complaints

Case 1: Twenty-nine persons representing seventeen residences signed a petition against this source.

Case 2: Eighteen persons signed a petition against this source.

Miscellaneous cases: Ten persons complained to CARB.

Type of odors

All kinds of chemical odors depending upon what type of chemicals the specific manufacturing company was producing.

Comments

Case 1: A nuisance violation was issued by local district.

Case 2: The company installed control equipment which alleviated the nuisance without enforcement action.

Miscellaneous cases: These sources, after investigation by the local districts, were not found to be in violation of the nuisance law.

Plastic manufacturing plants

Number of complaints

Case 1: Twelve persons complained to the CARB. A petition with eighty signatures was received by the local district. People complained of great discomfort and illness.

Miscellaneous cases: Seven persons complained to CARB.

Type of Odor

Styrene.

Comments

Case 1: The company cut production, no enforcement action was taken.

Miscellaneous cases: These sources, after investigation by the local districts, were not found to be in violation of the nuisance law.

Kraft pulp mills

Number of complaints

Case 1: One person complained to CARB.

Case 2: Two hundred and thirty people signed a petition against this source.

Miscellaneous cases: Five persons.

Type of odors

Total reduced sulfur compounds—terpene.

Comments

Case 1: The mill was issued a conditional abatement order from the local district. The company was ordered to install equipment to alleviate the condition according to an achievement schedule. The conditional abatement order, rather than requiring the mill's closure, would allow the company to continue operating subject to certain conditions which will ultimately bring its emissions into compliance with applicable regulations.

TABLE C-3 (*Continued*)

Case 2: An abatement order was suspended until the company had completed a three-phase improvement plan. This plan has been completed; however, the CARB is still receiving complaints indicating the public nuisance still exists. An abatement order is pending against this source.

Miscellaneous cases: These sources, after investigation by the local districts, were not found to be in violation of the nuisance law.

Agricultural odors

Number of complaints

We have received a significant number of complaints regarding odors from cattle feedlots, turkey farms, and chicken farms. Any operation involved in raising food is exempt from the California nuisance law, unless the fecal matter is being used in a separate process to produce fertilizer.

TABLE C-4 California Bay Area—Complaint Log—Summary of IBM Printout

1976	2,307
1977	2,425
1978 to 4/21/78	1,026

65% of the complaints are odor, odor and smoke, or other. IBM printout of each complaint is available.

The printout lists: Date
Time
Type of complaint—odor, other
Named source
City

TABLE C-5 San Diego, California

Summary of Major Complaints—1978

Chicken manure processing	380
Fish processing	19
Rendering plant	377

A. *Summary of test results—Chicken manure processing*
IITRE dynamic triangle olfactometer

		ACPM	ED_{50}	Odor Units/Minute
3/7/78	inlet	21,400	920	19,700,000
3/7/78	outlet	19,000	45	853,000
3/7/78	outlet	19,000	25	475,000

Re: Ceilcote report March 1978—C. W. Ritzrow
Scrubber *without* chlorine
Results are hard to believe—should follow up to determine if complaints were reduced.

B. *Tests conducted on fish processing exhaust equipped with Duall odor control scrubbing system*
Used static syringe method—drawing samples in syringes
Test 4— <2 ou/scf outlet
5— <6 ou/scf outlet
Inlet not tested—results impossible to believe—should obtain further details on scrubber, test procedure validation, and effect on complaints.

TABLE C-6 California, South Coast Air Quality Management District—1977 Summary of Complaints

Type of Source	Smoke	Odors	Dust	Fumes	Soot	Other	Total
Power plants	15	9	3	—	2	30	59
Ref.-Chem.	30	259	17	2	—	9	317
Rendering	—	1	—	—	—	—	1
Industrial	52	262	199	6	—	50	569
Vehicles	4	3	22	—	—	—	29
Ships	11	—	1	—	5	—	17
Sanitation	—	19	—	—	—	—	19
Open fire	7	—	—	—	—	1	8
Miscellaneous	74	252	189	1	1	66	583
Unknown	59	638	82	2	1	62	844
TOTALS	252	1,443	513	11	9	218	2,446

TABLE C-7 Dade County, Florida

Summary of Complaints 1977-78

Source Category or Odor Type	Approximate Number of Complaints Annually
Volatile solvents	50
Oil combustion fumes[a]	50
Waste water plants and lift stations	40
Landfills, garbage	30
Groundwater (lawn sprinkling, cooling water, etc., H_2S)	30
Unknown, undetermined	30
Blue-green algae (*Anabena*)	20
Pesticide sprays	20
Smoke (wild fires), incinerators, etc.	20
Industrial (galvanizing, chemicals, etc.)	20
Mercaptan (natural and LP gas)	10
Dog feces (domestic)	10
Food preparation and processing	20
Roofing asphalt	10
Diesel exhaust, mobile sources	10
Textile operations	10
TOTAL	380

Summary of Court Cases

Source Category	Odor Type	Number of Complaints	Court Action
Fiberglass molding	Volatile solvents	12	$300 fine in Circuit Court. Mfg. operation moved to an industrial park away from a nearby residential area.
Clothing mfg.	Fabric sizing	70	$300 fine in Circuit Court. Ordered to install control equipment to correct problem.
Coffee roaster	Burnt coffee odors	217	$1,500 fine in Circuit Court. Ordered to install control equipment (afterburner).
Rendering plant	Decayed animal odor	75	Civil case. Plant ordered to move to new location away from populated areas.
Fiberglass molding	Volatile solvents	63	$4,250 fine in Circuit Court. Operation moved out of Dade County.
Galvanizing plant	Acid, ammonia	52	Compliance plan currently being implemented. 4-4-78.

[a] The vast majoruty of oil burner fume complaints originate in Miami Beach, where there is a high density of old apartment buildings and residential hotels using oil-fired burners.

TABLE C-8 Illinois

Recent survey—240 air-pollution-related/month
140 non-health-related (referred to other agencies)
Nuisance complaints defined as non-health-related.

Examples:

Health-related
1. Ammonia leak adjacdnt to inhabited area—causes severe eye irritation, coughing and caustic action on skin.
2. Hydrogen sulfide emissions—extremely obnoxious odor with possible dizziness and burning sensation in eyes and mouth.
3. Asbestos particles from an insulation manufacturer—causes irritation in nostrils and possible shortness of breath.
4. Sulfuric acid mist from a chemical manufacturing operation—extremely irritating to skin areas, mouth, and throat.

Non-health-related
1. Garbage burning complaint by adjacent neighbor.
2. Odors from adjacent rendering plant.
3. Leaf burning.

TABLE C-9 Chicago, Illinois

Method of Evaluation

The enforcement inspector stops at about 16 locations in the vicinity of Lake Calumet to detect the type of odor, intensity, and wind direction. This is done on a daily basis for 5 working days each week or about 20 per month. If a source is detected once, this is about a 5% frequency for the month. The wind direction at the time and location is noted by him by observing the stock plumes in the vicinity.

The daily odors are summarized each month on the Odor Evaluation Summary Sheet by an engineer. For each location and date, the wind direction, intensity, and type of odor are placed in the respective box. For example, N/2C means an odor of intensity 2 and type C was detected while a north wind was blowing.

From the Summary Sheet the sources of significant odor problems with an intensity of 3 or higher are pinpointed as to intensity, type of odor and frequency. The intensity scale is as follows: 0—No Odor; 1—Very Faint; 2—Faint; 3—Noticeable; 4—Strong; 5—Overpowering.

Maps are plotted of each type of odor that causes significant problems. The arrows shown are self-explanatory for direction from which the odor is coming to the point of detection. The length of the arrow indicates the strength of an odor with a scale of 0.5 in. for each odor unit. A length of 2.5 in. is then an overpowering odor of intensity 5. The number of feathers indicates the frequency for the month.

Total Odor Intensity

In the case of a very substantial odor problem, the total odor intensity for the latest month is compared with the average monthly total odor intensity for the previous year. The monthly total odor intensity of the location due to a specific source is calculated as follows:

3X (number of occurrences of intensity 3) +
4X (number of occurrences of intensity 4) +
5X (number of occurrences of intensity 5) =
monthly total odor intensity.

Code of Common Odors

A	Ammonialike	O	Musklike
B	Aromatic	P	Oily, fatty
C	Burnt, smoky, tar	Q	Paintlike
D	Cooked vegetables	R	Petroleum, solventlike
E	Decayed	S	Putrid, foul, garbage
F	Dusty, earthy	T	Swamp odors
G	Fecal (like manure)	U	Sharp, pungent, disinfectant
H	Fishy	V	Sulphur
I	Fruit (citrus)	W	Sickening
J	Garlic, onion	X	Sour, acid, rancid
K	Grass cut	Y	Spicy
L	Meaty (cooked)	Z	Musty, mouldy, grainlike
M	Metallic	Δ	Soap
N	Minty, peppermint	π	Potato chip

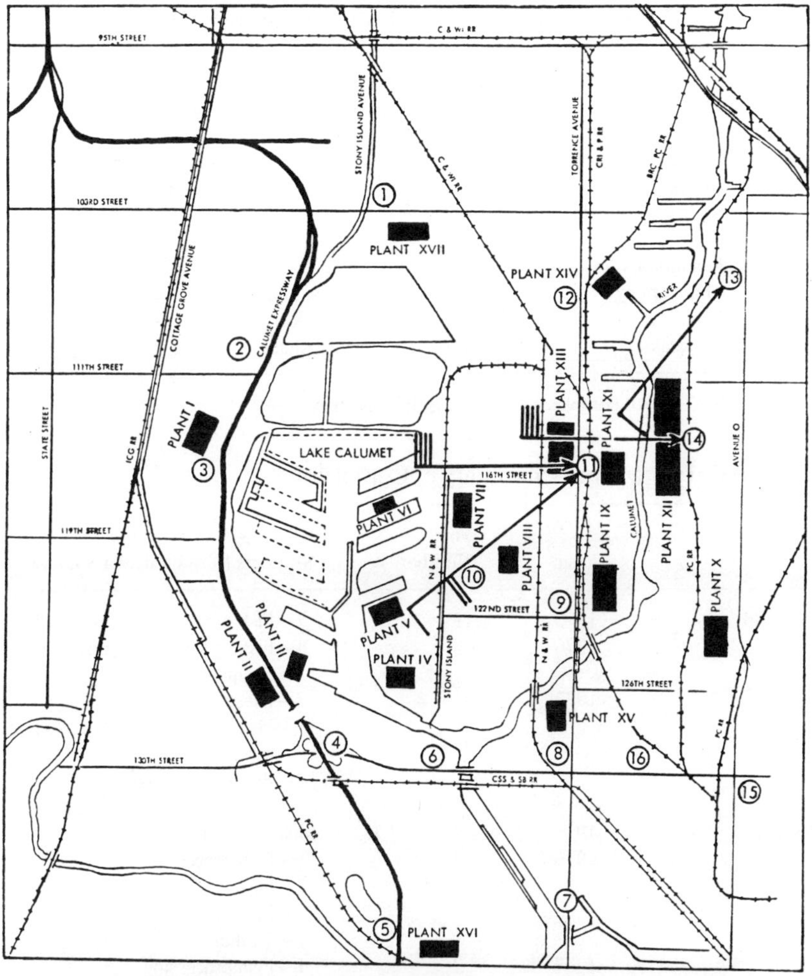

City of Chicago, Department of Environmental Control
Odor Source Evaluation of Worst Problems, Lake Calumet Region
December 1977 No. of Observations *18*

Source	Location	Type	Date and Wind Direction
Intensity of 3 Noticeable:			
Plant I	115th and Expressway	Paint	1st,W; 2nd,SW; 7th,SW; 9th,W; 20th,W; 21st,SW
Plant XIII	114th and Torrence	Tar	1st,SW; 2nd,SW; 9th,W; 14th,W; 20th,W; 27th,W
Plant XII	106th and Avenue "O"	Tar	2nd,SW
Plant XII	114th and Avenue "O"	Tar	9th,W; 14th,W; 20th,W; 27th,W
Intensity of 4 Strong:			
Plant XIII	114th and Torrence	Tar	12th,SW

TABLE C-10 Will County Health Department, Illinois

Summary of Complaints—July 1978

43 complaints in 2½ years for these types of odors prior to July 1978:

Sewage	Sulfur
Chemicals	Perfume
Ammonia	Vomit
Hydrogen Sulfide	Others

Populations impacted:

Channahom	2,700	Crest Hill	8,300
Joliet	74,000	Lockport	9,900
Rockdale	2,000	Plainsfield	2,900

NOTE: All complaints referred to Illinois EPA.

TABLE C-11 Maine—Summary of Complaints

	Year	Total Number	Numbers from Particular Sources
Portland	1975	19	7—Fish processing—1 company
	1976	8	4—Fish processing—same company
	1977	36	8—Diesel
			11—Fish, animal processing—same company
			3—Paper mill—1 company
Bangor	1975	4	Miscellaneous
	1976	5	1—Fish processing
			2—Paper mill
			1—Incinerator
			1—Tanker
	1977	2	1—Fish processing
			1—Burning trees
Central Maine	1975	13	2—Sanitary district
			3—Paper mill
			3—Chicken
	1976	19	2—Incinerator
			7—Chicken
			1—Paper mill
	1977	25	2—Diesel
			13—Paper mill
			2—Dump

State agency believes major odor sources are:

1. Pulp and paper
2. Rendering with chicken raising and packing
3. Agricultural operations—miscellaneous

TABLE C-12 Maryland—Summary of Complaints from One Chemical Company

Year	Number of Complaints	Number of Persons Complaining	Number of Days Referred to in Complaints
1961	1	1	1
1962	1	1	1
1963	—	—	—
1964	—	—	—
1965	1	1	N.S.
1966	7	7	3
1967	8	7	4
1968	94	37	49
1969	261	24	105
1970	113	23	70
1971	106	16	64
1972	28	26	6
1973	12	3	11
1974	32	11	27
1975	19	7	16
1976	2	2	2
TOTAL	685	166	359

NOTE: Not more than 75 people in 42 households.
REFERENCE: Final Report on Studies Conducted in Little Elk Creek Valley during 1974-1976, Maryland State Department of Health—November 1977.

TABLE C-13 Towson, Baltimore County, Maryland—Summary of Complaints—Number of People Affected—1973-1978

Source Type	People Affected
Acrylic polymerization	More than 50
Grain drying (spent brewing grain)	40
Sanitary landfill	35
Waste water treatment plant	30
Paint manufacture	25
Paint spray operations	15
Gasoline stations	5
Dry cleaners	3
Restaurant	1

TABLE C-14 Wayne County, Michigan

Odor Source	Type of Odor	Number and History of Complaints[a]	Type of Control Equipment	Date Installed	Odor Test Before Control Equipment Installation (date)	Odor Test After Control Equipment Installation (date)
1. Packing co.	Rendering cooker	1971—0, 1972—0, 1973—12, 1974—4, 1975—2, 1976—5, 1977—0	Fume burner incinerator	1976	440 ou/ft³ and 835,000 ou/min (1973)	115 ou/ft³ (1976)
2. Potato chip corp.	Corn chip fryer	1970—0, 1971—6, 1972—18, 1973—10, 1974—6, 1975—1, 1976—1, 1977—2, 1978—1	Wet scrubber followed by packed tower	1975	131,000 ou/ft³ (1971)	90 ou/ft³ (1977)
	Potato chip fryer	(as above)	Wet scrubber followed by fume burner	1974	131,000 ou/ft³ (1971)	None
3. Soap co.	Total rendering process	1974—65, 1975—21, 1976—29[b], 1977—42[b]	Wet chemical scrubber (hypochlorite) serves combined effluent of 2 stacks	1975	Stack #1—870 ou/ft³ and 29,000,000 ou/min (1975)	92 ou/ft³
					Stack #2—730 ou/ft³ and 12,000,000 ou/min	92 ou/ft³
4. Cooling condenser mfgr.	Paint solvent odor	1969—15, 1970—34, 1971—4, 1972—0, 1973—10, 1974—0, 1975—5, 1976—0, 1977—0, 1978—2	Fume incinerator	1971	4,800 ou/ft³ (1968)	None

5. Process co.	Amyl mercaptan	1969—2[c] 1970—0[c] 1971—7[c] 1972—3[c] 1973—4[c]	1974—4[c] 1975—1 1976—1 1977—4	Charcoal scrubber serves both stacks	1975	East stack—1,020 ou/ft³ (1973) West stack—462 ou/ft³ (1973)	21 ou/ft³ (1976)
6. Rendering co.	General rendering, process air	1969—19 1970—13 1971—95 1972—35 1973—1	1974—5 1975—7 1976—4 1977—25[d]	Wet chemical scrubber	1972	East cooling tower—30 ou/ft³ and 990,000 ou/min West cooling tower—180 ou/ft³ and 7,270,000 ou/min (1971)	East tower—43 ou/ft³ and 675,000 ou/min West cooling tower—15 ou/ft³ and 93,000 ou/min (1973)
7. Foundry	Three mold line	1971—90 1972—137 1973—168 1974—65	1975—21 1976—18 1977—45	Charcoal filter system manifold serves 3 stacks	1972	#1, 365 ou/ft³ (1971) #2, 72,000 ou/ft³ (1972) #3, 1,150 ou/ft³ (1972)	#1, #2, #3 combine 5.8 ou/ft³ (1975)
8. Fish and chips co.	Cooking oil heater stack	1975—65 (names on a petition) 1976—0 1977—0 1978—0		Wet chemical scrubber (hypochlorite)	1975	260 ou/ft³ (1975)	None

[a] Number of people affected will be significantly higher than the actual number of complaints registered. This could range anywhere from 10 to over 100 persons affected per complaint actually registered.

[b] Attributable to nonprocess related waste disposal problem.

[c] Before 1975 many complaints (100's) directed to Mich. Consolid. Gas Co. in addition to those received by WCAPC.

[d] Twenty-four complaints from one citizen.

TABLE C-15 Nebraska—Summary of Odor Source Categories—1978

Agriculture related
Feedlots
Meat packing plants
Herbicide and pesticide transfer and spraying
Ammonia storage
Dehydrated onion operation
Manure dryers
Petroleum related
Natural gas pipelines and pump stations
H_2S from storage of Canadian crude
Miscellaneous
Open dumps
Sewage treatment plants
Open burning
Hatcheries
Hydrocarbon sources (painting, degreasing, lithographic printing, etc.)
Foundries

TABLE C-16 New Hampshire—Summary of Major Complaints to 1978

Source Category	People Affected
Tannery	1,500
Underground fire	4,000+
Fiberglass tube manufacturing	50
Fuel oil and gasoline	Several

TABLE C-17 New Mexico—Summary List of Categories—Odor Complaints

Diesel Exhausts	Nitrogen Plants
Petroleum Facilities	Tire Retreading Facilities
Asphalt Facilities	Incinerators
Laboratory Fumes	

TABLE C-18 North Carolina—Summary List of Major Odorous Emissions—1978[a]

Source Category	No. Sources	Source Category	No. Sources
Pulp mill	6	Textiles	2
Rendering	5		2
	1		2
Fish meal	4	Metallurgical	1
Chemical	1	Furniture	1
	2	Graphite	1
	1	Ceramics	1
	1	Fiberglass	1
	1	Sludge treatment	
	1	Animal farms	
Meat packers	1	Barbecue houses	
Fertilizer	1	Spray painting	
		Tobacco processing	

[a] Not complete—not necessarily complaints.

TABLE C-19 North Dakota—Summary of Major Complaints—Source Categories—1978

Sugar beet refining—waste water odor, rotting beets
Charcoal briquet mfg.—heating lignite
Petroleum and petrochemical—pump stations, refining, natural gas processing, fertilizer mfg., and fertilizer application
Potato storage and processing—warehouse, potato frying, waste water treatment, and on-land disposal
Livestock—feed lot, slaughterhouse, rendering, smoke house, and waste treatment
Cheese processing—disposal of whey
Municipal sewage treatment plant—after spring thaw
Restaurants—charcoal broiling
Power plants—sulfur furnaces to reduce alkalinity
Metal plating shops

TABLE C-20 Ohio—Summary of Memorable Odor Complaints, 1977-1978[a]

Incineration of chemical wastes	Glove mfg.
Margarine mfg.	Sugar mfg.
Glue mfg.	Stone container mfg.
Glass company	Sewage treatment plant

[a] The state office received 132 odor complaints from April 1977 through April 1978. This represents only a small portion of the odor complaints in the state; most of the complaints are sent directly to local offcces.
NOTE: Particulate complaint frequently accompanies odor complaint.

TABLE C-21 Willamette Valley Region, Salem, Oregon—Summary of Complaints and Corrective Action

Relative Ranking	Type of Source	No. of Complaints/Year	Odor Data Available	Control Strategy
1	Pulp and paper	Total of 30 to 40 on 5 plants	No	Reduced levels of emissions.
2	Agricultural (animal waste, mushroom compost and mink and chicken farms)	25 to 35	Yes, Scentometer No. 1 to 2 on mushroom and chicken farms	These are exempt from rules by statute. We try to gain voluntary cooperation.
3	Rendering plants	Total of 5 per year on 2 plants	No	Adequate incineration or other treatment of emissions.
4	Resins, paints and dry cleaning solvent odors	5 to 10	No	Install stack or divert away from nearby occupied areas.
5	Incineration equipment	2 to 3	No	Adequate incineration or other treatment of emissions.
6	Zirconium	40+	Not typical of odor sources	

TABLE C-22 Philadelphia, Pennsylvania—Update of Odor Problems—Fiscal Year July 1, 1977-June 30, 1978

Category	Complaints	Violations
Petroleum refining	8	1
Web offset printers	12	4
Chemicals	23	2
Rendering	11	2
Metal smelting	9	0
Incineration	86	38
Sewage treatment	1	1
Metal fabricating	31	12
Restaurants	75	4

NOTE: Agency attempted to draft a specific odor regulation but it was discarded as being unworkable. Agency found the general nuisance provisions of the Air Management Code completely adequate for enforcement and problem solving.

Essentially all legal action taken against odor violators has been in lower court where fines are imposed. For fiscal year 1978, 16 cases were heard, resulting in $1,600 fines and $81 costs, compared to an average of 25 for the previous 5 years.

In 1970, Agency was completely successful in obtaining an injunction against a rendering plant for creating odor nuisances. The plant closed later that year when it became clear the court order could not be obeyed.

TABLE C-23 Rhode Island—Summary of Complaints and Actions—1978

Source Category	Population Impact	No. of Complaints	Action
Refining vegetable oil	R.I. 3 miles	100/year for 10 years	Court twice $250,000 for control
Surface coating—bookbinding	1/2 mile radius	20-30/year	Court $500,000 to go to water base
Fish meal	—	numerous	closed
Fish processing	—	3-4 per week in summer	?
Textiles, dying and stripping	12 factories	sulfur oxide odors	?

TABLE C-24 South Carolina—Summary of Major Odor Investigations—1977

Source Category	Number of Investigations
Industrial—Chemical companies	80
Cotton mills, Asphalt plants, Paper mill, Refinery, Rendering	57
General—Hospitals, Grocery stores	15
Domestic—not classified	75
Dumps, landfills, storage	25
Agricultural	9
Sewers/waste treatment	8

TABLE C-25 Chattanooga, Tennessee—Summary of Odor Complaints[a]

	Number of Complaints	
Source Category	1976	1977
Chemical	87	40
Foundry	2	31
Tar	3	
Gas, diesel fuel	6	12
Sewers	7	5
Paint	5	7
Rendering plants		5
Food processing	5	2
Open burning (smoke)	7	3
Nonindustrial (miscellaneous)	2	2
Waste disposal (sewage treatment plant)	17	20
Unknown	29	33
TOTAL	170	160

Data are not sufficient to give the total number of people affected by these odors.

NOTE: Mr. Rarr observes in his opinion citizens complain about odors as frequently *or* more frequently than any other air pollution problem.

[a] In 1976 the Bureau received 170 odor complaints; in 1977 the Bureau received 160 odor complaints; the above is a breakdown of the amount for each category and year.

TABLE C-26 Texas—Summary of Complaints Received September-December 1976

Region 1 (Abilene)
1 Construction
2 Feedlot
1 Crude oil production
1 Crude oil storage

Region 2 (Lubbock)
1 Oil refinery
2 Feedlot

Region 3 (Waco)
1 Farm (dead cow)
1 Chemical
1 Municipal landfill
2 Poultry rendering
1 Agricultural supply
1 Turkey farm

Region 4 (Harlingen)
1 Brush disposal
1 Incinerator
1 Ammonia terminal
1 Polyethylene extrusion

Region 5 (Corpus Christi)
3 Oil refinery
1 Grocery store incinerator
1 Metal reclamation
1 Beef processing

Region 6 (Odessa)
1 Tank car repair
1 Crude oil storage
8 Hog feeding operation
1 Oil well
1 Petroleum production
4 Hide processing
1 Pipeline leak
1 Petrochemical

Region 7 (Houston)
7 General odors
5 Chemical plants
3 Oil refinery
1 Pipe yard
3 Domestic garbage
1 Cereal manufacturing
1 Animal pen
1 Seafood processing
1 Grain drying
1 Brewery
1 Industrial waste disposal

Region 8 (Fort Worth)
2 Paint solvent
1 Fiberglass spraying
1 Burning tires
2 Slaughterhouse (blood)
2 Hog farm
2 Paint fumes
1 Organic vapors
1 Paint shop
1 Metal reclamation
1 Oil reclamation
4 Petrochemical
1 Feed mill
1 Manure composting
1 Unknown

Region 9 (San Antonio)
2 Yeast drying
1 Restaurant dumpster
3 Packing plant
4 Egg farm
1 Iron and metal company
1 Cement company
1 Paint fumes
1 Sewage treatment

Region 10 (Beaumont)
2 Oil refinery
1 Car fumes
9 Chemical plant
2 Rendering plant
3 Petrochemical
11 Unknown
1 Railroad fumes

Region 11 (El Paso)
6 Smelter
13 Acid plant
1 Refinery
1 Unknown

Region 12 (Tyler)
1 Petrochemical
1 Zoo
1 Slaughterhouse

Central Office
1 Fiberglass insulation mfg.
1 Aluminum smelter
1 Dead cow
1 Metal reclamation
1 Slaughterhouse (blood)
1 Hog pen
1 Gas well
1 Fertilizer plant
1 Cattle auction
2 Petrochemical
1 Egg farm
1 Resin curing

TABLE C-27 Vermont—Summary of Major Odor Complaints

Source Category	No. of Sources
Tanning—settling lagoons	1
Diary—whey settling lagoon	3
Poultry—chicken manure driers	1
Open dumps	several
Textiles—fabric drying unit	1
Agriculture—cow manure spreading	several

TABLE C-28 Wisconsin—Summary of All Complaints—6 Regions—January-June 1978

Month	Dust	Noise	Odor	Smoke	Misc.	Total
January	14	1	38	26	3	60
February	23	1	26	33	3	73
March	24	—	59	35	—	102
April	27	—	59	46	5	120
May	40	1	98	98	7	211
June	43	1	120	56	6	194
TOTAL	171	4	400	294	24	760